COMPUTATIONAL TECHNIQUES

The Intext Series in

Circuits, Systems, Communications, and Computers

Consulting Editor

S. C. Gupta

Chirlian – SIGNALS, SYSTEMS AND THE COMPUTER
Gupta, Bayless, –CIRCUIT ANALYSIS: WITH COMPUTER
  and Peikari   APPLICATION TO PROBLEM SOLVING
Ingels – INFORMATION AND CODING THEORY
Lathi – SIGNALS, SYSTEMS, AND CONTROLS
Matsch – ELECTROMAGNETIC AND ELECTROMECHANICAL
        MACHINES
Mickle and Sze – OPTIMIZATION IN SYSTEMS ENGINEERING
Nanavati – SEMICONDUCTOR DEVICE THEORY: BJTS, FETS,
        MOSFETS, AND INTEGRATED CIRCUITS
Sheng – INTRODUCTION TO SWITCHING LOGIC

Also by ALLEN DURLING:

AN INTRODUCTION TO ELECTRIAL ENGINEERING

A DISCRETE–TIME APPROACH FOR SYSTEM ANALYSIS
  (with M. Cuénod)

COMPUTATIONAL TECHNIQUES:
analog, digital and hybrid systems

ALLEN DURLING
University of Florida

Intext Educational Publishers
New York and London

Library of Congress Catalog Card Number: 73-21069
ISBN 0-7002-2454-8

Intext Educational Publishers
257 Park Avenue South
New York, New York 10010

Dedicated to: Neale Hamilton,
friend of the
educational process

preface

This book presents an introduction to computation and simulation techniques for use on analog, digital and hybrid computation and to put in perspective the relation between the analog computer, digital computer, digital simulation languages, straight-forward digital computation and hybrid computers.

Traditionally analog and digital computation has been taught in disjoint and often competing courses. Since both analog and digital computation may be performed in the analysis of a single problem it seems appropriate to treat the two simultaneously. The organizational basis of this book is the parallel discussion of analog and digital techniques in a complementary manner for comparison of solutions, accuracy and ease of programming. The reader develops his own digital simulation language as progressing through the book, and as new topics are covered they are incorporated into the program. The use of state variable description of systems reinforces the understanding of analog methods and its relationship to the bookkeeping problems in the numerical integration of high order differential equations.

I have assumed that the reader has a knowledge of FORTRAN and calculus through differential equations. Ideally a differential equations course would be taken concurrently with the use of this book and the material coordinated with the computation course in a complementary manner. Thus the computation illustrates and reinforces the relationships that the equations describe.

This book has evolved from class notes developed for a four quarter hour junior level course at the University of Florida for Electrical Engineering. The Computer Science department uses this course as corresponding to the ACM Curriculum 68 course A3-Analog and Hybrid Computing in the CIS program.

If only one course in computation can be taken it is important that it include an exposure to both continuous and discrete systems and methods. If more than one course is taken it is even more important to keep in perspective that the methods under study are from the many methods available or possible.

This treatment is certainly not exhaustive. The greatest difficulty I have had is deciding what to leave out and what material not covered in the text can be introduced in the problems. My decisions for coverage were based on presenting that material which would be most useful for the student's applications of computational techniques in following courses and professional activity, and a desire to emphasize the relationship between the methods.

I gratefully acknowledge the encouragement of my students over the past four years as the notes developed into a manuscript. Their scrutiny of the orgainzation and presentation has been most helpful. Many students were of great help in preparing examples, simulations and programs, particularly R. Gwin, K. Swan, R. Schall, N. Wenri, L. Jones, and D. Sekac. The encouragement and interest of Dr. Wayne H. Chen and the Department of Electrical Engineering of the University of Florida are gratefully acknowledged. Thanks go also to Mrs. Linda Smith Mihalcik and Mrs. Cathy Rushing Worsham for their efficient typing of the many versions of the manuscript. A. A. Arroyo and T. E. Bullock have been particularly helpful throughout the entire project.

contents

COMPUTATIONAL TECHNIQUES

1 introduction to computational techniques

1.1 INTRODUCTION

Engineers and scientists use computers to gain insight into their problems and analyses. Analog and digital computers are similar in that they are valuable tools for the analysis of systems which can be characterized by mathematical models. However, it is the dissimilar operating principles of these two devices that make them complementary rather than competititve in system analysis. This complementary role is emphasized by the continuing development of hybrid systems which combine the features of the analog and digital computers.

Mathematical modeling is the description of a physical situation in terms of mathematical equations or relations. The mathematical model can be analyzed on an analog computer by continuous analogs or on a digital computer as a discrete process. The modeling and representation of a physical or conceptual situation is called *SIMULATION*. The simulation is another system (the computer) that is programmed to behave like the system under study and the computer

1

thus becomes a working model of the system. The
initial values of the system variables and parameter
values become numbers in the digital computer program
and potentiometer settings on the analog computer.
Changing these numbers or potentiometer settings
provides a new situation and the computer provides the
corresponding response. The relationship between the
physical variables and the (simulated) system response
gives an intuition into system analysis and problem
solution.

The general problem-to-solution procedure is
outlined in the following steps:
1. Statement of the problem
2. Formulation of a mathematical model
3. Methods of solution (computation formulas, etc.)
4. Construct flow chart or block diagram
5. Write computer program
6. Debug program (correct mistakes)
7. Evaluate results

After the mathematical model has been formulated it
can be analyzed on the analog computer, the digital
computer, or both. To determine which approach to use
for each problem requires an understanding of both
types. In this book we use both analog and digital
computational techniques to solve various problems and
to gain an understanding of these techniques. Since
the emphasis is on techniques and their application,
little consideration is given to the internal
operation of the computer.

ANALOG AND DIGITAL COMPUTERS

The physical characteristics and operation of
analog and digital computers are markedly different.
An analog computer consists of amplifiers, potentiometers, and various electrical circuits, all of
which can be interconnected so that each variable and
element on the computer has a correspondence to the
physical system. The interconnection of the components
of the analog computer to make up a system that
satisfies the same mathematical equations as the
system under study is accomplished by plugging patchcords (wires) into the *PATCHBOARD* of the computer.

This interconnection constitutes the programming of the analog computer. When the analog computer program is complete, the operation of the computer circuits approximately satisfies equations that are analogs of the equation's of the system under study, making the computer a continuous analog of the system. Since each variable and element on the analog computer corresponds to a physical variable or element, the analog variables can be observed to get a "feel" for the system response. This can also be accomplished in digital simulations when the programmer can sit at the computer console, make changes in the program variables, and observe the corresponding response. However, in large installations the programmer generally does not have this opportunity.

Perhaps the strongest argument for analog simulation is that the analog computer is a "hands-on" simulation and the programmer runs his own program and personally obtains the records and plots of the output variables. If the plots do not look right the program can be corrected immediately and the simulation rerun. Furthermore, the results may suggest productive changes in the system variables which can be investigated immediately. The value of this intimate contact with the simulation cannot be overstated.

While the analog computer operates with continuous signals that are analogous to the system variables, the digital computer uses discrete numbers in a step-by-step fashion through sequential operations to obtain an approximate solution to the system equations. The digital computer is basically a very fast calculating machine with the facility for automatically sequencing through a predetermined set of arithmetic and logical operations called the *PROGRAM*. By this logical capability alternative paths of action in the program can be followed, depending on the value of numbers that can be the result of calculations within the program. Although the digital computer performs these operations at rates exceeding millions per second they must be completed sequentially, and although the analog computer circuits operate more slowly the computer completes all operations simultaneously. Thus analog and digital simulation operation times may be comparable. Of course the operation time depends on the problem and on the computers used in the simulation.

Time sharing digital computer systems give the appearance of parallel operation since the computer can sequence through the tasks demanded by each user on the system. Thus current time sharing systems operate in sequence just as in individual operation: only a small portion of the first program is executed, then a portion of the next program, and so on through all the users. Then the users are sequenced through again. Of course priorities are set regarding the precedence of operations and users, but the operation is sequential and not parallel.

The application of a sequential machine to a parallel type problem is discussed in considerable detail in this text with regard to a digital-analog simulation program. Parallel processing digital machines have been used and simulated and are discussed in some detail in Chapter 5.

Digital computer facilities can run many programs in a short period of time, allowing multiple users. However, this operation of the digital computer reduces the intimacy between the programmer and the operation of his program. The problem of allowing only one user at a time on the analog computer is reduced somewhat by having multiple patchboards that can be patched and checked while removed from the computer and replaced in the computer only when completed. Thus the analog facility is tied up only while it is actually in use. This is equivalent to having many card punch machines for punching digital computer programs: each program is completed before it is put on the computer, obviating the need for each user to type his program into the console.

Digital computers perform well on numerical problems such as the solution of algebraic and transcendental equations. Programs that logically fall into the class of sequential numerical procedures are easily worked with digital computers whereas these problems are generally quite difficult on the analog computer. The solution of differential equations, by contrast, is particularly well suited for analog computers.

The primary element of an analog computer is the *INTEGRATOR*. A patchcord connected to an input terminal of an integrator yields a response voltage on the output terminal which is the integral of the voltage applied to the input. With these integrators the analog computer is very suitable for the continuous solution of systems of differential equations. It is frequently desirable to have the analog computer alter the program (simulation) based on the program variables. To accomplish this, most analog computers have some logic and decision elements. Similarly, since it is frequently desirable to obtain continuous signals from the digital computer, perhaps as plots of the output variables, digital facilities usually contain some forms of continuous signal equipment.

With appropriate interface equipment an analog computer and a digital computer can be connected to each other, creating the ability to utilize the most efficient aspects of the two while freely transfering variables through analog-to-digital converters and digital-to-analog converters that change the variables from continuous signals to sequences of numbers and vice-versa. This kind of system, which can be used as either analog or digital computer separately or as a hybrid system is becoming increasingly popular.

NOTATION AND BASIC CONCEPTS

The basic analog computer element is the operational amplifier which can be connected as a *SUMMER* or *INTEGRATOR*. Figure 1.1 shows the symbols for these two elements and the mathematical representation of their operation.

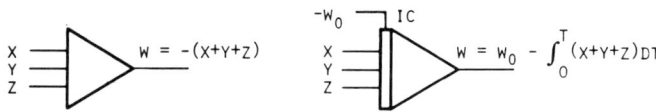

$$W = -(X+Y+Z)$$

$$W = W_0 - \int_0^T (X+Y+Z)\,DT$$

FIGURE 1.1 ANALOG COMPUTER SUMMER AND INTEGRATOR.

The output of the summer is the negative of the sum of the variables connected to the input terminals. The integrator output is minus the integral of the sum of the variables connected to the input terminals. The initial value of the integrator output can be set by connecting a value into the initial condition (IC) terminal of the integrator. Note that this initial condition also experiences a sign change through the integrator.

The procedure for the analog computer simulation of a syseem described by a differential equation is illustrated by the following short example. General analog computer procedures are presented in detail in the next chapter.

EXAMPLE 1.1

Obtain an analog computer block diagram for the analog computer solution of the differential equation

$$\frac{dy(t)}{dt} + y(t) = 0 \qquad (1.1)$$

with the initial condition $y(0) = 1$.

SOLUTION:

Equation (1.1) can be written[†]

$$\dot{y} + y = 0 \qquad y(0) = 1$$

[†] Throughout this text we will use the notation

$$\dot{y} = \frac{dy}{dt} \qquad\qquad y^{(3)} = \frac{d^3y}{dt^3}$$

$$\ddot{y} = \frac{d^2y}{dt^2} \qquad\qquad y^{(n)} = \frac{d^ny}{dt^n}$$

1. Solve for the highest derivative of the dependent variable in the given equation.

$$\dot{y} = -y$$

2. Assume this derivative exists as a signal on a line in the diagram and integrate it to obtain the signal -y neglecting the initial condition.

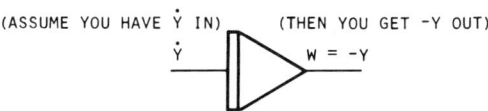

(ASSUME YOU HAVE $\dot{Y}$ IN)   (THEN YOU GET -Y OUT)

3. Take the resulting integrator output and use Equation (1.1) to obtain $\dot{y}$ = -y.

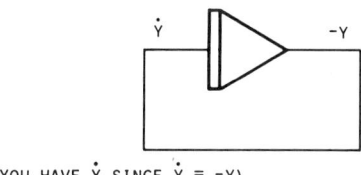

(YOU HAVE $\dot{Y}$ SINCE $\dot{Y}$ = -Y)

4. Now connect the initial condition y(0) = 1 into the integrator by setting IC equal to the negative of the desired integrator output at t = 0.

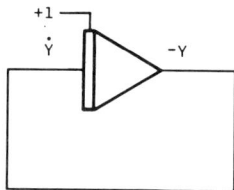

The integrator output at t = 0 is −1 thus −y(0) = −1 or y(0) = 1 which is the desired initial condition.

If this resulting block diagram is patched on an analog computer and the computer placed into operation a plot of the variable y(t) traces out the solution to Equation (1.1) since the equation describing this diagram or program is also Equation (1.1). Figure 1.2 shows the resultant response of this setup which is the solution to Equation (1.1)† (Verify that y(t) = e⁻ᵗ is the solution to 1.1 with the given initial condition.) ▼††

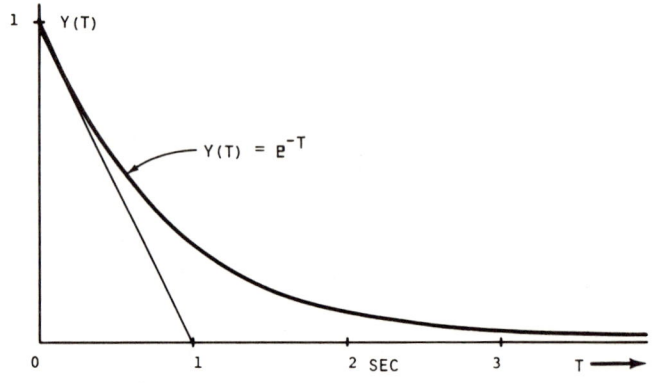

FIGURE 1.2 SOLUTION TO THE EQUATION Ẏ + Y = o SUBJECT
TO THE INITIAL CONDITION Y(o) = 1.

Since digital computation involves sequential operations with the variables of the computational algorithm it is convenient to give the following definitions of terms to be used throughout the text.

---

† Actually, only −y(t) is available on the diagram. The sign change can be accomplished with the plotter or another summer with unity gain.
†† The symbol ▼ indicates the end of a solution or example.

INDEPENDENT VARIABLE SEQUENCE: A set of values of the independent variables denoted $\{t_n\}$ consisting of the sequence elements t_n ($n=0, 1, 2, \ldots$).

INDEPENDENT VARIABLE INCREMENT: The distance between the t_n's is $\tau_n = (t_n - t_{n-1})$. If $(t_n - t_{n-1})$ is the same for all n we denote the independent variable increment $\tau = t_n - t_{n-1}$. In this case of equally spaced independent variable increments $t_n = n\tau$ ($n=0, 1, \ldots$).

DEPENDENT VARIABLE SEQUENCE: A set of values of a dependent variable denoted $\{y_n\}$ such that $y_n = y(t_n)$. If the independent variable sequence is equally spaced $y_n = y(t_n) = y(n\tau)$. Values of a sequence can be written in the form

$$\{y_n\} = [y_0, y_1, y_2, \ldots, y_n, \ldots]$$

EXAMPLE 1.2

If $y(t) = t + \cos t$ and $t_n = n\tau$, then $y_n = y(n\tau) = (n\tau) + \cos n\tau$. ▼

RECURSION RELATION: A recursion relation is a rule by which the value of a sequence element can be obtained from those elements with lower index (called past values) of that sequence and values of other sequences. A typical recursion relation is

$$y_{n+1} = 0.9y_n + 0.1x_n$$

EXAMPLE 1.3

Show that

a. $x_n = n\tau$ can be written as the recursion relation $x_n = x_{n-1} + \tau$ with the starting value $x_n = 0$.

b. $y_n = \tau x_0 + \tau x_1 + \ldots + \tau x_{n-1} = \sum_{k=0}^{n-1} \tau x_k$ can be written as the recursion relation $y_n = y_{n-1} + \tau x_{n-1}$ with the starting value $y_0 = 0$.

SOLUTION:

(a) $x_{n-1} = (n-1)\tau = n\tau - \tau$

$x_n = x_{n-1} + \tau = (n\tau - \tau) + \tau = n\tau$

(b) $y_{n-1} = \tau \sum_{k=0}^{n-2} x_k$ so

$y_n = \tau \sum_{k=0}^{n-2} x_k + \tau x_{n-1} = \tau \sum_{k=0}^{n-1} x_k$

For the digital simulation of differential systems consider an approximate solution at a discrete sequence of the independent variable. In general the independent variable increment τ depends on the system under study and the computational algorithm.

An approximation for the derivative of a function which illustrates the procedures of digital computation is the so-called *FORWARD DIFFERENCE*

$$\dot{y}_n \approx \frac{y_{n+1} - y_n}{\tau} \qquad (1.2)$$

Figure 1.3 shows the approximation of the derivative (dy/dt) at t = nτ by the forward difference Equation (1.2). Clearly for small τ this is a good approximation; in fact, as $\tau \to 0$, Equation (1.2) becomes exact.

EXAMPLE 1.4

Give a digital computer program (recursion relation) for the solution of differential equation

$$\dot{y} + y = 0 \qquad (1.3)$$

with the initial condition y(0) = 1.

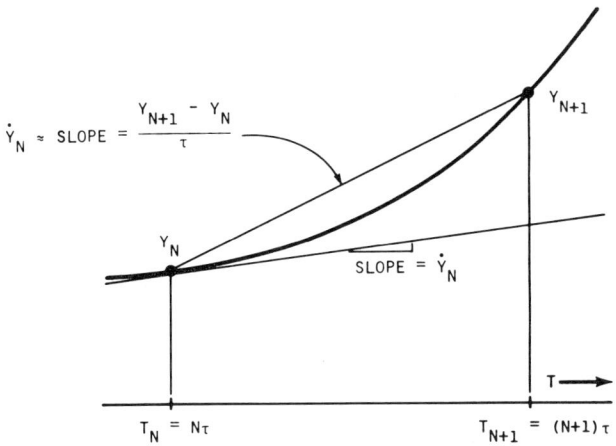

FIGURE 1.3 FORWARD DIFFERENCE APPROXIMATION OF $\dot{y}_N$.

SOLUTION:

With uniform increments of the independent variable we have

$$\dot{y}_n + y_n = 0$$

Substituting the forward difference for $\dot{y}_n$ we have

$$\frac{y_{n+1} - y_n}{\tau} + y_n = 0$$

or

$$y_{n+1} = (1-\tau) y_n \qquad\qquad (1.4)$$

with the starting value $y_0 = 1$.

The result of calculations using Equation (1.4) with the starting value $y_0 = 1$ yields different sequences $\{y_n\}$ depending on τ. For $\tau = 1$, Equation (1.4) gives $y_{n+1} = 0$ and

$$[y_n] = [y_0, y_1, \ldots, y_n, \ldots]$$

$$[y_n] = [1, 0, 0, \ldots]$$

for $\tau = 0.5$, $y_{n+1} = 0.5y_n$ and

$$[y_n] = [1, 0.5, 0.25, 0.125, \ldots]$$

for $\tau = 0.1$, $y_{n+1} = 0.9y_n$ which gives

$$[y_n] = [1, 0.9, 0.81, 0.729, \ldots]$$

Figure 1.4 shows the calculated values obtained from Equation (1.4) for various values of τ. Note the convergence of the sequence of solutions to the known exact solution $y(t) = e^{-t}$. ▼

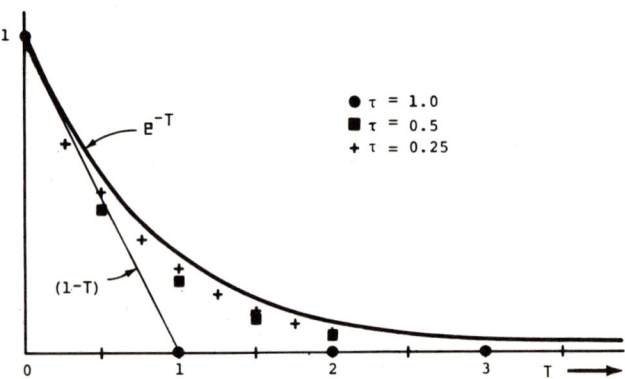

FIGURE 1.4   EXAMPLE 1.4 APPROXIMATE SOLUTIONS TO
$\dot{Y} + Y = 0$ FOR $Y_0 = 1.0$.

1.2 ANALOG AND DIGITAL COMPUTING ELEMENTS

One of the first steps in analog computation is the construction of a block diagram showing how the various units of the computer are interconnected to form the analog system. The block diagram is made up of symbols representing analog computer operations. The symbols are interconnected by lines representing the signal flow in the actual computer setup. To implement the block diagram the elements of the computer are wired together by plugging patchcords corresponding to the signal flow lines on the diagram into the patchboard of the computer.

Standard † symbols for basic analog computer elements are shown in Figure 1.5 together with their functions and some explanatory notes. Included with the functional specification is a FORTRAN statement that performs equivalent digital operation. With the exception of the integrator these are exact operations. Below is a description of each element and its digital counterpart. With these digital equivalents of the analog operations and some rules given in the next chapter for ordering the operations, we can go directly from an analog computer block diagram to a digital computer program. To some extent this corresponds to using the block diagram as a flow chart for the digital program.

The FORTRAN statements representing operations equivalent to the analog computer elements are trivial applications of the powerful programming flexibility of the digital computer. However, the interconnection of these statements provides a systematic technique for programming the digital simulation of large systems as seen in Chapter 2.

The actual positioning and representation of analog computer elements on the patchboard of different computers depends upon the specific computer. Thus you must consult the users manual for patching instructions for your particular equipment.

† "Uniform Graphics for Simulation," *Simulation,* December, 1967.

POTENTIOMETER: The potentiometer attenuates the signal by an adjustable constant factor. Usually, potentiometer settings are between 0.01 and 1.0. However, the multiplicative factor P in the equivalent FORTRAN statement X = P*A can take on any value in the range of floating point numbers in the digital computer (usually about 10^{75}). The potentiometer setting (value) is written next to the symbol as indicated by 'p' in Figure 1.5. On an analog computer

ELEMENT NAME	SYMBOL	FUNCTION AND FORTRAN STATEMENT	NOTES
POTENTIOMETER	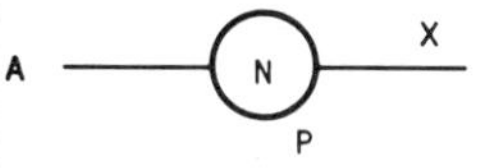 A ─(N)─ X P	X = PA X = P*A	N INDICATES WHERE THE NUMBER OF THE POT SHOULD BE PLACED.
SUMMER	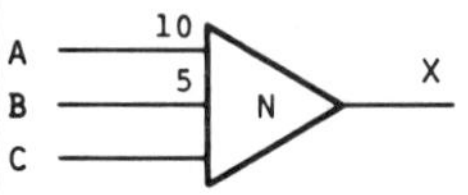 A ─10─ B ─5─ ▷N─ X C	X = -(10A+5B+C) X = -(10*A+5*B+C)	NO ARROWS SINCE THE SHAPE INDICATES THE DIRECTION OF FLOW. N INDICATES WHERE THE NUMBER OF THE SUMMER SHOULD BE PLACED.
INVERTER	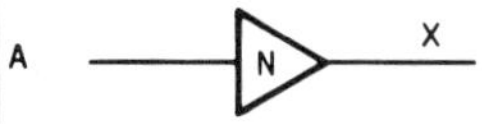 A ─▷N─ X	X = -A X = -A	
INTEGRATOR	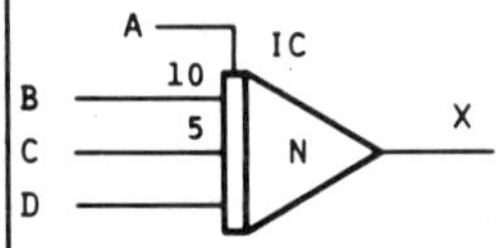 A ─ IC B ─10─ C ─5─ ▷N─ X D	$X = -A - \int_0^T (10B+5C+D)\,DT$ X = -A X = X-T*(10*B+5*C+D) = X+T(-(10B+5C+D))	N INDICATES WHERE THE NUMBER OF THE INTEGRATOR SHOULD BE PLACED. $X_0 = -A$ RECTANGULAR INTEGRATION T IS THE INTERVAL SIZE.
MULTIPLIER	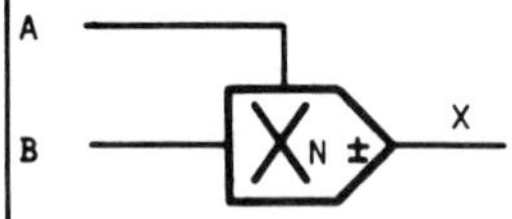 A ─┐ ⟨X N ±⟩─ X B ─	X = +AB OR X = -AB X = A*B	BECAUSE EQUIPMENT DIFFERS THE SIGN OF THE OUTPUT SHOULD BE GIVEN ON THE DIAGRAM.

FIGURE 1.5 BASIC ANALOG COMPUTER BLOCK DIAGRAM SYMBOLS.

the potentiometers are set after the circuit is connected by placing a reference voltage across the POT and measuring the output voltage while varying the POT setting.

SUMMER: A summer has an output equal to the negative of the sum of the inputs each multiplied by the number written above the line going into the summer as shown in Figure 1.5. If no number is indicated, unity gain is assumed. The multiplicative factors available are given in the users manual of the computer and are usually indicated on the patchboard. The constants given in the FORTRAN statement $X = -(10.*A+5.*B+C)$ can of course be chosen as any desired numbers. The sign reversal through the summer results from the design of summers from operational amplifiers and causes no programming difficulty.

INVERTER: The inverter is a single input, unity gain summer which merely changes the sign of the signal. Inverters are frequently necessary to obtain a signal with the desired polarity. The symbol for an inverter is usually drawn smaller than a summer symbol.

EXAMPLE 1.5

Give an analog computer block diagram and FORTRAN Program to perform the following operations assuming analog summer gains of 1, 5 and 10. The variables on the right side of the expressions are available numbers or signals.

(a) $x = a + 10b$ (b) $x = 0.4y$

(c) $x = 0.40$ (d) $w = 0.4y - 6x$
    (assume ± 1 is available)

SOLUTION:

Figure 1.6 gives the block diagrams and FORTRAN statements. ▼

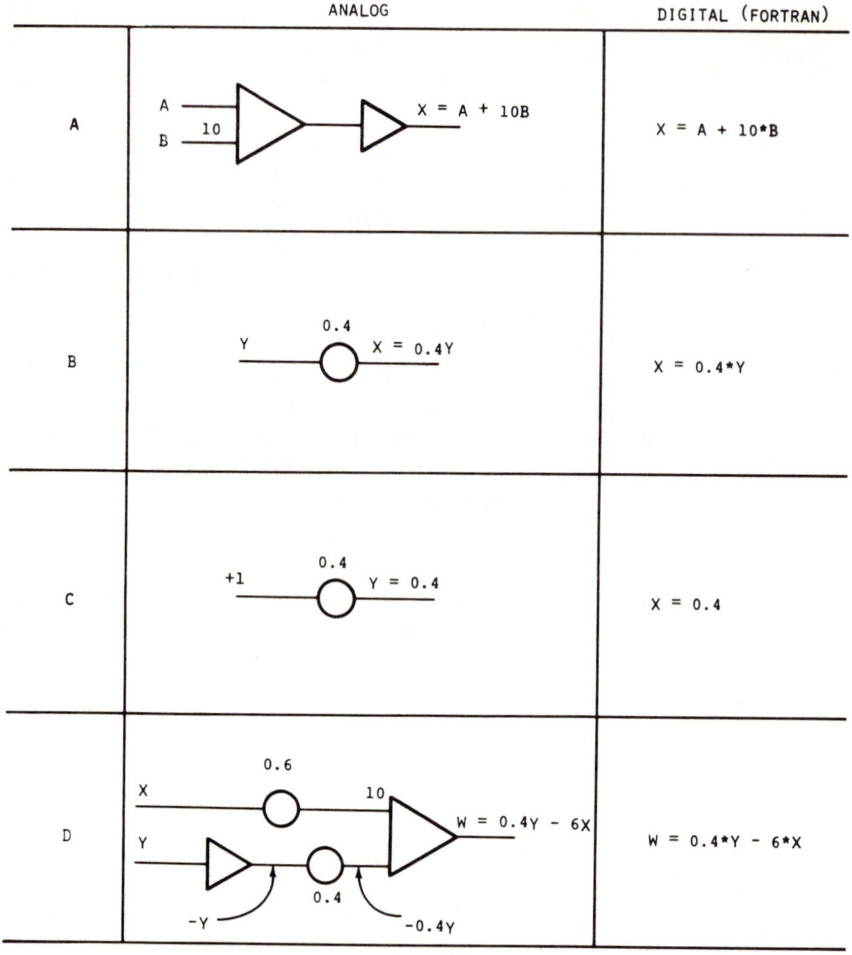

	ANALOG	DIGITAL (FORTRAN)
A	A ——▷—▷ X = A + 10B, B —— 10	X = A + 10*B
B	Y ——(0.4)—— X = 0.4Y	X = 0.4*Y
C	+1 ——(0.4)—— Y = 0.4	X = 0.4
D	X ——(0.6)—— 10 ▷ W = 0.4Y – 6X, Y ▷ (0.4), −Y, −0.4Y	W = 0.4*Y – 6*X

FIGURE 1.6 EXAMPLE 1.5 SOLUTIONS.

INTEGRATOR: The integrator performs summation with appropriate gains as a summer and integrates the result. The negative of the initial value of the integrator output is connected into the IC junction as shown in Figure 1.5.

A digital computer program cannot integrate exactly so an approximation must be used. Most approximate

integration schemes are based upon an assumed
representation of the function being integrated, and
several of these are presented in Chapter 3. For our
present purposes we consider only rectangular
integration which assumes that the function being
integrated is constant between sample values. Figure
1.7 shows this rectangular approximation to the
integral.

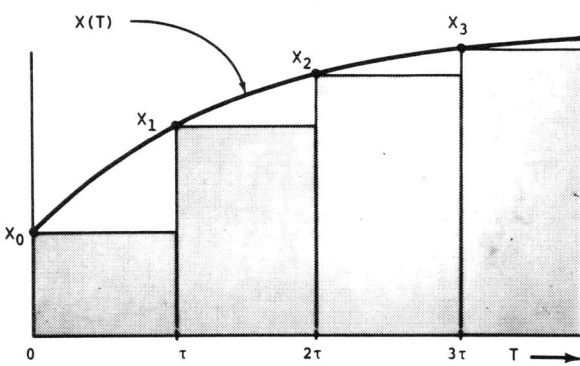

FIGURE 1.7 RECTANGULAR INTEGRATION.

The integral over the first interval is
approximately given by

$$\int_0^\tau x(t)\,dt \approx \tau x_0$$

and the integral over n intervals is approximately

$$\int_0^{n\tau} x(t)\,dt \approx \tau x_0 + \tau x_1 + \ldots + \tau x_{n-1}$$

$$= \tau \sum_{k=0}^{n-1} x_k \tag{1.5}$$

Equation (1.5) can be written as a recursion relation by taking $y(n\tau) = \int_0^{n\tau} x(t)dt$. Then

$$y_n = \tau \sum_{k=0}^{n-1} x_k$$

$$= \tau \sum_{k=0}^{n-2} x_k + \tau x_{n-1}$$

$$= y_{n-1} + \tau x_{n-1} \qquad (1.6)$$

Equation (1.6) indicates that the integral to the nth increment t is the value of the integral at the (n-1)th increment plus the area of the rentangle τx_{n-1}.

An alternative approach to obtain y_n by Equation (1.6) is obtained by taking $y = \dot{x}$ and approximating $\dot{x}$ by the forward difference. (See Problem 1.8.)

The initial value of this rectangular integrator is obtained by merely setting x_0 equal to the desired initial value.

EXAMPLE 1.6

Give a FORTRAN Program to integrate a function x(t) from t = 0 to t = 10 using rectangular integration.

If XI is the integral of X and TAU the step (or integration step) size, then XI at t=nτ is given by Equation (1.6) as

```
XI=XI+TAU*X(N-1)
```

A suitable program incorporating this statement, assuming X(N) is read into the computer or can be calculated, is given below. †

---

† The initial index is increased by 1 to start the FORTRAN subscript on x at 1. Thus $X(1) = x_0$ and $X(N) = x_{n-1}$.

```
        TAU=0.1
        N=10./TAU
        XI(1)=0.
        M=N+1
        DO 100 J=2,M
100  XI(J)=XI(J-1)+TAU*X(J-1)
```

EXAMPLE 1.7

Draw an analog computer block diagram and give a FORTRAN program to perform the following operations. Assume that the variables on the right side of the equations and the constants +1.0 and −1.0 are available.

$$\text{(a)} \quad y = -\int_0^t (a+5b)\,dt$$

$$\text{(b)} \quad w = \int_0^t (x+6y)\,dt$$

$$\text{(c)} \quad y = -\int_0^t (5x)\,dt + 0.2$$

$$\text{(d)} \quad x = \int_0^t (y-2w)\,dt - 0.1$$

SOLUTION:

Figure 1.8 shows the block diagrams and FORTRAN Programs. Note that the initial conditions on the integrators can be obtained from passing the constant voltage available on the computer through an appropriate *POTENTIOMETER*. Initial conditions on the digital integrators are obtained by a statement that sets the integrator output equal to the desired value.

The integration interval selection is very important as seen in Example 1.4 of the last section. Unfortunately there are no specific rules for

	ANALOG	DIGITAL (FORTRAN)
A	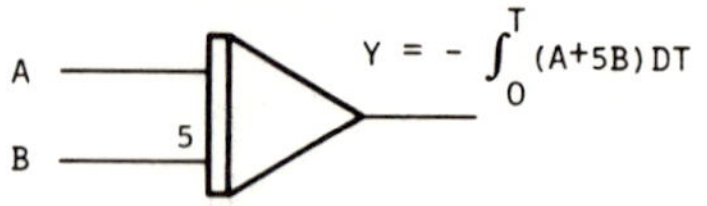 A ───── ▷ 5 ── $Y = -\int_0^T (A+5B)\,DT$ B ─────	$Y(1) = 0$ $Y(N) = Y(N-1) - T*(A+5*B)$
B	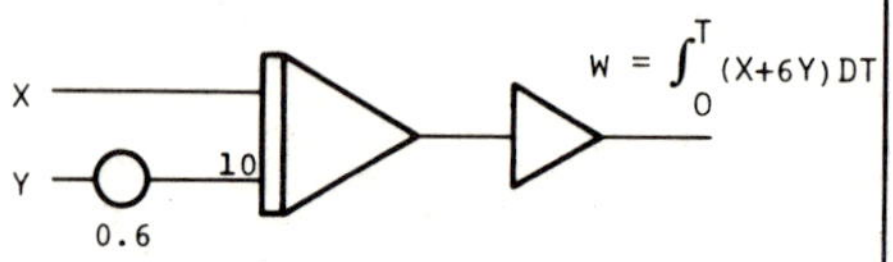 X ───── ▷ 10 ── ▷ $W = \int_0^T (X+6Y)\,DT$ Y ─○── 0.6	$W(1) = 0$ $W(N) = W(N-1)$ $+ T*(X(N-1+6*Y(N-1)))$
C	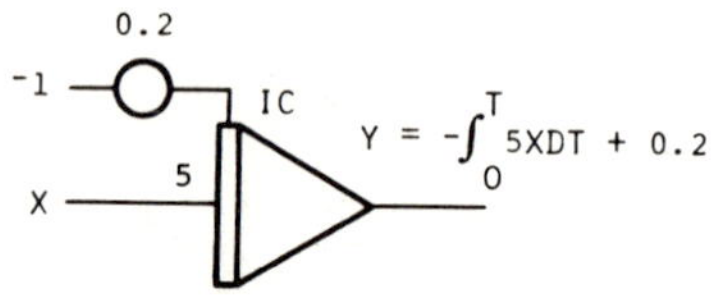 0.2 -1 ─○── IC $Y = -\int_0^T 5X\,DT + 0.2$ X ───── 5 ▷	$Y(1) = 0.2$ $Y(N) = Y(N-1) - 5*T*X(N-1)$
D	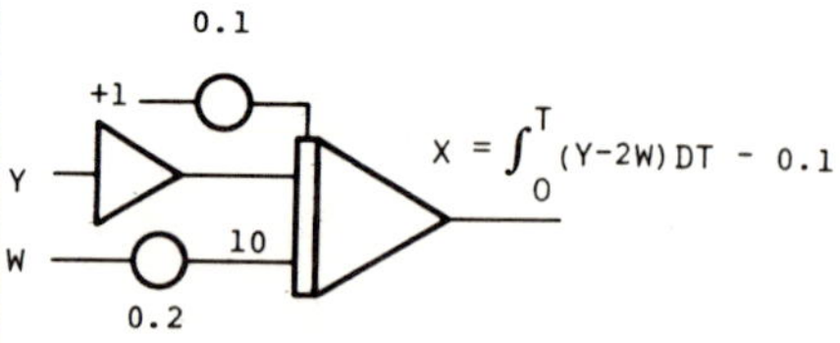 0.1 +1 ─○── $X = \int_0^T (Y-2W)\,DT - 0.1$ Y ── ▷ W ── ○ 10 ▷ 0.2	$X(1) = -0.1$ $X(N) = X(N-1)$ $+ T*(Y(N-1)-2*W(N-1))$

FIGURE 1.8 EXAMPLE 1.7 SOLUTIONS.

selecting the increment size so it is generally chosen on an intuitive basis depending upon the required accuracy, the permissible computation time, and the functions being integrated. A frequently used procedure for checking results is to run a computation twice, with the second run using half the increment size of the first. The conclusion is that if the answers are essentially the same, the increment is sufficiently small. However, this gives no indication that the increment might be much smaller than necessary resulting the use of excessive computation time.

Chapter 3 contains a discussion of several schemes for systems with variable increment sizes in which the increment is changed as a function of the variables being integrated.

Some indication of the performance of the integrator with different increment sizes is shown in the following example.

EXAMPLE 1.8

Evaluate the integral $\int_0^1 e^{-t} dt$ by rectangular integration with various increment sizes.

SOLUTION:

From Equation (1.6) we have the program.†

```
      DIMENSION N(7)
      READ(5,31)N
      DO 63 I=1,7
      TAU=1./N(I)
      X=0.
      WRITE(6,97)X
      IN=N(I)
      DO 63 J=1,IN
      A=J-1
      B=-A*TAU
      X=X+TAU*EXP(B)
   63 WRITE(6,97)X
      STOP
   31 FORMAT(7I4)
   97 FORMAT(F9.6)
      END
```

---

† This is not an efficient program because it does not print out the increment size τ nor the number of increments N, and the output is a single column of each value of the integral in succession. Making this program more efficient and informative is left as an exercise for the reader. (See Problem 1.12.) Hereafter mixed mode arithmetic will be assumed.

The results of the calculation are shown in Table 1.1. The integration is correct to 2 places when integrated with the increment equal to 1/100 the time constant of the integrand.▼

More efficient integration schemes are presented in Chapter 3. Initially we use a rough approximation of rectangular integration to illustrate the simulation techniques; then we adapt our program for more sophisticated problems.

TABLE 1.1 EXAMPLE 1.8 SOLUTIONS.

TIME	N	1	2	5	10	20*	100*	1000***	EXACT
	TAU	1.0	0.5	0.2	0.1	0.05	0.01	0.001	$1-e^{-T}$
0.0		0.0	0.0	0.0	0.0	0.0	0.0	0.0	0.0
0.1					0.1	0.0976	0.0956	0.0952	0.0952
0.2				0.2	0.1905	0.1858	0.1822	0.1814	0.1813
0.3					0.2724	0.2657	0.2605	0.2593	0.2592
0.4				0.3637	0.3464	0.3380	0.3313	0.3298	0.3297
0.5			0.5		0.4135	0.4034	0.3954	0.3937	0.3935
0.6				0.4978	0.4741	0.4626	0.4534	0.4514	0.4512
0.7					0.5290	0.5161	0.5059	0.5036	0.5034
0.8				0.6067	0.5787	0.5646	0.5534	0.5509	0.5507
0.9					0.6236	0.6084	0.5964	0.5937	0.5934
1.0		1.0	0.8032	0.6974	0.6643	0.6480	0.6353	0.6324	0.6321

*   EVERY 2ND CALCULATED VALUE
**   EVERY 10TH CALCULATED VALUE
*** EVERY 100TH CALCULATED VALUE

MULTIPLIER: Potentiometers and summers can be used to obtain multiplication by a constant. However, the multiplier gives the product of two inputs, each of which can be variable in a computer setup. Depending on the equipment and the patching, the output may be either the product of the two inputs or the inverted product, so the sign of the output is shown on the diagram.

EXAMPLE 1.9

Given the signals x and y, give an analog computer block diagram and FORTRAN statements to perform the following operations.

$$\text{(a)} \quad w = xy + x$$

$$\text{(b)} \quad z = x^2y$$

SOLUTION:

(See Figure 1.9)

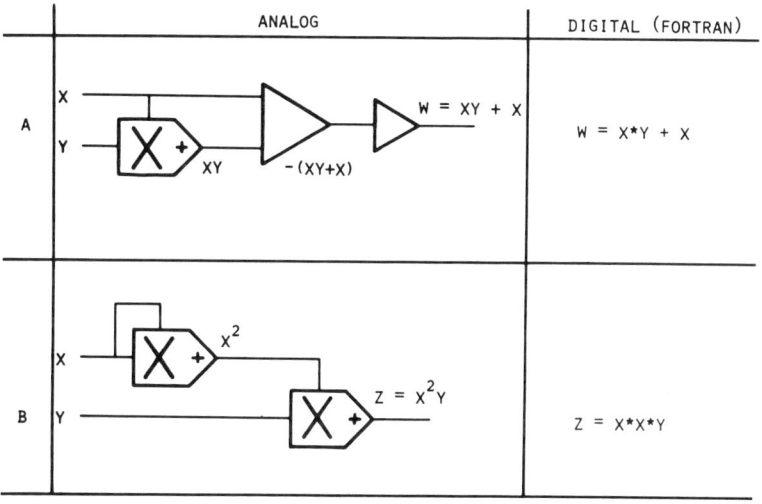

FIGURE 1.9 EXAMPLE 1.9

Each of the analog computer elements in the block diagram is identified with a number corresponding to the number of the amplifier or potentiometer on the equipment used for that element. The recommended positions of these identifying numbers are shown in Figure 1.5 by "N".

1.3 ERRORS IN ANALOG AND DIGITAL COMPUTATION

Analog and digital techniques seldom result in exact answers or solutions. The accuracy of the approximate solution depends upon the accuracy of the input data (which contains measurement errors), the computation algorithm, the equipment, and the problem being analyzed. Accuracy can in some cases be increased by decreasing the increment size or increasing the complexity of the computational scheme. In this section we discuss several sources of errors in analog and digital computation, their effects on results, and some techniques for error detection and control.

MAXIMUM AND MINIMUM SIGNAL LEVELS, NOISE AND ROUND OFF ERRORS

Electronic analog computation equipment generates noise signals which limit the minimum magnitude of useful signals. Usually this noise is less than 10mV on a 100V computer or one part in 10^4. To minimize the effects of this error it is desirable to maintain the signal levels as high as possible. The maximum signal level is limited by the dynamic range of the equipment, usually between +100V and -100V. Thus problems must be scaled so that the computer variables are scaled to the system variables in such a way that they will not exceed these limits. Of course, if the limits of the computer are exceeded the results are not valid. Scaling procedures are discussed in detail in Chapter 2.

In the early 1950's digital computers did not have floating point arithmetic, so problems for them had to be scaled to the dynamic range of the computer in the same way as those for analog computation. If 6 decimal places were used, any operation that resulted in a number greater than 10^6 gave an overflow invalidating the results. These calculations were also corrupted by ROUND OFF ERROR. If a calculation, such as division, gave results for more places than were allotted, some digits had to be discarded and this resulted in a round off error. The random nature of the round off error can be considered a form of noise corrupting the calculated values.

Figure 1.10 shows an analogy between digital round off error and analog noise. For this figure the analog signal is of a magnitude comparable to the noise level and the digital signal is shown correct to 1 decimal place. The analogy between analog noise and round off error effects can be analyzed as an additive random disturbance.

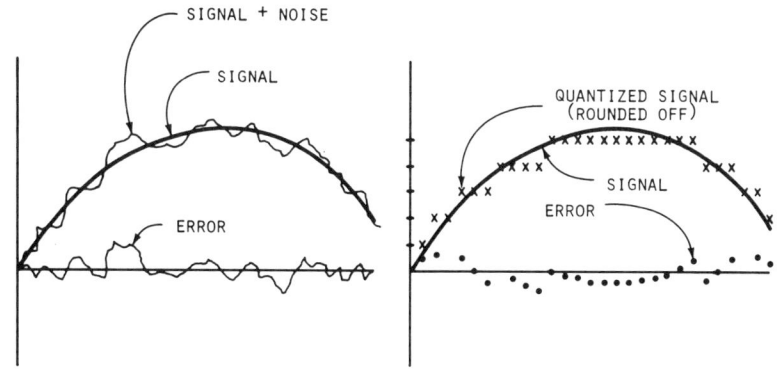

(A) ANALOG SIGNAL AND NOISE   (B) DIGITAL SIGNAL AND ROUND OFF ERROR

FIGURE 1.10 ANALOG NOISE AND DIGITAL ROUND OFF ERROR.

With the development of floating point arithmetic in the mid 1950s the problems of overflow in numerical calculations were virtually eliminated. At present, most computers have a dynamic range of at least 10^{-75} to 10^{+75} which allows problem solution in most cases without scaling. However, the problem of round off errors has not been eliminated, although variables can take on values as small as 10^{-75}. The accuracy of the calculation depends upon the number of digits carried in the calculation. The round off error still occurs because of the representation of numbers by a finite number of decimal (or binary) digits.

With floating point arithmetic the round off error has different magnitudes depending upon the number being rounded off. For example, consider rounding off numbers to 4 significant figures in floating point representation: $A = 0.77777 \times 10^8$ is rounded to 0.7778×10^8, giving an error of 3×10^3, and $B = 0.77777 \times 10^{-20}$ is rounded to 0.7778×10^{-20} giving an error of 3×10^{-25}. Clearly what is important here is not the *ABSOLUTE ERROR* but the error relative to the magnitude of the number. We define the *RELATIVE ERROR* (ε_R) of a number A which approximates A as the asbsolute error (ε_A) divided by the number A. Or

$$\varepsilon_R = \frac{\varepsilon_A}{A} = \frac{A - \tilde{A}}{A} \approx \frac{A - \tilde{A}}{\tilde{A}}$$

In fixed point arithmetic the *ABSOLUTE* round off error is limited by the number of places carried in the operations, while in floating point arithmetic it is the *RELATIVE* round off error that is limited by the number of places used in the representation of the numbers.

In practice an indication of the effects of round off error on a particular program can be obtained by running the program twice, once in single precision and again in double precision. The difference between the solutions is then an estimate of the round off error. Of course this assumes that the round off error is negligible in double precision. Most practical programs use double precision arithmetic for those portions of the computation that are particularly sensitive to round off error, such as repeated operations on small numbers and differences between large and almost equal numbers.

Also, the ordering of the operation frequently contributes to the sensitivity to round off error, as seen in the following example.

EXAMPLE 1.10

Use 8 digit arithmetic to perform the operations indicated.

$$x = a(b-c)$$

$$x = 10001(1.0002-1.0001)$$

SOLUTION:

(a) First take b - c then multiply by a,

$$b - c = 1.0002 - 1.0001 = 0.0001$$

$$a(b-c) = 1.0001$$

(b) First multiply ab and ac, then take

$$x = ab - ac$$

$$ab = 1.0003.000 \quad \boxed{2} \text{ Round off}$$

$$ac = 1.0002.000 \quad \boxed{1} \text{ Round off}$$

$$x = ab - ac = 1.000$$

For any particular program or sequence of operations the round off error can be reduced only by carrying more digits in the computations.

TRUNCATION ERROR: Using an inexact representation of a mathematical expression results in truncation errors. The most common type of truncation error results when a finite approximation is taken to a limiting process. Estimates of the truncation error for specific approximations can be obtained and are discussed in great detail in books on numerical analysis. Usually error estimates in numerical integration or differentiation schemes are developed from an analysis of the Taylor series representation of the function.

EXAMPLE 1.11

Estimate the error in approximating the derivative $\dot{y}$ by the forward difference $(y_{n+1} - y_n)/\tau$.

SOLUTION:

Expanding y in a Taylor series about y_n gives

$$y_{n+1} = y_n + \tau \dot{y}_n + \frac{\tau^2}{2!} \ddot{y}_n + \frac{\tau^3}{3!} y_n^{(3)} + \ldots$$

which can be solved for $\dot{y}_n$ to give the exact expression

$$\dot{y}_n = \frac{1}{\tau} (y_{n+1} - y_n) - \left(\frac{\tau}{2} \ddot{y}_n + \frac{\tau^2}{6} y_n^{(3)} + \ldots \right)$$

Clearly the last term on the right is the truncation error in taking

$$\dot{y}_n = \frac{1}{\tau} (y_{n+1} - y_n)$$

A reasonable estimate of this truncation error for small τ is the first term $\tau/2\, \ddot{y}_n$. ▼

ERROR BUILDUP:
    The propagation of errors in numerical calculation and the buildup of errors in analog computation are topics of utmost importance. Unfortunately very little is known about these phenomena. When doing a numerical calculation that has truncation errors, round off errors, and perhaps inaccurate data one obtains a sequence of calculated values $\{y_n^*\}$ as an approximation to the exact sequence $\{y_n\}$. Each successive calculation uses these inexact values to continue the solution. This use of inexact past values in the calculation introduces additional errors. For some problems, and some increment sizes, this effect is negligible after a few calculations. However, for other problems these effects grow with each calculation and result in *ERROR BUILDUP*. Error buildup can be considered to be new input errors at each step of calculation. An appreciation for this problem can be obtained from the following example.

EXAMPLE 1.12

Suppose an error E is introduced at some point of a numerical calculation and the error is increased by 1 percent with each additional calculation. Estimate the error in the result of 1000 calculations due to this error alone.

SOLUTION:

$$E_0 = E$$

$$E_{n+1} = 1.01E_n$$

Thus

$$E_{1000} = (1.01)^{1000} E > 2 \times 10^4 E$$

(after 100 calculations, $E_{100} = (1.01)^{100} E > 2.7E$) ▼

In most numerical schemes the truncation error goes to zero as the interval becomes small, but a small increment size implies many calculations which may increase the sensitivity to the propagation of round off errors. Perhaps the greatest error control device is an understanding of the system under study and a knowledge of what to expect from your computer analysis. If the results are not as expected, then either the solution is in error or the understanding was not correct. But it is important to decide what results to expect before seeing the output or simulation response. Printed output has a sacred appearance to all 16 decimal places even if not a single digit is correct, and invalid results frequently bias one's judgement. Of course, a system simulation by both analog and digital computation serves as a cross check on the accuracy of each.

Analog computation is subject to errors from several sources. Although components having accuracy to 0.01 percent are available, their value may change with changes in the temperature. Therefore, most critical elements for which accuracy is critical are frequently maintained in a constant-temperature oven. Furthermore, at high frequencies (fast computing speeds) resistors and other components have dynamic effects that cause phase shift and resultant errors.

The operational amplifiers of the analog computer are velocity limiting. That is, they have limitations on the rate at which the output can change. Clearly this limitation is directly related to the highest frequency the amplifier can pass without distortion in magnitude or phase so it limits the bandwidth of operation of the amplifier. (See Problem 1.19.) As in digital calculations it is usually difficult to determine the overall accuracy of an analog computer solution. However, in analog computation the problem can be scaled to obtain an indication if high frequency effects are causing errors.

Operational amplifiers also have voltage and current offsets, which are adjustable, as well as drift, which is a slow change in offset. The offset voltage is typically less than 100mV, and the drift changes the offset by less than 100 percent in a few hours. Thus with proper adjustment drift and offset are negligible.

Finally, the equations derived for analog computer integrators assures an infinite gain amplifier with infinite input impedance and zero output impedance. These values are usually sufficient to obtain satisfactory operation. Typical values are a gain in excess of 10^7 and input and output impedances of greater than $10^5\Omega$ and less than 1Ω respectively.

PROBLEMS

1.1 Show that the 5-point moving average

$$y_n = 1/5 (x_n + x_{n-1} + x_{n-2} + x_{n-3} + x_{n-4})$$

can be written as the recursion relation

$$y_n = y_{n-1} + 1/5 (x_n - x_{n-5})$$

for $n \geq 5$.

1.2 The Taylor series expansion

$$y(x+h) = y(x) + h\dot{y}(x) + \frac{h^2}{2!}\ddot{y}_n + \ldots$$

$$= \sum_{k=0}^{\infty} \frac{h^k}{k!} y^{(k)}(x)$$

can be written

$$y_{n+1} = y_n + \tau\dot{y}_n + \frac{\tau^2}{2!}\ddot{y}_n + \ldots$$

$$= \sum_{k=0}^{\infty} \frac{\tau^k}{k!} y_n^{(k)}$$

If τ is small the high order terms are negligible and

$$y_{n+1} \approx y_n + \tau\dot{y}_n$$

Draw a sketch of Figure 1.3 identifying the points y_n, y_{n+1}, and $y_n + \tau\dot{y}_n$. Is $y_n + \tau\dot{y}_n$ a good approximation for y_{n+1}?

1.3 The expression obtained in Problem 1.2 and Equation (1.2) in the text

$$y_{n+1} \approx y_n + \tau\dot{y}_n \qquad \text{and} \qquad \dot{y}_n \approx \frac{y_{n+1} - y_n}{\tau}$$

although similar are neither identical nor equivalent. Briefly explain how these expressions differ.

1.4 Use the Taylor series in Problem 1.2 with $\dot{y}_n = -y_n$ and neglect the high order terms to obtain a recursion relation to solve Equation (1.3).

1.5 Write a digital computer program for the simulation of Equation (1.3) using Equation (1.4). Plot the error in the computed value of $y(0.1)$, $y(0.2)$, ...$y(1)$ as a function of the independent variable increment τ (do the first few calculations by hand).

1.6 Replace the zero on the right side of Equation (1.3) by the function $x(t)$ and plot the response to this excitation for $\tau = 0.1$ and $y(0) = 1$.

(a) $x(t) = \sin t$

(b) $x(t) = \sin 20t$.

1.7 (a) Use the program for Problem 1.6(a) to calculate the value of $y(2)$ for various values of the increment τ. Plot $y(2)$ as a function of τ.

(b) Repeat part (a) for $x(t) = \sin 10t$.

(c) Why must τ be smaller in 1.7(b) than in 1.7(a) for "good" results?

1.8 Show that replacing $\dot{x}$ by the forward difference in the equation $y = \dot{x}$ gives the same result as Equation (1.6).

1.9 Write a program to integrate the following functions, for various values of the independent variable increment, from $t = 0$ to $t = 1$

(a) $\sin t$ (b) $t + \cos t$

(c) t^2 (d) te^{-t}

1.10 Give an analog computer block diagram and FORTRAN

program to perform the following operations.

(a) $x = 4a + 3b$ (b) $x = 0.5y$

(c) $x = 3y + 2$ (d) $x = y - x$

1.11 Give an analog computer block diagram and FORTRAN program to perform the following operations.

(a) $y = \int_0^t xdt + 0.5$ (b) $y + \int_0^t (x-y)dt$

(c) $y = .x^2 + x$ (d) $y = \int_0^t x^2dt$

1.12 For programming practice, complete the program in Example 1.8 to give the output in a table as shown in Table 1.1.

1.13 Plot the truncation error that results from using rectangular integration in Example 1.8 as a function of the independent variable increment τ. (Use the results in Table 1.1.)   Supplement the table with values of τ sufficiently small to obtain round off error effects.

1.14 An indication of the effects of round off error can be obtained by intentionally adding a small amount to the variables at each calculation. In the program in Example 1.8 add 0.0001 to the integrand   [X = X + TAU*EXP(B) + 0.0001]   and obtain a plot of the error as a function of the independent variable increment τ. Compare your plot with the results of Problem 1.13.

1.15 Suppose a numerical calculation is carried out with increment sizes τ_0 and $2\tau_0$ yielding solutions that differ by E. Assume that no error occurs for $\tau = 0$ and that the error varies linearly with τ. Determine the error in the two solutions in terms of E.

1.16 Give the relationship between the input and output variables in the analog computer diagrams shown in Figure 1.11.

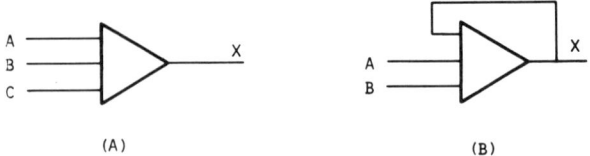

(A)   (B)

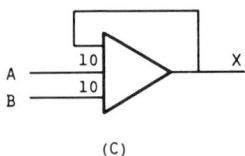

(C)

FIGURE 1.11   PROBLEM 1.16.

1.17 A diode is a nonlinear device that acts as a short circuit or open circuit depending on the potentials in the circuit containing it. The diode shown in Figure 1.12(a) is a short circuit for $x > 0$ and an open circuit for $x < 0$, which can be depicted by the nonlinear characteristic shown in Figure 1.12(b). Determine the characteristic of the circuit shown in Figure 1.12(c).

1.18 Give block diagrams that will have the following characteristics

(a) $y = -x$ for $x > 0$

   $y = -x/2$ for $x < 0$

(b) $y = x$ for $x > 0$

   $y = -x$ for $x < 0$

(c) $y = 1$ for $x > 0$

   $y = 0$ for $x < 0$

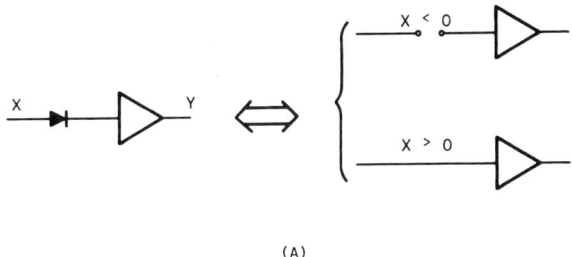

(A)

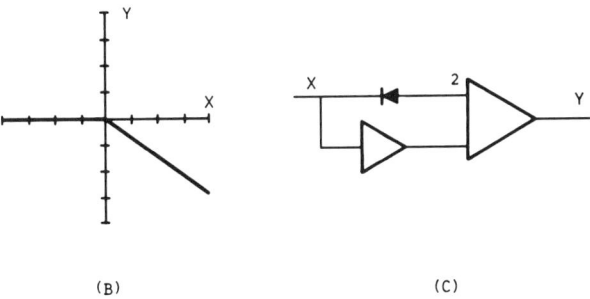

(B)   (C)

FIGURE 1.12   PROBLEM 1.17.

1.19 The differential equation $\ddot{y} + \omega^2 y = 0$ can be implemented on an analog computer by the diagram shown in Figure 1.13. Set up this block diagram on an analog computer and observe the output y on an oscilloscope for several values of frequency ω. Increase ω to obtain errors due to frequency limitations of the amplifiers.

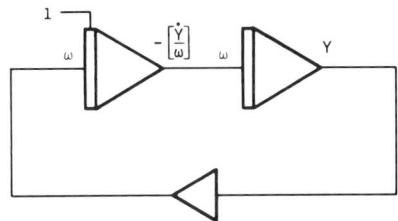

FIGURE 1.13   PROBLEM 1.19.

1.20 In digital simulations it is essential to sample analog signals sufficiently fast not to lose the detail of the signal. Suppose you have a sample frequency $\omega_s = 2\pi f_s = 2\pi/\tau$ so the function $y(t) = \cos(\omega_s + \omega_1)t/2$ is sampled at the times $t_n = n\tau = (2\pi\tau/\omega_s)$. Show that the sample values are

$$y(t_n) = y_n = \cos\left(\pi\tau + \frac{\pi\omega_1}{\omega_s}\tau\right)$$

$$= \cos\left(\pi\tau - \frac{\pi\omega_1}{\omega_s}\tau\right)$$

Thus to obtain the same sample values for y if ω_1 is positive as if ω_1 is negative. If you sample at least twice the highest frequency in the signal then all frequencies are less than $\omega_s/2$ and this eliminates the ambiguity. Actually you must sample many times the highest frequency, 10 to 100 times as fast, for reasons of accuracy and stability.

2 basic analog computation

In this chapter we connect analog computer elements to solve differential equations and introduce a Digital-Analog Simulation using floating arithmetic which does not require the magnitude scaling encountered in analog computation. A distinct advantage of digital simulation for classroom use is the availability of batch processing of digital programs. Unless many analog computers are available for student use there are complications due to competition for the machines. So, in Section 2.2 we combine the digital counterparts of the analog computer elements into a digital-analog simulation of systems from their analog computer block diagram. With this capability problems can be checked on the digital computer before an analog simulation, to get accurate estimates of the maximum values of the variables for scaling purposes. Furthermore, after an analog simulation gives initial estimates of design parameters, the digital simulation can be used to obtain accurate results.

It is frequently desirable to sample a continuous variable and hold its value contant during a portion of the calculation or to use a computed result to automatically adjust the initial conditions or a parameter for another run. This is accomplished by what is called *MODE CONTROL* on the analog computer. Mode control is discussed in Section 2.3.

Some complications arise unless the problem variables are scaled so that the computer variables are maintained below the reference level. Section 2.4 is devoted to a discussion of amplitude scaling.

2.1 ANALOG COMPUTER SIMULATION OF DYNAMIC SYSTEMS

The general procedure for the analog computer solution of differential equations is no more complicated than that discussed in Example 1.1. The basic steps are outlined below.

1. Solve the differential equation for the highest derivative of the dependent variable present.

2. Assume this derivative exists as a signal on a line in the block diagram. (Just draw a line on the left side of the paper and say "this is $\dot{x}$" or whatever the highest derivative is.) Then connect this signal to the input of an integrator to obtain the negative of the next lower derivative. Continue passing the signal through enough integrators to obtain the independent variable.

3. Now the equation from step 1 gives the highest derivative in terms of the lower derivatives so feed back (connect) these signals through the appropriate POTS, gains and nonlinear elements to obtain the signal you originally assumed you had.

4. Connect the proper initial conditions on all integrators.

5. Patch up the diagram on the computer.

EXAMPLE 1.1 (revisited)

a. Obtain an analog computer block diagram for the analog computer solution of the differential equation

$$\dot{y} + y = 0$$

with the initial condition $y(0) = 1$.

SOLUTION:

Assume the highest derivative ($\dot{y}$) exists and integrate to obtain $-y$.

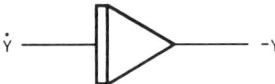

Now $\dot{y} = -y$ from (1.1) so the diagram is

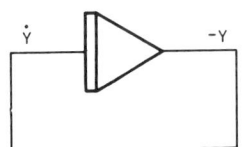

Finally connect in the initial condition

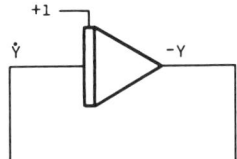

b. In this example suppose the equation were $\dot{y} + 2y = 0$, or $\dot{y} = -2y$. The block diagram is obtained by feeding back $-y$ with a gain of 2 which can be obtained by a summer with gain 10 and a POT set at 0.2, but the summer changes the sign. The easiest method here is to use a 10 gain on the integrator giving the diagram shown in Figure 2.1. Note now that $\dot{y}$ does not appear in the diagram. ▼

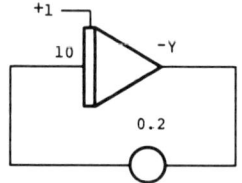

FIGURE 2.1 BLOCK DIAGRAM FOR THE EQUATION $\dot{Y}$ + 2Y = 0, Y(0) = 1.

For higher order systems the procedure is the same. The only change is that more integrators are required.

EXAMPLE 2.1

Give the analog computer block diagram which when implemented will model the differential equation

$$\ddot{x} + 0.5\dot{x} + x = 0$$

having the initial condition $x(0) = 0.1$ and $\dot{x}(0) = 0.5$ (Assume REF = +1).

SOLUTION:

First assume we have $\ddot{x}$ and integrate it twice to obtain $-\dot{x}$ and $+x$.

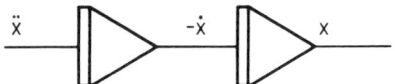

Now $\ddot{x} = -0.5\dot{x} - x$ can be obtained from the signals available. We can take $-\dot{x}$ through a POT set at 0.5 and x through an inverter giving the diagram shown in Figure 2.2. The initial conditions are now connected. (Check the sign on the initial conditions.) ▼

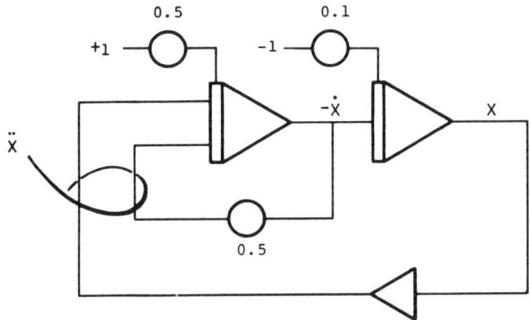

FIGURE 2.2 EXAMPLE 2.1 BLOCK DIAGRAM FOR THE EQUATION
$\ddot{x} + 0.5\dot{x} + x = 0$, $x(0) = 0.1$, $\dot{x}(0) = 0.5$,

EXAMPLE 2.2

Draw an analog computer block diagram to generate the signal $x(t) = \cos 2t$.

SOLUTION:

In this example we are given the solution rather than the differential equation. Assume this signal $x(t) = \cos 2t$ is the output of an integrator. Then the input to that integrator is x = 2 sin 2t.

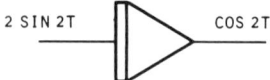

If 2 sin 2t is the output of another integrator we have

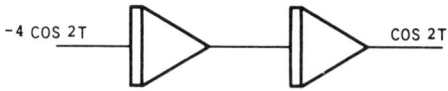

The input signal −4 cos 2t can be obtained by passing x(t) = cos 2t through an inverter and a gain of 4. Including   initial   conditions,   we have the block diagram in Figure 2.3.▼

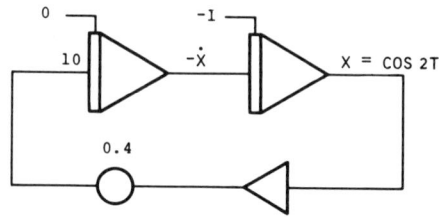

FIGURE 2.3 EXAMPLE 2.2 BLOCK DIAGRAM TO GENERATE x(T) = cos 2T.

    In the above example we constructed a signal generator to generate the function cos 2t by assuming that it is the output of cascaded integrators. This procedure is satisfactory for generating functions that are exponentials or sinusoids. However, it is usually easiest if you know which differential equation the signal satisfies. It is easy to verify that $y(t) = Ae^{-\alpha t}$ is the solution to the differential equation $\dot{y} + \alpha y = 0$ with the initial condition $y(0) =$ A. (Just add together αy and $\dot{y}$.) Also $y(t) =$ A cos ωt + B sin ωt † is the solution to the differential equation $\ddot{y} + \omega^2 y = 0$ with initial conditions $y(0) = A$ and $\dot{y}(0) = B\omega$. Thus to obtain a sinusoidal signal for an analog computer program we can use the diagram shown in Figure 2.4.

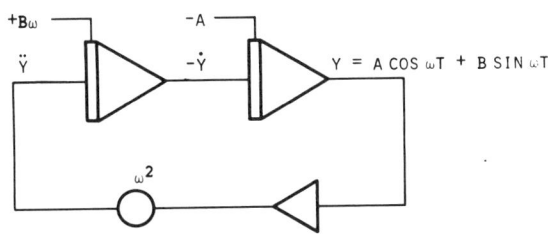

FIGURE 2.4 SINEWAVE GENERATOR $(\ddot{Y} + Y = 0)$.

In Example 2.1 we develop an analog computer block diagram which will give the solution to a second order differential equation. This solution is the natural (or force-free) response of the system. A mechanical system which is described by a second order differential equation is shown in Figure 2.5.

HOMOGENEOUS EQUATIONS for which the right hand side is zero have response due only to the initial conditions. If the mass M in Figure 2.5 is held over to the right so x = 1 and $\dot{x} = 0$, and released at t = 0, the mass will oscillate back and forth with decreasing amplitude as the energy is dissipated by the frictional damping. On the analog computer we set the initial conditions $x(0) = 1$ and $\dot{x}(0) = 0$ on the

---

† Of course, A cos ωt + B sin ωt = C cos (ωt+θ).

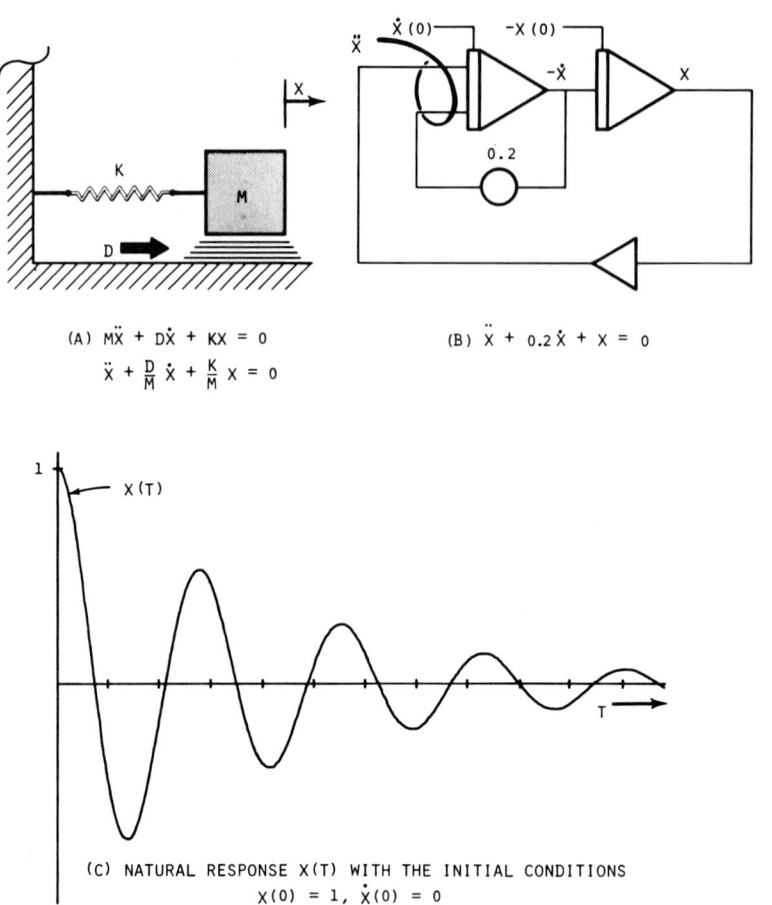

(A) $M\ddot{X} + D\dot{X} + KX = 0$

$\ddot{X} + \frac{D}{M}\dot{X} + \frac{K}{M}X = 0$

(B) $\ddot{X} + 0.2\dot{X} + X = 0$

(C) NATURAL RESPONSE X(T) WITH THE INITIAL CONDITIONS
$X(0) = 1, \dot{X}(0) = 0$

FIGURE 2.5 SECOND ORDER MECHANICAL SYSTEM NATURAL RESPONSE.

setup shown in Figure 2.5(b). Figure 2.5(c) shows a typical response for the differential equation $\ddot{x} + 0.2\dot{x} + x = 0$, [D/M = 0.2 and K/M = 1]. Problem 2.3 discusses the natural response of this second order system for various ratios of D/M and K/M.

Now, suppose you want to wiggle the mass with an applied force and measure the response. Specifically, to get the response of this system to an excitation y(t) we need only write the equation

$\ddot{x} + 0.2x + x = y(t)$ and   the block diagram merely has an additional input $y(t)$ to the first integrator.

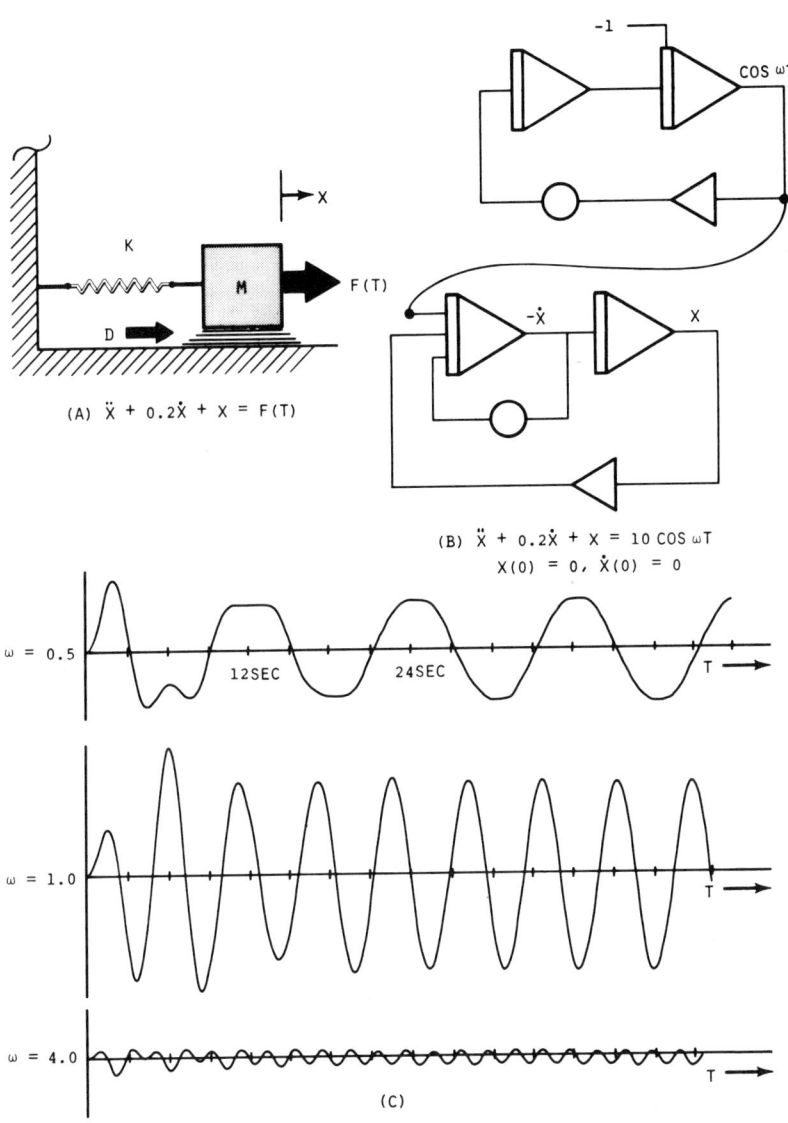

(A) $\ddot{X} + 0.2\dot{X} + X = F(T)$

(B) $\ddot{X} + 0.2\dot{X} + X = 10 \cos \omega T$
$X(0) = 0, \dot{X}(0) = 0$

(C)

FIGURE 2.6 EXAMPLE 2.3 SYSTEM SIMULATION AND FORCED RESPONSE.

EXAMPLE 2.3

Give an analog computer diagram to obtain the response of the system shown in Figure 2.6(a) to a sinusoidal excitation force $f(t) = \cos \omega t$; for several values of ω.

SOLUTION:

The block diagram to generate $\sin \omega t$ is given in Figure 2.4. This signal is connected as an input to the system simulation as shown in Figure 2.6(b). Figure 2.6(c) shows the resultant system simulation response for various frequencies of excitation.▼
The addition of nonlinear elements to analog computer block diagrams is straightforward as demonstrated in the following example.

EXAMPLE 2.4

Give a block diagram to simulate a system described by the equation

$$\ddot{y} + 2y\dot{y} = 0$$

$$y(0) = 0$$
$$\dot{y}(0) = 1$$

SOLUTION:

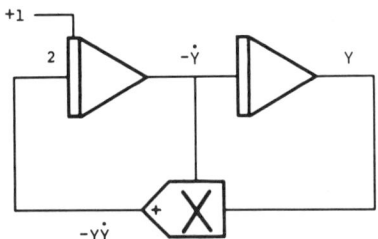

FIGURE 2.7   EXAMPLE 2.4 BLOCK DIAGRAM FOR THE EQUATION
$\ddot{Y} + 2Y\dot{Y} = 0$, Y(0) = 0, $\dot{Y}$(0) = 1,

The block diagram is given in Figure 2.7. (Check the diagram and initial conditions.)

2.2 DIGITAL-ANALOG SIMULATION

The examples in the last section had particularly "nice" numbers for the coefficients of the differential equations. In real problems the equations must first be scaled so the gains on the summers and integrators, and the POT settings are reasonable for analog equipment. For example, if you try to program the differential equation $\ddot{y} + 10^6 y = 0$ without scaling you need an integrator gain of 10^6. So even for this trivial equation we encounter difficulties. In Section 2.3 we discuss time and amplitude scaling for analog computation.

This section deals with another approach to the problem - the digital simulation of analog computation. This digital-analog simulation has the computational ability of the digital computer with its essentially unlimited variable range, as well as the structural organization of the analog computer. The procedure is to devise an analog computer block diagram for the simulation and replace the analog elements with their digital equivalents. In this section we write continuous system simulation programs in FORTRAN based on the digital equivalents given in Chapter 1.

There are two difficulties in this type of digital simulation. First, real systems and their analog simulations have parallel devices and concurrent events. Since the digital computer must perform operations one at a time in sequence, the operations must be ordered so that for each value of the independent variable, the output of each device is not computed until the values of all the inputs to that device have been computed. Because of this sequential processing on the digital computer the digital simulation of high order systems frequently runs considerably slower than "real time".

The second difficulty is the selection of the independent variable increment, τ, which must be specified in the digital simulation. Here we choose the increment size on the basis of the experience gained from the last chapter and from Problems 1.5, 7, 9, and 13. It was observed that rectangular integration of exponential functions is accurate to about two decimal places if the increment size is 1/100th of the time constant. Comparable accuracy is obtained when integrating a sine wave with about 400 increments per period, while accuracy to one decimal place is obtained by taking about 40 increments per period. For these initial studies we choose τ consistent with these observations, the system time constants, and the frequencies of the excitation. In Chapter 3 we make our simulation procedures more sophisticated by including the facility for our program to automatically adjust the independent variable increment, and utilize more sophisticated integration formulas.

The basic ideas of digital-analog simulation (DAS) are demonstrated in the following example.

EXAMPLE 2.5

From the analog computer block diagram in Figure 2.7, for the solution of the differential equation $\dot{y} + y = 0$, write a digital computer program to simulate the analog setup (Figure 2.8), where $y(0) = 1$.

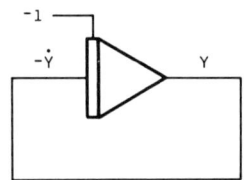

FIGURE 2.8 BLOCK DIAGRAM FOR THE EQUATION $\dot{y} + y = 0$, $y(0) = 1$.

SOLUTION:

First we set the duration of the simulation, say 5 seconds [5 time constants]

$$FINTIM=5.0$$

Then we set the increment size, say 1/100th of the time constant, to obtain about two decimal places accuracy

$$TAU=0.01$$

The number of steps in the calculation is

$$NSTEPS=FINTIM/TAU$$

Set the initial condition

$$Y=1.0$$

The input to the integrator is from the block diagram

$$X=Y$$

Finally the output of the integrator $[y_{n+1} = y_n - \tau (x_n)]$

$$Y=Y-TAU*X$$

x is the input to the integrator and $x_n = -y_n = y_n$ by the system equation.   Note that this is the same recursion relation obtained in the last chapter $[y_{n+1} = (1-\tau)y_n]$ for this differential equation.

Putting these together with some input output statements results in the program below

```
FINTIM=5.0 - - - - - - - - - - - - (FINAL TIME)
TAU=0.01 - - - - - - - - - - - - - (INCREMENT SIZE)
NSTEPS=FINTIM/TAU - - - - - - - - (NUMBER OF INCREMENTS)
Y=1.0 - - - - - - - - - - - - - - (INITIAL CONDITION)
TIME=0. - - - - - - - - - - - - - (TIME = 0)
WRITE(6,20) - - - - - - - - - - - (OUTPUT HEADING)
WRITE(6,30)TIME,Y - - - - - - - - (PRINT INITIAL VALUES)
DO 10 J=1,NSTEPS - - - - - - - - - (DO N STEPS FROM 0 TO FINTIM)
X=Y
Y=Y-TAU*X - - - - - - - - - - - - (DEPENDENT VARIABLE SEQUENCE)
TIME=J*TAU - - - - - - - - - - - - (INDEPENDENT VARIABLE SEQUENCE)
10 WRITE(6,30)TIME,Y - - - - - - - (PRINT TIME AND OUTPUT)
STOP
20 FORMAT(' TIME      OUTPUT')
30 FORMAT(1X,F5.2,4X,F8.4)
END
```

The above program results in 500 lines of output.

To get a more reasonable output record it is probably sufficient to print the output at every 10th increment or 10 values per time constant. This and several other formalities of digital-analog simulation are presented in the following paragraphs. However it is important to note that printing out every tenth value is not the same as calculating with a larger increment size.

THE SIMULATION LANGUAGE

The statements of our digital-analog simulation language are divided into the following three general categories.

INITIAL STATEMENTS: Those statements of the program which set the duration, the increment size and the initial values of *ALL* of the variables are called *INITIAL STATEMENTS*.

REPRESENTATION STATEMENTS: The representation of the analog elements in FORTRAN is accomplished by the *REPRESENTATION STATEMENTS*.

CONTROL STATEMENTS: The operating procedures for the program and the format of the input/output data are provided by the *CONTROL STATEMENTS*.

Notation for the outputs of each of the analog elements is given in Table 2.1.

TABLE 2.1 NOTATION FOR OUTPUT VARIABLES IN
DIGITAL-ANALOG SIMULATION LANGUAGE DAS.

ELEMENT	OUTPUT VARIABLE NAME
POTENTIOMETER 'N' (POT NO. 5)	PN (P5)
SUMMER 'N' (SUMMER NO. 8)	SN (S8)
INTEGRATOR 'N' (INTEGRATOR NO. 3)	IN (I3)

The values of potentiometer settings are given the names POT1, POT2, and so on. Inverters are treated as single input summers. The integrator outputs I1, I2, I3, ... IK must be specified as real (floating point) variables. With this notation we have a name for the signal on each line of the block diagram.

EXAMPLE 2.6

Use the above notation to assign names to the variables in the block diagram in Figure 2.9(a).

SOLUTION:

See Figure 2.9(b).

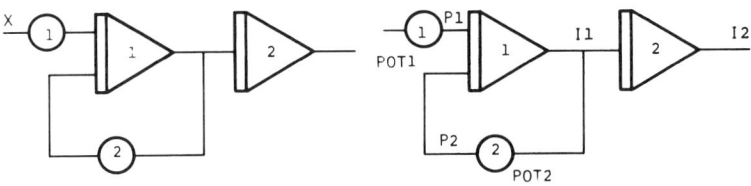

(A) ANALOG COMPUTER BLOCK DIAGRAM   (B) VARIABLE NAME ASSIGNMENT
                                               FOR DIGITAL PROGRAM

FIGURE 2.9 NOTATION AND VARIABLE NAME ASSIGNMENT
            FOR DIGITAL ANALOG SIMULATION PROGRAM.

It is convenient to write the integration of our digital analog-simulation language as a real function. Taking the function INTGRL(A,B) where A is the output and B is the input to the analog integrator we have

```
REAL FUNCTION INTGRL(A,B)
COMMON TAU
INTGRL=A-TAU*B †
RETURN
END
```

---

† NOTE: INTGRL = A-TAU*B = A + TAU*(-B). Or the integrator changes the sign of the input and integrates it.

This real function program designates the function name INTGRL as a floating point function and uses rectangular integration with the independent variable increment TAU common to the main simulation program. The first argument of the function, A, is the output of the integrator and the second argument, B, is an arithmetic statement representing the integrator input.

The statement INTGRL = A-TAU*B takes the output of the integrator at the last time increment and subtracts off the area TAU*B to give the new output as shown in Figure 2.10. In most digital simulation programs the statement would be A+TAU*B. However, we retain the negative sign here to have a one-to-one correspondence between the inverting analog integrator and the digital program.

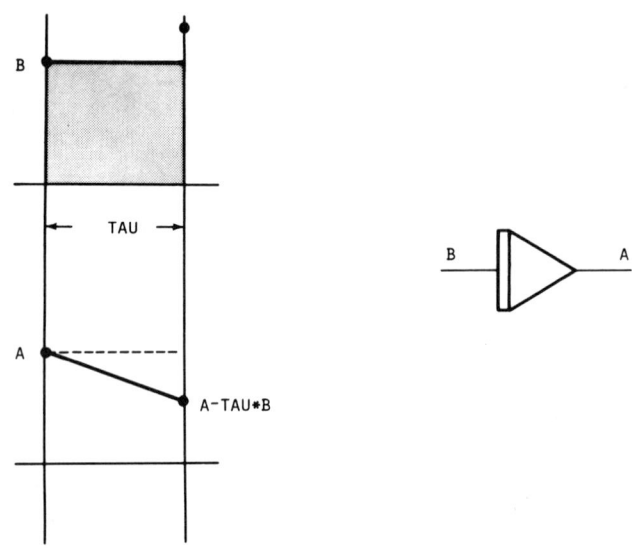

FIGURE 2.10 INVERTING RECTANGULAR INTEGRATION.

EXAMPLE 2.7

Use the INTGRL function given above to approximately represent the analog integrator shown in Figure 2.11

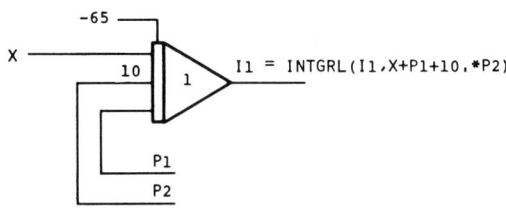

FIGURE 2.11 EXAMPLE OF REAL FUNCTION INTGRL (A,B) FOR EXAMPLE 2.7

SOLUTION:

(See Figure 2.11)

```
I1=65.0
DO 50 K=1,N
I1=INTGRL(I1,X+P1+10.*P2)
```

(other representation statements) ▼

At this point in the development of our simulation language we consider only the rectangular integration function. In Chapter 3 we discuss higher order integration schemes having more complex formulas and methods for implementating them into the program.

It is desirable to control the output so that the output increment can be different than the computational increment TAU. For instance, we call the output increment OUTDEL and define the variable TIMOUT as the time at which the next output is to be printed. Since TIME is an accumulation of the independent increments [that is, TIME = K*TAU], we can print the simulated variable values when TIME = TIMOUT by the following output statements.

```
      TIMOUT=OUTDEL
      DO 100 K=1,NSTEPS
```

(simulation program)

```
      IF(TIME.LT.TIMOUT) GO TO 100
      WRITE(6,19)TIME,I2
      TIMOUT=TIMOUT+OUTDEL
  100 CONTINUE
```

The first output occurs at TIME = TIMOUT = OUTDEL, so TIMOUT is increased by OUTDEL to the next output time. Generally, about 50 output values are sufficient to obtain a rough idea of system response and the data output can be conveniently cut to standard 8 1/2 by 11 inch size for inclusion in reports. Furthermore, since the above program does not print out the initial conditions of the system variables, it is convenient to print these and the headings on the output right after the initial conditions are set. This is also an aid in debugging programs since it indicates whether the initial conditions have been set properly.

Occasionally the comparison (TIME – TIMOUT) does not check when it is zero due to round off errors, since TIME and TIMOUT are calculated differently, thus causing the output values to be shifted by the increment TAU. This problem is eliminated by starting TIMOUT at TIMOUT = OUTDEL – TAU/2. This results in TIME exceeding TIMOUT by half the increment size when the next output is to be printed. Alternatively we could leave TIMOUT as shown and change the comparison statement to IF(TIME – TIMOUT + TAU/2). (Draw a line representing the independent variable and mark off points TIME, TIMOUT and TIMOUT + TAU/2 to verify the above discussion.) †

---

† T. E. Bullock has suggested that for this program it is more efficient to insert the statement NPRT = OUTDEL/DEL + 0.5 before the DO loop, and replace the three lines before statement 100 with the single statement

```
      IF (K/NPRT*NPRT).EQ.K) WRITE(6,19)TIME,I2
```

If you can figure out how this works, you understand integer operations in FORTRAN pretty well.

With these portions of the simulation programming completed we look at the second order system given in Figure 2.3.

A SECOND ORDER SYSTEM
The system shown in Figure 2.3(a) is described by the differential equation $\ddot{x} + A\dot{x} + Bx = 0$ with the initial conditions $x(0) = X0$ and $\dot{x}(0) = XODOT$. Figure 2.12 shows the assignment of variable names to the block diagram of the system.

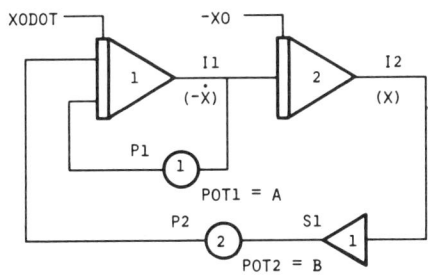

FIGURE 2.12 DIGITAL PROGRAM NOTATION FOR THE EQUATION
$\ddot{x} + A\dot{x} + Bx = 0$, $x(0) = X0$, $\dot{x}(0) = XODOT$.

To write the program we need *INITIAL STATEMENTS* to set TAU, FINTIM and the number of output increments NUMOUT. Then

        OUTDEL=FINTIM/NUMOUT

(Of course we could choose OUTDEL and calculate NUMOUT.) We also need as initial statements the POT settings

        POT1=A

        POT2=B

and the initial conditions

        I1=D=-XODOT

        I2=C=XO

To write the *REPRESENTATION STATEMENTS* we calculate all inputs to all integrators before calculating the integrator outputs. Following this rule we have

```
S1=-I2
P2=POT2*S1
P1=POT1*I1
I2=INTGRL(I2,I1)
I1=INTGRL(I1,P1+P2)
```

At the first calculation after the initial conditions are set, TIME = TAU, we must calculate the output of each integrator based on the inputs to the integrators at TIME = 0. This is accomplished by the above ordering of the statements. If the last two statements are reversed in order, I1 at TIME = TAU is calculated and then I2 is calculated. I1 at TIME = 0 should be used to calculate I2, but we have already changed I1 to its new value. Thus the ordering of the integration should be from right to left on the diagram.

The above ideas and techniques are incorporated into the program in the following example.

EXAMPLE 2.8

Give a program to simulate a system described by the differential equation $\ddot{x} + 0.2\dot{x} + x = 0$ with the initial conditions $x(0) = 10$, $\dot{x}(0) = 0$

SOLUTION:

The undamped natural frequency is $\omega_n = 1$ (where $\ddot{x} + 2\zeta\omega_n\dot{x} + \omega_n^2 x = 0$) so we pick FINTIM = 25 to get about 4 periods of the solution. Pick NUMOUT = 50 to get 50 output values, and TAU = 0.01. (We look at other values of TAU below.)

Figure 2.13 gives the output and a plot of the system response. Compare this with Figure 2.3. ▼

```
      REAL I1,I2,INTGRL - - - - - (FLOATING POINT VARIABLES)
      COMMON TAU - - - - - - - - (SAME TAU IN PROGRAM AND
C                                    SUBPROGRAM FUNCTION)
C     INITIALIZATION STATEMENTS.
C
      TIME=0. - - - - - - - - - - (SET INITIAL VALUE OF TIME)
      FINTIM=25.0 - - - - - - - - (SET FINAL TIME)
      TAU=0.01 - - - - - - - - - (SET COMPUTATIONAL INCREMENT)
      NUMOUT=50 - - - - - - - - - (NUMBER OF OUTPUT VALUES)
      NSTEPS=FINTIM/TAU - - - - - (NUMBER OF COMPUTATIONS)
      OUTDEL=FINTIM/NUMOUT  - - - (OUTPUT INCREMENT SIZE)
      TIMOUT=OUTDEL-0.5*TAU - - - (TIME FOR NEXT OUTPUT)
C
C     SET POTS.
C
      POT1=0.2
      POT2=1.0
C
C     SET THE INITIAL CONDITIONS ON THE INTEGRATORS.
C
      I1=0.0
      I2=10.0
C
C     WRITE OUT THE HEADINGS, INCLUDING INITIAL VALUES.
C
      WRITE(6,1)
    1 FORMAT(///' TIME    OUTPUT')
      WRITE(6,2)TIME,I2
    2 FORMAT(1X,F4.1,2X,F8.4)
C
C     SIMULATION LOOP...TO STATEMENT #100.
C
      NN=1
      DO 100 K=1,NSTEPS - - - - - (DO N STEPS)
      S1=-I2
      P2=POT2*S1
      P1=POT1*I1  - - - - - - - - (REPRESENTATION STATEMENTS)
      TIME=K*TAU  - - - - - - - - (TIME AT THIS CALCULATION)
      I2=INTGRL(I2,I1)
      I1=INTGRL(I1,P1+P2)
C
C     IS IT TIME TO WRITE THE OUTPUT VALUE?
C
      IF(TIME.LT.TIMOUT) GO TO 100
      NN=NN+1
      WRITE(6,2)TIME,I2 - - - - - (PRINT OUT IF TIME = TIMOUT)
      TIMOUT=NN*OUTDEL-0.5*TAU  - (SET TIMOUT TO NEXT OUTPUT TIME)
  100 CONTINUE
      STOP
      END


      REAL FUNCTION INTGRL(A,B)
      COMMON TAU - - - - - - - - - (SAME TAU AS IN MAIN PROGRAM)
      INTGRL=A-TAU*B - - - - - - - (RECTANGULAR INTEGRATION WITH
      RETURN                           SIGN INVERSION)
      END
```

The order of the statements must allow the calculation of the inputs to each integrator at the current increment before the next integrator output is calculated.

TIME	OUTPUT
0.0	10.0000
0.5	8.8353
1.0	5.7108
1.5	1.5347
2.0	-2.6213
2.5	-5.7875
3.0	-7.3136
3.5	-6.9914
4.0	-5.0641
4.5	-2.1278
5.0	1.0396
5.5	3.6766
6.0	5.2149
6.5	5.3946
7.0	4.2968
7.5	2.2941
8.0	-0.0632
8.5	-2.1918
9.0	-3.6150
9.5	-4.0640
10.0	-3.5221
10.5	-2.2061
11.0	-0.4937
11.5	1.1787
12.0	2.4239
12.5	2.9894
13.0	2.8048
13.5	1.9813
14.0	0.7693
14.5	-0.5126
15.0	-1.5583
15.5	-2.1449
16.0	-2.1768
16.5	-1.6970
17.0	-0.8640
17.5	0.0955
18.0	0.9455
18.5	1.4974
19.0	1.6492
19.5	1.4016
20.0	0.8489
20.5	0.1479
21.0	-0.5241
21.5	-1.0126
22.0.	-1.2200
22.5	-1.1235
23.0	-0.7730
23.5	-0.2736
24.0	0.2444
24.5	0.6583
25.0	0.8808

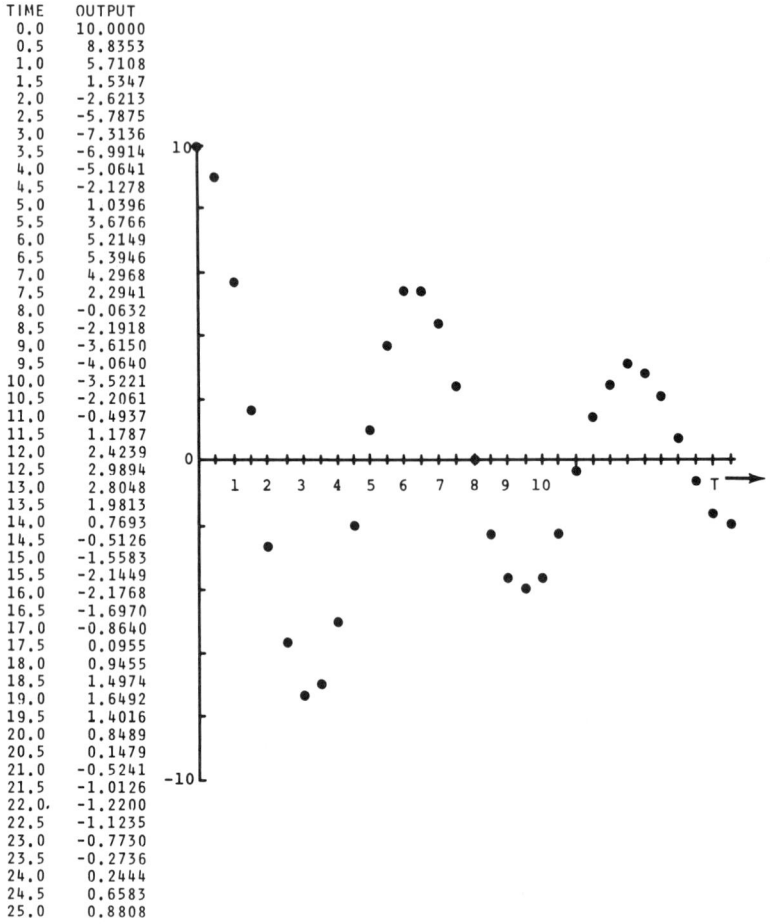

FIGURE 2.13 EXAMPLE 2.8 DIGITAL SIMULATION RESPONSE
OF THE SYSTEM $\ddot{x} + 0.2\dot{x} + x = 0$.

The program given in Example 2.8 calculates S1, P2, and P1 for the first time inside the DO loop. Therefore, at statement 100 these variables correspond to TIME = $(K-1)*$TAU, and I1 and I2 correspond to TIME = $K*$TAU. This causes no problems unless it is desired to output S1, P2 or P1. If this is desired we can take the alternative program

```
C
C    INITIAL STATEMENTS.
C
     I1=0.0
     I2=10.0
     S1=-I2
     P2=POT2*S1
     P1=POT1*I1
     DO 100 K=1,NSTEPS
C
C    REPRESENTATION STATEMENTS.
C
     TIME=K*TAU
     I2=INTGRL(I2,I1)
     I1=INTGRL(I1,P1+P2)
     S1=-I2
     P2=POT2*S1
     P1=POT1*I1
100  CONTINUE
```

This change requires the addition of initial state-ments for the variables other than integrators and the reordering of the representation statements. If you understand the equivalence of this and the previous program you can probably program any analog block diagram into a digital simulation without mixing up the order of the statements. (Good luck!)

A general procedure, which preserves proper order of the representation statements, is to draw the analog computer block diagram with all integrators in a single line and number them from left to right as in Example 2.8. Then the integration statements begin with the last integrator and proceed to integrator number 1. After all integrator outputs are evaluated, all other variables are updated for the next calculation. (See Problem 2.24)

The digital simulation program corresponds more closely to the analog computer functions with the addition of the function subprograms given in Figure 2.14. With the functions SUM, POT and INTGRL the program becomes a list of the wires patched onto the analog computer for analog simulation.

A program is presented in the appendix which includes an output subprogram, called PLOT, that gives a plot of desired output variables. The PLOT program determines the maximum and minimum values of the output variables and scales the output plot to fit on an 8 1/2 by 11 inch sheet. The PLOT program is discussed in the appendix and used hereafter in the text without further explanation.

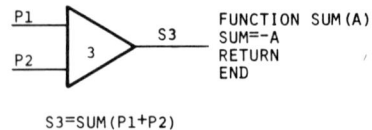

S3=SUM(P1+P2)

(A) SUMMER

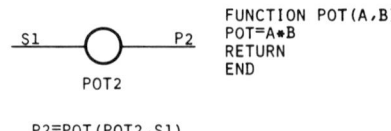

P2=POT(POT2,S1)

(B) POTENTIOMETER

FIGURE 2.14   ADDITIONAL FUNCTION SUBPROGRAMS FOR THE
DIGITAL-ANALOG SIMULATION LANGUAGE.

2.3 MODE CONTROL

Each analog computer has several knobs or buttons that determine the disposition of the computer. These controls have different names on different machines and even those machines having the same names for the controls do not necessarily operate identically. Thus it is imperative to study the instruction manual for the machine being used before beginning any computation or simulation activity. There are controls with names like STANDBY, POTSET and REPOP. Problem 2.12 requests that you go to your instruction manual and determine what the equivalent operations are on your computer, and how they affect its operation.

There are several operational modes for typical analog computers. In the *OPERATE* or *COMPUTE* mode the computer has all voltages applied and the solution to the simulated equations, with the applied initial conditions, is traced out by the voltages which represent the system variables. If these conditions and the patching of the program correspond to the system under study then this is the solution desired. Of course the system model, the block diagram and the patching must be correct to obtain correct results, but it is important to remember that the computer does not know if you have made a mistake. If you have made a mistake, you get the results to some other problem.

After the program, including initial conditions, is patched up on the patchboard, the POTs all set to their correct values, and the computer warmed up and ready for operation, the RESET button is depressed (or the mode control switch is placed in the RESET position). The *RESET* mode sets the initial conditions on all the integrators to whatever value is plugged into their initial condition jacks. This value is usually obtained from the reference voltage through a POT. The computer can remain in the RESET mode indefinitely but, depending on the machine, only a fraction of a millisecond (typically 100 μ sec) is required for the RESET operation. In RESET the output of the integrators should be equal to the desired initial conditions. They can be addressed through the switches provided to check their outputs. If all integrator outputs are correct you can address each summer and check whether the output is correct and corresponds to the constant inputs. This procedure is called a *STATIC CHECK* since all variable voltages remain constant and equal to their initial value in RESET. The static check frequently detects all patching errors.

After the initial conditions are set in RESET the program can be run by depressing the OPERATE button and changing the computer from RESET to OPERATE. If the variable you wish to observe is connected to a pen recorder or oscilloscope its output will display the solution to the equation.

Now suppose that you have run your program and the results are not what you expected. In that case, you guessed the results incorrectly, you made a mistake in the program, or something is wrong with the patchboard setup. When the error is corrected, run the program again. To do this, depress the RESET button, then the OPERATE button, and then observe the new solution. It is important that the RESET button be depressed for each new run to insure that you start off with the desired initial conditions.

After a successful run is obtained and recorded, any of the system initial conditions or parameter values represented by the various POTs can be changed and the corresponding system response observed. On the basis of these results an intuition is obtained into the action of the system under different conditions and the POTs can be changed to obtain the desired response. This close contact, with the programmer actually running the program, provides a "feel" for the system that is not obtained from large digital installations where the programmer does not operate the machine. With good estimates for parameter values the system can be digitally simulated to obtain more accurate results if necessary.

The remaining mode is the *HOLD* operation. When the computer is in HOLD the inputs to the integrators are set at zero. If an integrator has zero input then the output does not change, so in HOLD all signals remain constant. Putting the computer into HOLD will stop the operation without changing the solution variables and allow the operator to make measurements on the simulation. When the measurements are completed the COMPUTE button can be depressed to continue the solution. In general, the computer will hold values constant for a few minutes, but if you put the computer into HOLD and go to lunch do not expect the same values when you come back.

EXAMPLE 2.9

The block diagram shown in Figure 2.15(a) is

patched up on an analog computer and the computer then switched between COMPUTE and HOLD and RESET during the intervals shown. Sketch the integrator output x(t).

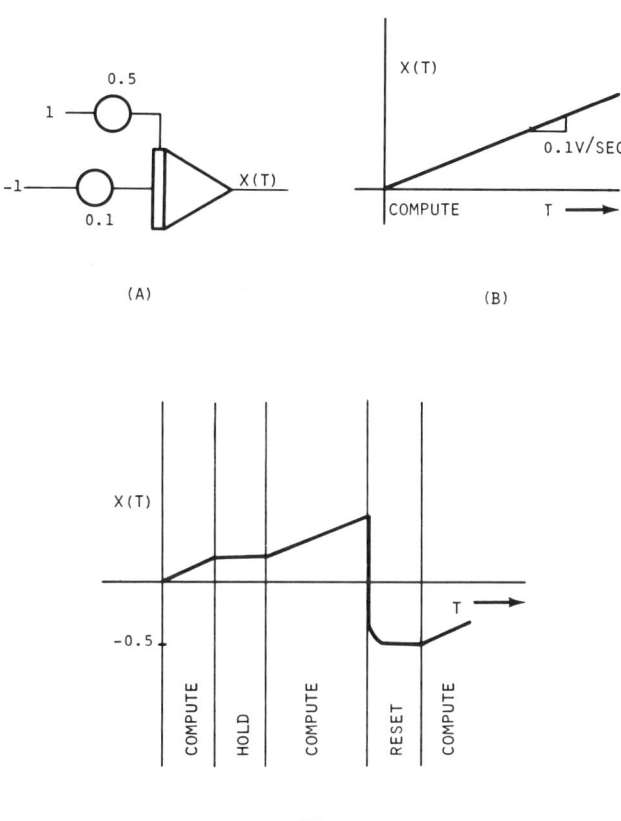

(A)   (B)

(C)

FIGURE 2.15 EXAMPLE 2.9 MODE CONTROL OF AN INTEGRATOR.

SOLUTION:

   Figure 2.15(b) shows the response for COMPUTE only. Figure 2.15(c) shows the response to a sequence of RESET, COMPUTE and HOLD. During RESET the integrator resets to the initial value x = 0.5, during COMPUTE, x(t) is the integral of the inverted input, and during HOLD the output x(t) remains constant.

If the system variables are plotted on an X-Y plotter with the horizontal input, or time base, generated on the computer going to HOLD also holds the time base and the pen does not move during the HOLD period. The resultant output waveform is the same as if the HOLD regions were cut out and the other sections taped together.▼

It is occasionally desirable to have a circuit that keeps the output constant while other variables change. This is accomplished by patching the integrators into the desired mode during each step of the simulation. The instruction manuals give patching instructions for this type of operation. In a typical simulation some integrators may be in HOLD while others are in COMPUTE and still others are in RESET.

Many computers have mechanical or electronic switches which will sequentially switch the mode control between two states. Thus the computer can automatically sequence, with the mode of each integrator at each stage controlled according to the patching of its mode control. This automatic switching of the modes is called *REPETITIVE OPERATION* or *REPOP*. Some computers do not have the REPOP operation and some have more than two states in REPOP.

NONLINEAR FUNCTION GENERATORS

The symbol and operating characteristics on analog computer multipliers is given in Section 1.2 and Figure 1.5 as a basic analog computer block diagram symbol. The shape of the multiplier symbol is used for all nonlinear functions and the internal symbol (X.for the multiplier) depicts the function. The procedure for patching the multipliers or any nonlinear function may differ with each particular computer and is outlined in the instruction manual for the computer. Therefore we use the standard symbol shown in Figure 2.16 for an arbitrary nonlinear function. When the function being generated can be represented by a standard mathematical symbol, the f

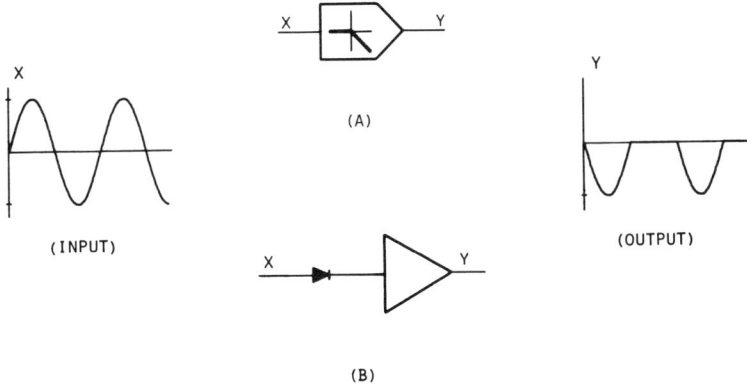

(A)

(INPUT)

(B)

(OUTPUT)

FIGURE 2.16   HALF WAVE RECTIFIERS FOR ANALOG COMPUTATION.

is replaced by the appropriate symbol, such as the square root, the multiplier and the divider as shown in Table 2.2. The nonlinear functions not available on the computer can be generated with diode function generators. The procedure for setting these elements

TABLE 2.2   NONLINEAR FUNCTION ELEMENTS FOR ANALOG COMPUTATION.

TYPICAL FUNCTION SYMBOLS F

SQUARE ROOT	$\sqrt{}$
MULTIPLIER	X
DIVIDER	÷
ABSOLUTE VALUE	⤓

to a desired function are also given in the computer instruction manual. Several problems at the end of this chapter deal with the generation of nonlinear elements using diodes and operational amplifiers.

There are many circuits that generate specific nonlinear elements and once the concept is understood most nonlinear functions can be constructed quite easily. Consider, for example, the half wave rectifier shown in Figure 2.16(a), which has an input-output relationship

$$y = \begin{cases} -X & \text{for } X > 0 \\ 0 & \text{for } X < 0 \end{cases}$$

A diode conducts current in only one direction, so when it is biased in one direction it acts like a short circuit. When biased in the other direction, it acts like an open circuit. Thus inserting a diode in the signal path of an amplifier allows signal flow in only one direction. Figure 2.16(b) shows this circuit corresponding to the element depicted in Figure 1.16(a). The input x gives the rectified output y as shown in the figure.

In practice it is better to place the diode in the feedback path of the operational amplifier when generating nonlinear functions. A comparison of several nonlinear functions is left to the problems.

SIMULATION OF NONLINEAR EQUATIONS

The concepts developed in the last section for the analog computer simulation of systems described by linear differential equations is easily extended to the solution of linear differential equations with time varying coefficients and nonlinear differential equations. Since a multiplier is available as a computing element, systems including multipliers are no more difficult to amplitude scale. If each input to a multiplier is scaled correctly then the output is scaled correctly.

The simulation of systems described by linear ordinary differential equations is useful in obtaining a knowledge of the operation and application of analog and digital computers. However, linear ordinary differential equations can be solved analytically to obtain the exact response of linear systems. It is generally difficult to determine analytically the response of nonlinear systems or of linear systems with time varying coefficients. Thus the application of analog and digital computational techniques is generally the only way to obtain the response of these systems.

Almost every book on analog computer techniques and nonlinear systems has a discussion of Equation (2.1), the van der Pol equation.

$$\ddot{y} = \varepsilon(1-y^2)\dot{y} + y = 0 \qquad (2.1)$$

Naming a nonlinear equation after the first man to have studied it emphasizes the difficulty of finding the solution. In fact, there are so few nonlinear differential equations with known solutions that they are generally referred to by the name of the first man to study them and not by the equation itself. For example, we have the Duffing equation, the Bernoulli equation, the Mathieu equation, and the van der Pol equation.

The van der Pol equation is a nonlinear ordinary differential equation that can quite easily be simulated on an analog computer. Following the procedures for the analog computer solution of differential equations, we solve the van der Pol equation for the highest derivative $\ddot{y}$.

$$\ddot{y} = \varepsilon(1-y^2)\dot{y} + y$$

Now, assuming that we have $\ddot{y}$, we pass the signal through an integrator to obtain $-\dot{y}$ which is in turn passed through a second integrator to obtain the dependent variable y as shown in Figure 2.17. To obtain the first term on the right side of the equation for $\ddot{y}$ we need the term y^2, which can be generated by connecting the signal y into both inputs of a multiplier resulting in the output y^2. If this y^2 signal is added in a summer to the constant -1 we obtain $1 - y^2$ which is fed into another multiplier to be multiplied by the signal $-\dot{y}$ which is in turn available out of the first integrator. The output of this multiplier $-(1-y^2)\dot{y}$ can be passed through a POT of value ε and an inverter to obtain the first term of the right hand side of the equation for $\ddot{y}$. The second term $-y$ is obtained by passing the signal y through an inverter into the input of the first integrator. This completes the analog computer block diagram for the solution of the van der Pol equation. The completed analog computer diagram is shown in Figure 2.17.

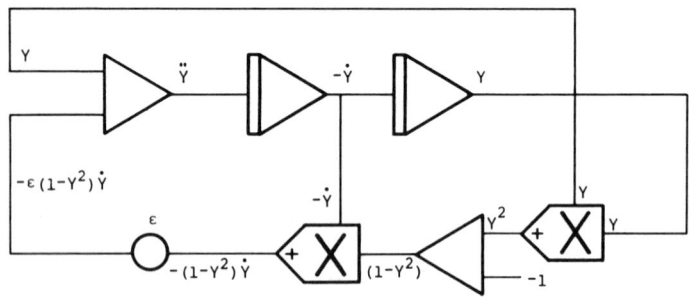

FIGURE 2.17 ANALOG COMPUTER SOLUTION OF THE VAN DER POL EQUATION.

Clearly the construction of the block diagram for the analog computer solution of differential equations with multipliers is essentially the same as presented in the last section.

Before obtaining the solution to the van der Pol equation on the analog or digital computer we try to estimate the character of its solution and the magnitude of the dependent variable y and its derivative from the equation. First, when y is small the equation is approximately

$$\ddot{y} + \varepsilon k \dot{y} + y = 0$$

which is a linear differential equation approximating the van der Pol equation for small y. The corresponding characteristic equation has roots with

positive real part for positive ε. Thus the response
of the system has the form of increasing exponentials
or exponentially increasing sinusoids. So if the
initial value of y is smaller than 1 the solution to
the van der Pol equation increases with time. Of
course, as y gets larger this linear approximation
loses its validity.

Now if y^2 is larger than 1 we have the approximate
equation

$$\ddot{y} + \varepsilon y^2 \dot{y} + y = 0$$

If we imagine momentarily that the y^2 term is constant
then we have the resultant linear approximation to the
van der Pol equation

$$\ddot{y} + \varepsilon k \dot{y} + y = 0$$

where y^2 is considered to be the positive constant k.
Now the characteristic equation of this linear
approximation has roots with negative real parts.
Therefore, the natural response of this system is
decreasing exponentials or exponentially decaying
sinusoids. Thus it appears that for large y the
solution of the van der Pol equation would tend to go
toward zero.

If for small y the solution to the van der Pol
equation increases and for large y the solution
decreases, then perhaps the solution is oscillatory in
nature, tending to increase when y < 1 and decrease
when y > 1. Therefore we might expect the maximum
magnitude of y to be in the order of magnitude of 1
and somewhat larger than 1, say 5 or 10.

EXAMPLE 2.10

Obtain the solution of the van der Pol equation
with the initial conditions $\dot{y}(0) = 0$, $y(0) = 1$, and
$\varepsilon = 5$ by the digital-analog simulation using
rectangular integration.

SOLUTION:

Figure 2.18 shows the block diagram for the van der Pol equation with the variable names for the digital analog-simulation. The simulation program outputs are presented after the program. (Check the representation statements of this program to be sure each variable is updated and in admissable order and check subroutine PLOT in the appendix for making time series plots.)

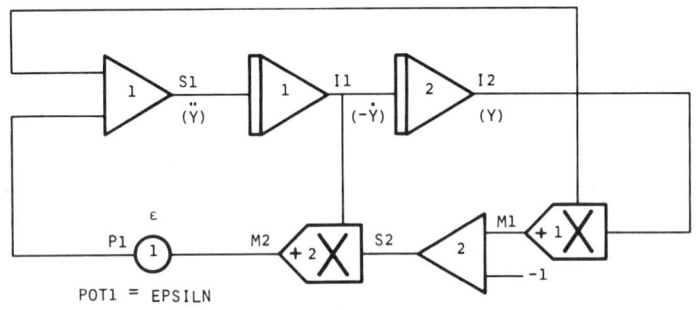

FIGURE 2.18 EXAMPLE 2.1 DIGITAL-ANALOG SIMULATION OF THE VAN DER POL EQUATION.

2.4 TIME AND AMPLITUDE SCALING

Perhaps the most important concept involved in analog computation is that of scaling the problem variables to the simulation variables which cannot exceed the reference voltage. Of course, the simulation variables must not be too small or excessive errors result. This problem does not exist on the digital computer since gains of 10^{-5}, 10^8 or almost any gain can appear in the same program. For this reason those who learn digital simulation first are generally reluctant to learn analog simulation because scaling seems difficult at first exposure. But almost everything useful and new is difficult on first contact.

```
      REAL I1,I2,INTGRL,M1,M2
      DIMENSION TIMEO(101),PLOTI1(101),PLOTI2(101)
      COMMON TAU
C
C     INITIALIZATION STATEMENTS.
C
      TIME=0.0
      FINTIM=15.0
      TAU=0.01
      NUMOUT=50
      NSTEPS=FINTIM/TAU
      OUTDEL=FINTIM/NUMOUT
      TIMOUT=OUTDEL-0.5*TAU
      EPSILN=2.0
C
C     SET POTS, INITIAL CONDITIONS AND PLOT ARRAYS.
C
      POT1=EPSILN
      I1=-1.0
      I2=0.0
      NN=1
      TIMEO(1)=TIME
      PLOTI1(1)=-I1
      PLOTI2(1)=I2
C
C     SIMULATION LOOP...TO STATEMENT # 100.
C
      DO 100 K=1,NSTEPS
      M1=I2*I2
      S2=-(M1-1.)
      M2=I1*S2
      P1=POT1*M2
      S1=-(P1+I2)
      TIME=K*TAU
      I2=INTGRL(I2,I1)
      I1=INTGRL(I1,S1)
C
C     IS IT TIME TO STORE OUTPUT VALUES?
C
      IF(TIME.LT.TIMOUT) GO TO 100
      NN=NN+1
      TIMEO(NN)=TIME
      PLOTI1(NN)=-I1
      PLOTI2(NN)=I2
      TIMOUT=NN*OUTDEL-0.5*TAU
      IF(NN.GE.(NUMOUT+1)) GO TO 999
  100 CONTINUE
  999 CALL PLOT(TIMEO,PLOTI2,NN)
      CALL PLOT(TIMEO,PLOTI1,NN)
      STOP
      END
```

```
MAXIMUM =    2.041566
MINIMUM =   -2.042158

                               (OUTPUT SCALED BY: 1.0E 00)
                                  <===== - 0 + =====>
                     -2.5  -2.0  -1.5  -1.0  -0.5   0.0   0.5   1.0   1.5   2.0   2.5
     TIME    OUTPUT  +----+----+----+----+----+----+----+----+----+----+----+
    0.0      0.0     |    |    |    |    |    +    |    |    |    |    |    |
    0.300    0.398   |    |    |    |    |    |****+|    |    |    |    |    |
    0.600    0.991   |    |    |    |    |    |***********+   |    |    |    |
    0.900    1.509   |    |    |    |    |    |****************+    |    |    |
    1.20     1.683   |    |    |    |    |    |*******************+ |    |    |
    1.50     1.650   |    |    |    |    |    |******************+  |    |    |
    1.80     1.544   |    |    |    |    |    |*****************+   |    |    |
    2.10     1.402   |    |    |    |    |    |****************+|   |    |    |
    2.40     1.224   |    |    |    |    |    |**************+  |   |    |    |
    2.70     0.995   |    |    |    |    |    |***********+    |    |    |    |
    3.00     0.672   |    |    |    |    |    |*******+   |    |    |    |    |
    3.30     0.152   |    |    |    |    |    |*+   |    |    |    |    |    |
    3.60    -0.760   |    |    |    |    +********|    |    |    |    |    |
    3.90    -1.764   |    |   +**************|    |    |    |    |    |    |
    4.20    -2.039   |   +**********************|    |    |    |    |    |
    4.50    -1.999   |    +*********************|    |    |    |    |    |
    4.80    -1.907   |    |+********************|    |    |    |    |    |
    5.10    -1.801   |    | +*******************|    |    |    |    |    |
    5.40    -1.683   |    |   +*****************|    |    |    |    |    |
    5.70    -1.551   |    |    +****************|    |    |    |    |    |
    6.00    -1.398   |    |    |+***************|    |    |    |    |    |
    6.30    -1.215   |    |    |  +*************|    |    |    |    |    |
    6.60    -0.979   |    |    |    +***********|    |    |    |    |    |
    6.90    -0.646   |    |    |    |  +*******|    |    |    |    |    |
    7.20    -0.105   |    |    |    |    |   +|    |    |    |    |    |
    7.50     0.839   |    |    |    |    |****+*****+   |    |    |    |    |
    7.80     1.808   |    |    |    |    |**********************+ |    |    |
    8.10     2.042   |    |    |    |    |************************+   |    |
    8.40     1.994   |    |    |    |    |***********************+    |    |
    8.70     1.901   |    |    |    |    |**********************+|    |    |
    9.00     1.794   |    |    |    |    |*********************+ |    |    |
    9.30     1.675   |    |    |    |    |******************+ |    |    |
    9.60     1.542   |    |    |    |    |*****************+   |    |    |
    9.90     1.388   |    |    |    |    |****************+|   |    |    |
   10.2      1.202   |    |    |    |    |**************+  |   |    |    |
   10.5      0.962   |    |    |    |    |***********+    |    |    |    |
   10.8      0.621   |    |    |    |    |******+    |    |    |    |    |
   11.1      0.061   |    |    |    |    |+   |    |    |    |    |    |
   11.4     -0.909   |    |    |    |+**********|    |    |    |    |    |
   11.7     -1.842   |    | +**************|    |    |    |    |    |    |
   12.0     -2.042   +***********************|    |    |    |    |    |
   12.3     -1.989   |+***********************|    |    |    |    |    |
   12.6     -1.895   ||+*********************|    |    |    |    |    |
   12.9     -1.787   |    +******************|    |    |    |    |    |
   13.2     -1.668   |    |+*****************|    |    |    |    |    |
   13.5     -1.533   |    |   +**************|    |    |    |    |    |
   13.8     -1.378   |    |   |+*************|    |    |    |    |    |
   14.1     -1.189   |    |    |  +**********|    |    |    |    |    |
   14.4     -0.945   |    |    |    |+*********|    |    |    |    |    |
   14.7     -0.595   |    |    |    | .  +******|    |    |    |    |    |
   15.0     -0.015   |    |    |    |    |    +    |    |    |    |    |    |
```

```
MAXIMUM =    3.820433
MINIMUM =   -3.842517

                              (OUTPUT SCALED BY: 1.0E 00)
                                  <===== - 0 + =====>
                    -5.0  -4.0  -3.0  -2.0  -1.0   0.0   1.0   2.0   3.0   4.0   5.0
   TIME   OUTPUT    +-----+-----+-----+-----+-----+-----+-----+-----+-----+-----+
   0.0     1.000    |     |     |     |     |     |*****+     |     |     |     |
   0.300   1.704    |     |     |     |     |     |*********+  |.   |     |     |
   0.600   2.098    |     |     |     |     |     |*************+   |     |     |
   0.900   1.135    |     |     |     |     |     |******+   |     |     |     |
   1.20    0.114    |     |     |     |     |     |+    |     |     |     |     |
   1.50   -0.273    |     |     |     |     |   +*|     |     |     |     |     |
   1.80   -0.423    |     |     |     |     |  +**|     |     |     |     |     |
   2.10   -0.531    |     |     |     |     |  +**|     |     |     |     |     |
   2.40   -0.666    |     |     |     |     | +***|     |     |     |     |     |
   2.70   -0.890    |     |     |     |    |+****|     |     |     |     |     |
   3.00   -1.330    |     |     |     |  +*******|     |     |     |     |     |
   3.30   -2.297    |     |     |  +***************|     |     |     |     |     |
   3.60   -3.770    |+************************|     |     |     |     |     |
   3.90   -2.107    |     |     |  +**************|     |     |     |     |     |
   4.20   -0.095    |     |     |     |     |   +|     |     |     |     |     |
   4.50    0.266    |     |     |     |     |     |*+   |     |     |     |     |
   4.80    0.335    |     |     |     |     |     |*+   |     |     |     |     |
   5.10    0.373    |     |     |     |     |     |*+   |     |     |     |     |
   5.40    0.415    |     |     |     |     |     |**+  |     |     |     |     |
   5.70    0.472    |     |     |     |     |     |**+  |     |     |     |     |
   6.00    0.554    |     |     |     |     |     |***+ |     |     |     |     |
   6.30    0.685    |     |     |     |     |     |****+|     |     |     |     |
   6.60    0.915    |     |     |     |     |     |*****+|     |     |     |     |
   6.90    1.377    |     |     |     |     |     |*******+    |     |     |     |
   7.20    2.397    |     |     |     |     |     |**************+    |     |     |
   7.50    3.820    |     |     |     |     |     |************************+
   7.80    1.887    |     |     |     |     |     |***********+|     |     |     |
   8.10    0.042    |     |     |     |     |     +     |     |     |     |     |
   8.40   -0.274    |     |     |     |     |   +*|     |     |     |     |     |
   8.70   -0.338    |     |     |     |     |   +*|     |     |     |     |     |
   9.00   -0.376    |     |     |     |     |   +*|     |     |     |     |     |
   9.30   -0.418    |     |     |     |     |  +**|     |     |     |     |     |
   9.60   -0.476    |     |     |     |     |  +**|     |     |     |     |     |
   9.90   -0.561    |     |     |     |     |  +**|     |     |     |     |     |
  10.2    -0.695    |     |     |     |     | +***|     |     |     |     |     |
  10.5    -0.934    |     |     |     |    |+*****|     |     |     |     |     |
  10.8    -1.419    |     |     |     |  +********|     |     |     |     |     |
  11.1    -2.486    |     |     |  +***************|     |     |     |     |     |
  11.4    -3.843    |+************************|     |     |     |     |     |
  11.7    -1.698    |     |     |   +**********|     |     |     |     |     |
  12.0    -0.001    |     |     |     |     |     +     |     |     |     |     |
  12.3     0.281    |     |     |     |     |     |*+   |     |     |     |     |
  12.6     0.340    |     |     |     |     |     |x*+  |     |     |     |     |
  12.9     0.378    |     |     |     |     |     |**+  |     |     |     |     |
  13.2     0.421    |     |     |     |     |     |**+  |     |     |     |     |
  13.5     0.480    |     |     |     |     |     |**+  |     |     |     |     |
  13.8     0.567    |     |     |     |     |     |***+ |     |     |     |     |
  14.1     0.706    |     |     |     |     |     |***+ |     |     |     |     |
  14.4     0.955    |     |     |     |     |     |*****+|     |     |     |     |
  14.7     1.463    |     |     |     |     |     |********+   |     |     |     |
  15.0     2.579    |     |     |     |     |     |***************+    |     |     |
                    |_____|_____|_____|_____|_____|_____|_____|_____|_____|_____|
```

The thought processes required for time and amplitude scaling of an analog simulation are valuable tools and give an insight into system behavior even if a simulation is not planned. The block diagram manipulation involved shows the relative effects of each term of a differential equation and indicates when some terms may be negligible. Furthermore, if a problem is posed incorrectly scaling may give an indication of the error.

TIME SCALING: If the time constant of a chemical reaction being simulated is in the order of milliseconds and the computer simulation output is to be plotted on a mechanical X-Y plotter, the time of the system reaction and the equivalent simulation must be different. Likewise, if the reaction takes a few days, the simulation time ought to be shorter than the reaction time.

To time scale a simulation, change the gain of all the integrators by the same factor. If all integrator gains are increased by a factor β then the simulation will run β times as fast. If $\beta > 1$ the simulation runs faster and if $\beta < 1$ the simulation runs slower. That's all there is to it. On the computer there is usually a switch that automatically changes the gain of all integrators by a factor of 10 (or some other value) thereby accomplishing time scaling of a simulation that is already patched up. If the computer can time scale with a single switch it must be easy. When changing the time scale only the integrator gains are changed. POT settings, multipliers, and initial conditions all remain the same.

EXAMPLE 2.11

Use a constant input to an integrator to demonstrate that an increased gain effectively increases the speed of the computation.

SOLUTION:

Figure 2.19 shows an integrator with a constant input of −1 and the corresponding response. Note that the output waveform goes from 0 to 5 in 5 seconds with an integrator gain of unity. With a gain of 5 the output reaches 5 in 1 second, so the system with the

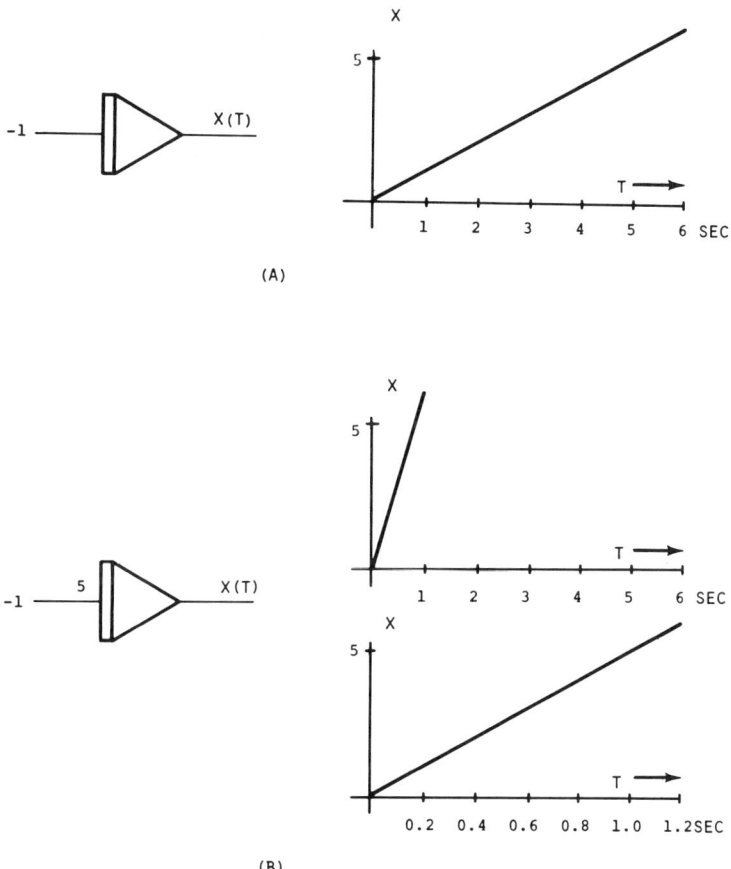

(A)

(B)

FIGURE 2.19 EXAMPLE 2.11 TIME SCALING BY CHANGING THE GAIN OF THE INTEGRATOR.

increased gain on the integrator is 5 times as fast. The solutions for the two systems are identical if drawn to time scales which differ by a factor of 5. ▼

When we say that a simulation is time scaled so it runs faster we mean that it takes less time for the simulation variables to trace out the solution. If the time base for output plots is generated by integrating a constant, the plots will be invariant with time scale provided that the time base integrator is scaled along with the rest of the system. This is demonstrated in the example below.

EXAMPLE 2.12

Give the block diagram for a sinewave generator and obtain plots of the output for several time scale factors.

SOLUTION:

Figure 2.20 gives the response for various time scale factors k. ▼

Note that in the above example the time base integrator is *NOT* scaled. If the time base integrator is also scaled by the gain factor k, all the plots are identical. Of course, there are limitations on how fast the mechanical plotter can move accurately, so if the gains are very large resulting in high speeds neither the computer nor the plotter can follow. But this is precisely why we want to time scale. When we have gains that are too large, like 10^4 or 10^{10}, we just time scale to get all integrator gains down to about 1 or 10.

While scaling a simulation it is convenient to draw an integrator, time scaled along with the rest, off to the side to keep track of the time scale factor since it frequently happens that in the process of scaling we might time scale several times.

The above examples demonstrate that time scaling works at least in those cases. In general, the input to each integrator in a simulation is a function of

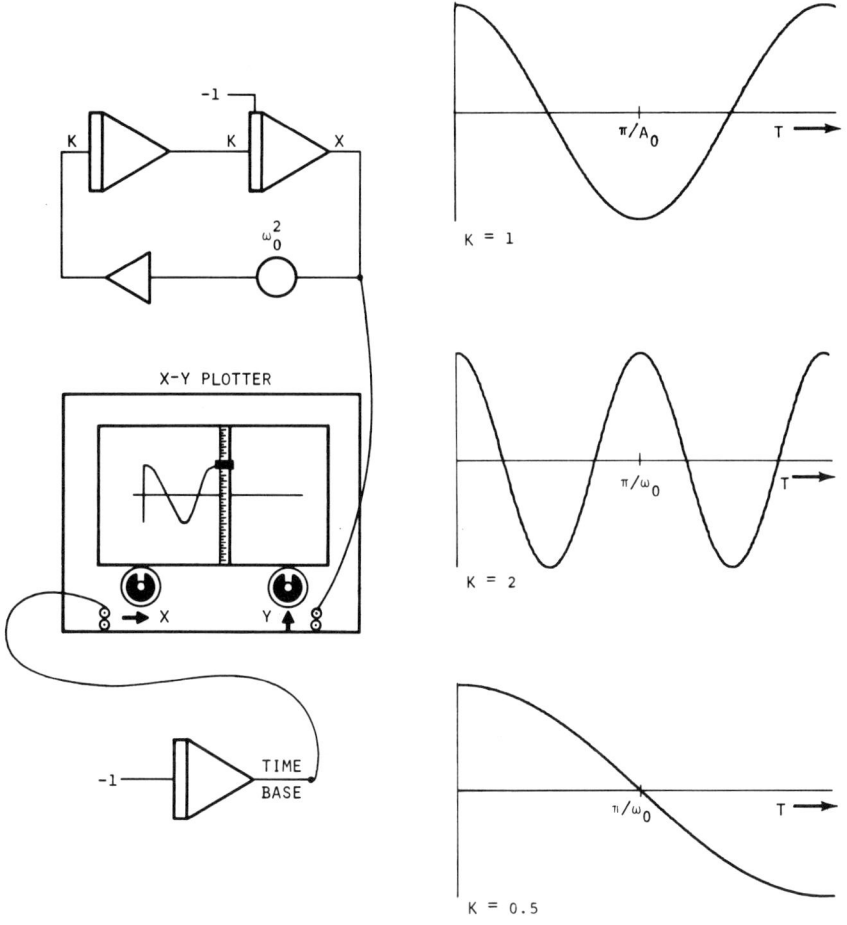

FIGURE 2.20 EXAMPLE 2.12 TIME SCALING BY CHANGING
THE GAIN OF ALL SYSTEM INTEGRATORS.

the other variables in the diagram.

Suppose the output of an integrator is y. Then the integrator input $\dot{y} = dy/dt$ is some combination of the other variables including perhaps y. We write the equation for one integrator as

$$\dot{y} = \frac{dy}{dt} = f$$

where f is a function of all the signals on the diagram. To time scale, we change the problem time t to computer time τ by some scale factor, say $\tau = \beta t$ or $t = \tau/\beta$, then

$$\frac{d}{dt} = \beta \frac{d}{d\tau}$$

and

$$\frac{dy}{dt} = f$$

becomes

$$\beta \frac{dy}{d\tau} = f$$

or

$$\frac{dy}{d\tau} = \frac{1}{\beta} f$$

So the new input to the integrator is $1/\beta$ times what it was before and we have time scaled that integrator by β. Of course this is true of all integrators, so to time scale a simulation we just change the gain of all the integrators in the simulation. If $\beta = 10$, then $t = 1$ seconds corresponds to 10 seconds of computer time and we make the simulation run slower by a factor of 0.1 by changing the integrator gains by $1/\beta = 0.1$.

If f is also a function of time, as in $f = y + \cos t$, then that t must also be scaled by $t = (1/\beta)\tau$ to obtain

$$\frac{dy}{d\tau} = \frac{1}{\beta} f\ (\ldots,\ \frac{1}{\beta}) = \frac{1}{\beta} [y + \cos \frac{1}{\beta} \tau]$$

This is equivalent to saying that if the simulation runs β times faster, then you must run the excitation β times faster also. For example, if the equation

$$\ddot{y} + 2\dot{y} + y^2 y = \cos 5t$$

is time scaled to run 10 times as fast you must apply
the excitation cos 50τ, where τ is the computer time
variable.

AMPLITUDE SCALING

The concept involved in *AMPLITUDE SCALING* is the
straightforward application of the properties of
linear computing elements and the desire to maintain
the simulation variables below the reference voltage.
The system variables are related to the simulation
variables by *SCALE FACTORS* which depend on the
reference voltage and the maximum values of the system
variables. It seems that we must know the character
of the system variables to determine their maximum
values, although if we knew how the system variables
changed we would not have to simulate the systems.
Well, this is not quite true because all we really
need is a guess of the maximum values from our limited
knowledge of the system and estimates from the system
equations. Usually, reasonable estimates for the
maximum values can be obtained by examining the
equations and the initial conditions. We know that
each variable has a maximum at least as large as its
initial value. Furthermore, a digital simulation
using rectangular integration gives a quick and
inexpensive estimate of the response that can be used
to scale the analog system. With estimates for the
maximum values of the system variables we can get down
to the business of amplitude scaling the equations.

It is desirable to have all variables extend over
the total range of operation of the computer, so for
the simulation we scale the signal magnitudes to the
reference voltage. To do this we can scale to a
maximum signal level assuming a reference of ± 1 for
all variables and obtain a block diagram that is
appropriate for patching on any analog machine.

Linear systems have the property of homogeneity:
If all inputs to a linear system are multiplied by a
factor α. then every variable of the system response
including the outputs is also multiplied by that
factor α. The waveshapes all remain the same and
their magnitudes are scaled by the factor α. In
particular, if we scale linear equations for a maximum

of ±1 on all variables, patch up the diagram, and multiply all inputs including initial conditions by REF, we get all variables multiplied by REF. Remember, *ALL* inputs must be multiplied by REF. But inputs and initial conditions can be obtained on the scaled diagram from the assumed reference of ±1 and then patched up to ±REF on the computer. Scaling for REF of 1 is called scaling to *MACHINE UNITS*. We see below that scaling to machine units is also valid for systems containing multipliers.

The scaling procedure is best presented by application to a specific equation. Suppose we wish to obtain a solution to the equation

$$\ddot{x} + 5\dot{x} + 20x = 15 \qquad\qquad (2.2)$$

with initial conditions $x(0) = 1$ and $\dot{x}(0) = 5$. Suppose also that we estimate the maximum values of x and $\dot{x}$ as $x_{max} = 5$ and $\dot{x}_{max} = 10$. These are "nice" numbers larger than the initial conditions. It is convenient to scale variables by numbers like 1, 2, 5 or powers of 10 of these.

To scale to machine units we must have the signals on the computer remain between −1 and +1. But we also want the signals to be as large as possible to minimize the errors and noise effects. So we scale the equations to obtain machine or computer variables that have a maximum of unity. Thus instead of the problem variables x and $\dot{x}$ on the block diagram we want the *COMPUTER VARIABLES* $[x/x_{max}] = [x/5]$ and $[\dot{x}/x_{max}] = [\dot{x}/10]$.

Now we rewrite Equation (2.2) as a function of the machine variables, $[\dot{x}/10]$ and $[x/5]$ as

$$\ddot{x} + (5)10\left[\frac{\dot{x}}{10}\right] + (20)5\left[\frac{x}{5}\right] = 15 \qquad (2.3)$$

Since $\ddot{x}$ is not going to appear as a signal on the diagram, it need not be scaled.

Solving for $\ddot{x}$ we have

$$\ddot{x} = -50\left[\frac{\dot{x}}{10}\right] - 100\left[\frac{x}{5}\right] + 15$$

Now we divide through the equation by an appropriate factor to give gains in the range 0.1.–.10 as reasonable POT settings and integrator gains. In this case we can divide by 100 to obtain

$$\frac{\ddot{x}}{100} = -.5\left[\frac{\dot{x}}{10}\right] - \left[\frac{x}{5}\right] + .15 \qquad (2.4)$$

which has coefficients on the right of .5, 1 and .15 – "nice" numbers for the computer.

While keeping the variables and their scale factors together and drawing the block diagram, we assume we have [$\ddot{x}$/100] which can be integrated once to obtain the next lower derivative signal, which is [$\dot{x}$/10]. However we also need a gain of 10 to get this signal. This additional gain in each integrator is the only change from the previous block diagrams. Proceeding through another integrator with a gain of 2, we obtain the block diagram shown in Figure 2.21.

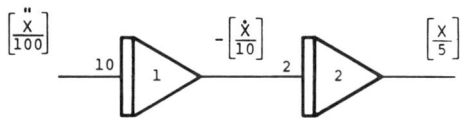

FIGURE 2.21 MAGNITUDE SCALED VARIABLES ON ANALOG BLOCK DIAGRAM.

Note that the lines on the diagram are labeled with the computer variables. It is important to keep these labels on the diagram and to keep them consistent with any changes you make in order to keep track of the relationship between the computer variables and the problem variables. The notation from the original equations should be maintained in the computer variable. That is, on the diagram the output of the integrator is [x/5], not a new variable such as y = [x/5], since for this new name we would have to keep track of the way computer variables related to the problem variables.

With the signals from Figure 2.21 scaled to the range of between −1 and +1, we can finish the block diagram corresponding to Equation (2.4) as shown in Figure 2.22.

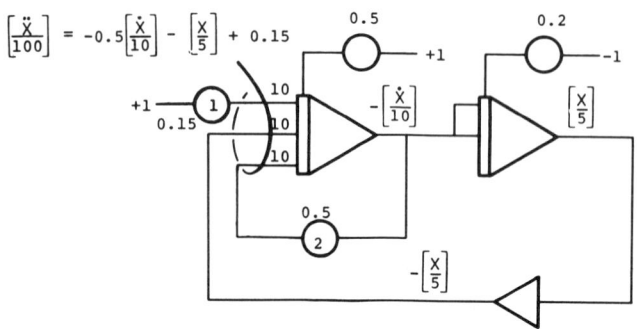

$$\left[\frac{\ddot{X}}{100}\right] = -0.5\left[\frac{\dot{X}}{10}\right] - \left[\frac{X}{5}\right] + 0.15$$

(A) AMPLITUDE SCALED VARIABLES

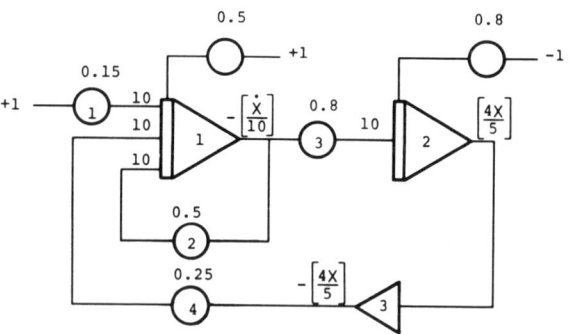

(B) SCALED BLOCK DIAGRAM

FIGURE 2.22   SCALED BLOCK DIAGRAMS FOR THE EQUATION
$\ddot{X} + 5\dot{X} + 20X = 15,\ X(0) = 1,\ \dot{X}(0) = 5.$

The gain of 10 on integrator 1 must be included on all inputs to the integrator. Also the gain of 2 on integrator 2 can be obtained by patching in two unity gains.

The initial conditions are added as before. Since $x(0) = 1$ the output of integrator 2 must start off at $[x(0)/5] = [1/5] = 0.2$ so we patch -0.2 into the IC jack. Similarly we need $+0.5$ into the IC jack of integrator 1. (Verify this.) Also the diagram can be simplified somewhat by combining some gains and POTs. (See if you can do it.)

When we patch up the block diagram we connect the reference voltage .REF to each point marked ± 1 on the diagram. Since the system is linear each of the signals is then multiplied by REF, and the output of integrator 1 is $[\dot{x}/10]$REF, the output of inverter 3 is $-[x/5]$REF, and so on.

Now suppose we press the RESET button to set the initial conditions, then press the OPERATE button and nothing overloads, but when we look at $[x(t)/5]$ on the meter, or plotter, we find that it has a maximum magnitude of about 25 volts and we are using a 100V computer. This means that our estimate of x was about 4 times too large. A maximum of 25 percent full scale is acceptable but we scale it to 100V to demonstrate the rescaling procedure. One way to rescale is to go back to the original equation and repeat the scaling with the newly obtained maximum

$$\left[\frac{x_{max}}{5}\right]\text{REF} = \left[\frac{x_{max}}{5}\right]100 = 25, \quad \text{or} \quad x_{max} = 1.25$$

the rescaled equation becomes

$$\ddot{x} + 50\left[\frac{\dot{x}}{10}\right] + \frac{100}{4}\left[\frac{4x}{5}\right] = 15$$

or

$$\ddot{x} + 50\left[\frac{\dot{x}}{10}\right] + 25\left[\frac{4x}{5}\right] = 15 \qquad (2.5)$$

Compare Equation (2.5) with the original scaled Equation (2.3). This rescaling is equivalent to multiplying the input to integrator 2 by 4 (a total input gain of 8) to obtain 4 times the previous output, or $[4x/5]$. This is obtained by inserting POT3 as shown in Figure 2.22(b). To have the same equation represented we must also divide the output somewhere by 4. This can be accomplished by inserting POT4. The resultant diagram is shown in Figure 2.22(b). Note that the labels and initial conditions must also be changed.

Example 2.13 demonstrates the application of time and amplitude scaling in the same problem.

EXAMPLE 2.13

Give an analog computer block diagram scaled to machine units for the solution of the differential equation

$$10^{-3}\ddot{y} + 10^{2}\dot{y} + 5 \times 10^{6}y = 90$$

$$y(0) = 5 \times 10^{-5} \qquad\qquad (2.6)$$

$$\dot{y}(0) = 0$$

SOLUTION:

The steady state solution (when $\dot{y}$ and $\ddot{y}$ are both zero) gives

$$5 \times 10^{9}y = 9 \times 10^{4}$$

or $y = 9/5 \times 10^{-5}$ and $y(0) = 5 \times 10^{-5}$ so we can guess that $y_{max} = 10^{-4}$ (twice the initial value).

Multiplying Equation (2.6) through by 10^{3} gives

$$\ddot{y} + 10^{5}\dot{y} + 5 \times 10^{9}y = 9 \times 10^{4} \qquad (2.7)$$

If $10^{5}\dot{y}$ and $5 \times 10^{9}y$ are about the same order of magnitude we can guess that

$$10^{5}\dot{y}_{max} \approx 5 \times 10^{9}y_{max}$$

which is the same as assuming that the gains of the two feedback lines are about the same. That is, both terms contribute significantly to the equation and neither is negligible.

So we can take $\dot{y}_{max} = 5 \times 10^4 y_{max} = 5$, or 10 to be on the safe side. We will probably have to rescale after running the problem anyway.

Rewriting Equation (2.7) in terms of the unity magnitude computer variables $[y/10^{-4}]$ and $[\dot{y}/10]$ we have

$$\ddot{y} + 10^5 \times 10 \left[\frac{\dot{y}}{10}\right] + 5 \times 10^9 \times 10^{-4} \left[\frac{y}{10^{-4}}\right] = 9 \times 10^4$$

or

$$\ddot{y} + 10^6 \left[\frac{\dot{y}}{10}\right] + 5 \times 10^5 \left[\frac{y}{10^{-4}}\right] = 9 \times 10^4 \qquad (2.8)$$

Divide through by a factor to give "nice"coefficients, 10^6 for instance.

$$\left[\frac{\dot{y}}{10^6}\right] + \left[\frac{\dot{y}}{10}\right] + 0.5 \left[\frac{y}{10^{-4}}\right] = 0.09 \qquad (2.9)$$

Assume we have $[\ddot{y}/10^6]$. Integrate with a gain of 10^5 to get $-[\dot{y}/10]$ which is integrated with a gain of 10^5 to get $[y/10^{-4}]$. Then connect the feedback consistent with Equation (2.9). The resultant diagram is given in Figure 2.23(a).

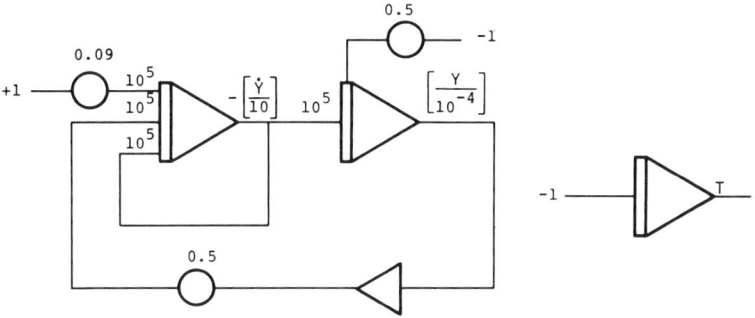

(A) AMPLITUDE SCALED VARIABLES

FIGURE 2.23 EXAMPLE 2.13 AMPLITUDE SCALING.

The initial condition $y(0) = 5 \times 10^{-5}$ gives an initial output for integrator 2 of $[5\times10^{-5}/10^{-4}] = 0.5$ as shown on the diagram.

Finally we can time scale by a factor 10^{-5} to get rid of the 10^5 integrator gains. Changing all of the gains on the integrators by 10^{-5} gives the diagram shown in Figure 2.23(b).

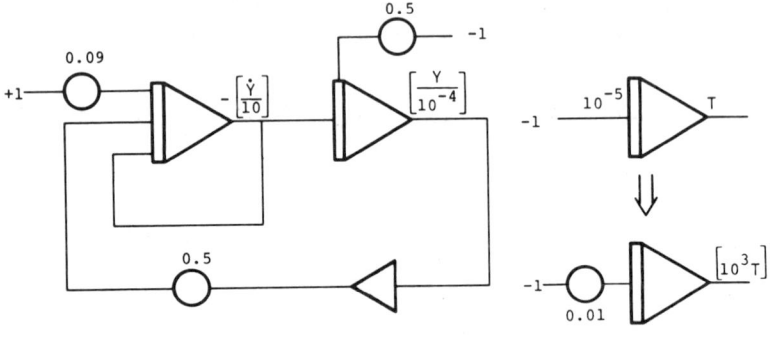

(B) SCALED BLOCK DIAGRAM

FIGURE 2.23 EXAMPLE 2.13

In the example above the time base generator ends up with a gain of 10^{-5} which cannot be realized on the computer, but it reminds us that we have scaled by 10^{-5}, so the simulation will run slower than the given system by a factor 10^{-5}. If the time base integrator is now magnitude scaled to have a reasonable gain of, say, 0.01, its output is $[10^3 t]$ which would overload in 100 seconds on the computer. This indicates that a reasonable running time for the simulation is 10 - 100 seconds of computer time up to $10^3 t = 1$ or $t = 10^{-3}$ seconds = 1 millisecond of the system time. On running the problem we find that the period of the undamped system is about 20 seconds of computer time, which is consistent with this estimate.

Suppose we want to run the solution faster, because we don't want to wait for 100 seconds to plot the solution most plotters can follow frequencies up to about one cycle per second. So, to make a run time of

10 seconds instead of 100 seconds, we can simply change the gains of all the integrators by 10 by patching each integrator input into a 10 gain rather than a 1 gain. Most computers have a switch that does this automatically.

Since the final step in scaling is changing the gains of all the integrators it is desirable to have the integrator gains all of about the same magnitude (within a factor of 10 of each other). An alternative approach to estimating the maximum of the derivative terms is to distribute the gain from the highest derivative to the output variable equally among the integrators. For the example above we have

$$\ddot{y} + 10^5 \dot{y} + 5\times10^5 \left[\frac{y}{10^{-4}}\right] = 9\times10^4 \qquad (2.10)$$

Dividing by 10^6 to get nice gains on $\dot{y}$ and $[y/10^{-4}]$ gives

$$\left[\frac{\ddot{y}}{10^6}\right] + 0.1\dot{y} + 0.5\left[\frac{y}{10^{-4}}\right] = 0.09 \qquad (2.11)$$

A gain of 10^{10} is required on the two integrators to go from $[\ddot{y}/10^6]$ to $[y/10^{-4}]$. If this is distributed between the two integrators each has a gain of 10^5, so the output of the second integrator is $-[\dot{y}/10]$ giving the maximum for $\dot{y}$ as given in the example and the scaled Equation (2.9).

Electronic analog computers with no mechanical elements, using an oscilloscope for output, can be time scaled to run at much higher frequencies and give thousands of solutions per second. This is accomplished in the same way as was the time scaling above - the integrator capacitors are decreased to increase the integrator gains to 100 or 1000.

Usually after completing scaling some gains and POTs can be combined to give better POT values (greater than 0.1).

In any problem it may be necessary to change the gain of the integrators for time scaling and then again for rescaling when the computer results show that rescaling is necessary. Since it is almost always necessary to do some rescaling when the problem is put on the machine it is frequently necessary, when rescaling, to put a POT where none existed before. During the rescaling of one or more elements (or variables) it is extremely important to label the

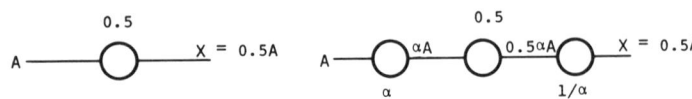

(A) POT

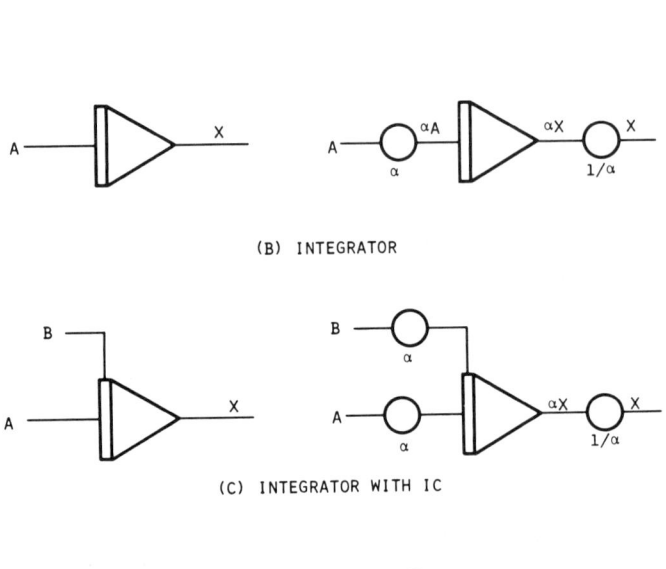

(B) INTEGRATOR

(C) INTEGRATOR WITH IC

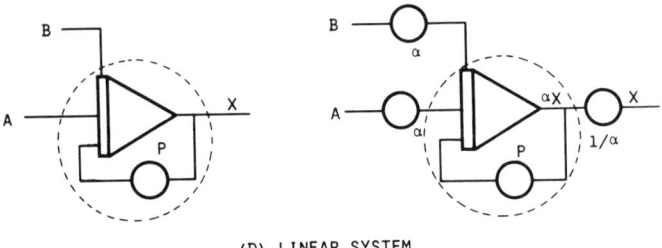

(D) LINEAR SYSTEM

FIGURE 2.24 EQUIVALENT LINEAR COMPUTING ELEMENTS.

diagram with the new variable name every time there is
a   change.   Furthermore,   when   obtaining plots   of
simulated   variables,   the   axes   and variables   are
labeled   in   terms   of   the   problem variables and the
problem independent variable,   not   the   computer
variables.   It is helpful to make a note on the corner
of the plot of what the computer variable and plotter
sensitivity are for that plot, for use during a later
check on the labels of the plotted variables if
necessary.

In the rescaling process the inputs to any set of
linear computing elements can be multiplied by a
constant α and the outputs multiplied by $1/\alpha$ without
changing the equations of the simulation.   Figure 2.24
shows some linear computing elements and equivalent
elements with preceding and following gains of α and
$1/\alpha$ respectively.

Now if an element is preceded by a gain of 10, we
can add a gain of $1/10$ in each input line and   10 in
each output line.   The effect is that of "pushing" the
original 10 gain through the linear elements since the
10 and $1/10$ cancel and leave just the 10 on the output
lines as shown in Figure 2.25.   Of   course   a   POT   or
gain can be pushed through a linear element in the
reverse direction also.

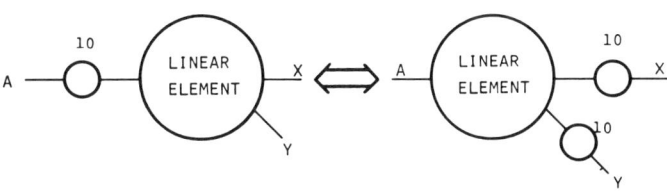

FIGURE 2.25 "PUSHING POTs" THROUGH LINEAR ELEMENTS.

The multiplier does not obey the rules for linear
elements,   but gains can be pushed   through   the
multiplier in a manner similar to that used with the

linear elements. If each of the inputs is multiplied by a constant α and the output multiplied by $1/\alpha^2$ the diagram is not changed. Alternatively, a gain can be pushed from one input to the output. Figure 2.26 shows some equivalent systems involving multipliers and pushing POTs.

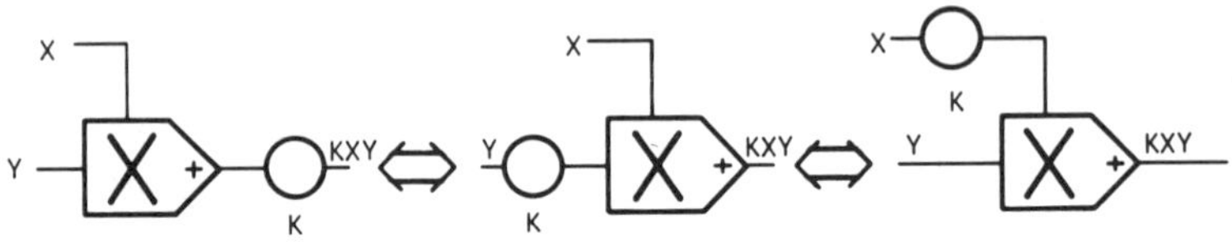

FIGURE 2.26 "PUSHING POTs" THROUGH MULTIPLIERS.

To scale block diagrams containing multipliers to machine units the variables are scaled in the same way as in the linear elements scaled above. Then if each of the inputs to the multiplier is scaled to less than one, the output of the multiplier is also scaled to a value less than one.

Suppose we have an analog computer block diagram containing linear computing elements and a multiplier having inputs x and y and output xy. If this block diagram has been scaled for machine units, and if each of the inputs is multiplied by REF, then x and y become REFx and REFy and the output of the multiplier is REF^2xy. Therefore, the block diagram no longer represents the system equations. Now consider the systems shown in Figure 2.27(a), which represents an analog computer block diagram containing a multiplier. The output of the multiplier xy is considered to be an input to the linear portion of the block diagram. If each of the inputs, including the input xy to the linear portion of the block diagram, is multiplied by REF, then each of the outputs of the block diagram is also multiplied by REF, and the output of the multiplier is multiplied by REF^2. Now, to obtain the

REFxy that we assumed we had, we need only to pass REF^{2}xy through a POT of value 1/REF as shown in Figure 2.27(b).

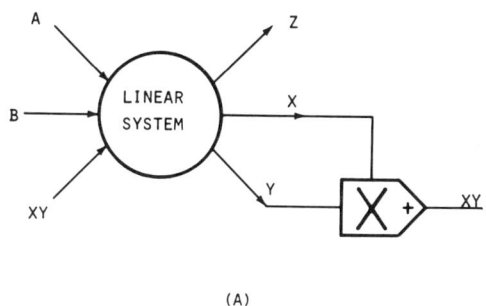

(A)

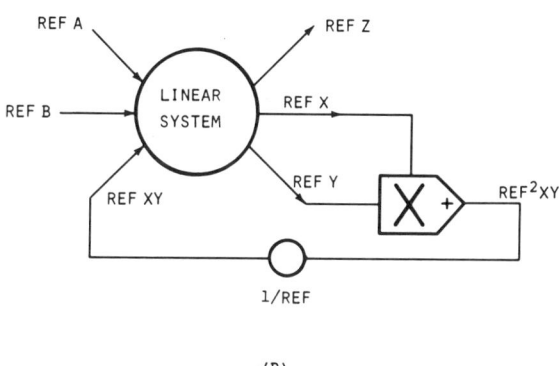

(B)

FIGURE 2.27   ADJUSTMENT OF MULTIPLIER OUTPUT FOR MACHINE UNIT SCALING.

Since we almost always need the factor 1/REF when setting up multipliers on the analog computer, most analog computers are wired to have the multiplier output equal to the product of the inputs divided by REF. This is reasonable since if x = REF and y = REF, then xy = REF^2 and this exceeds the output limitations of the computer. However, xy/REF = REF is okay.

Multipliers with this built in scale factor of 1/REF have an added advantage in that if we scale in machine units assuming REF = 1 then the multiplier with the output xy/REF has the homogeneity property for inputs multiplied by 1 or REF. Thus if each input to the overall block diagram is multiplied by REF, each variable is also multiplied by REF and the block diagram still represents the equations of the system. However, this homogeneity property for multipliers does not work for multiplication of all inputs by any number except 1 and REF.

Now we will scale the van der Pol equation to machine units.

EXAMPLE 2.14

Give a block diagram for the solution of the van der Pol equation scaled to machine units.

$$\ddot{y} - \epsilon(1-y^2)\dot{y} + y = 0 \qquad (2.12)$$

SOLUTION:

We know that $y_{max} > 1$, so for a first guess we can take $y_{max} = 5$. If $\epsilon = 0$, $y = \cos t$ so let us guess $\dot{y}_{max} = y_{max}$. However, since the equation is nonlinear this might not be a good estimate so we take $y_{max} = 10$ to be on the safe side. Then we can rescale if necessary after seeing some solutions. So our guesses for maximum values are $y_{max} = 5$ and $\dot{y}_{max} = 10$.

Rewriting the equation with all variables scaled to unity magnitude we have

$$\ddot{y} = \epsilon(1-25\left[\frac{y}{5}\right]^2)10\left[\frac{\dot{y}}{10}\right]-5\left[\frac{y}{5}\right]$$

Dividing through by 25 to get reasonable coefficients we have

$$\left[\frac{\ddot{y}}{25}\right] = \epsilon(0.4\left[\frac{\dot{y}}{10}\right] - 10\left[\frac{\dot{y}}{10}\right]\left[\frac{y}{5}\right]^2) - 0.2\left[\frac{y}{5}\right] (2.13)$$

Assume that we have $[\ddot{y}/25]$ and integrate twice to get $[\dot{y}/10]$ and $[y/5]$ as shown in Figure 2.28(a). Then combine the signals to obtain $[\ddot{y}/25]$ consistent with Equation (2.14).

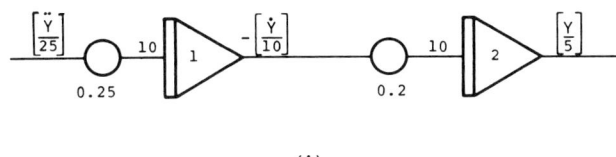

(A)

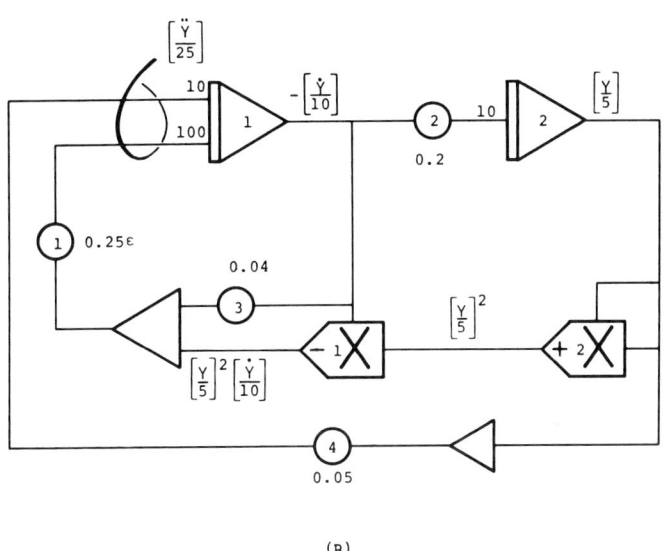

(B)

FIGURE 2.28 ANALOG COMPUTER SOLUTION TO THE VAN DER POL EQUATION.

Note that the sign of the multiplier outputs are specified so that Multiplier 1 has an output that is the negative of the product of its inputs and Multiplier 2 has an output with no sign reversal. This is patched up on the computer consistent with the block diagram. Thus the sign of the output is a portion of the symbol for a multiplier.

Now we can time scale by a factor of 10^{-1} by decreasing the gain of all integrators by 10^{-1} to eliminate the 100 gain. The resultant simulation will run one-tenth as fast as the "real time" solution. The final scaled block diagram is shown in Figure 2.29.

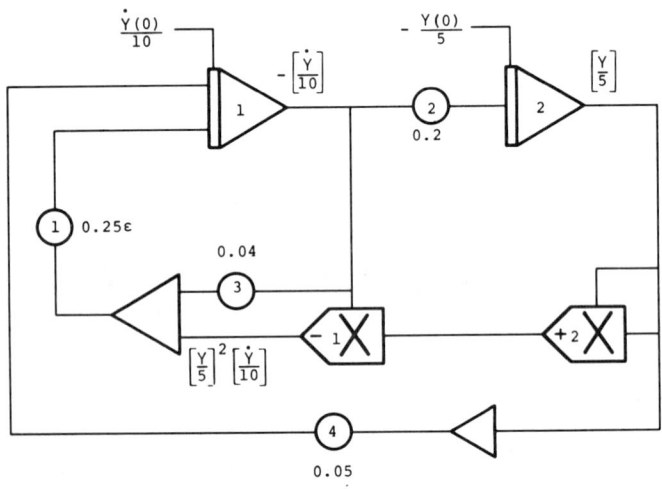

FIGURE 2.29   SCALED BLOCK DIAGRAM FOR THE SOLUTION
OF THE VAN DER POL EQUATION.

We have reasonable POT values and signal levels with the assumed values for y_{max} and $\dot{y}_{max}$. However, it turns out that these estimates of y_{max} and $\dot{y}_{max}$ are somewhat high, and after obtaining the response on the computer we can rescale the problem for more appropriate values. (See Problems 2.28 and 2.29.) ▼

In the above example and in Figure 2.29 every detail of the scaling is included for clarity. When actually scaling, POTs are crossed out and inserted on the diagram rather than redrawn on the diagram at each step. It is often necessary to draw a clear diagram after scaling the original diagram.

2.5 PHASE PLANE PLOTS

There are two basic ways to present analog data. One is with the use of time records as used thus far in this text and obtained from multichannel recorders. The other obtains data through the plotting of trajectories in the so-called *PHASE PLANE*. The phase plane plot is a continuous plot of one dependent variable versus another dependent variable. Thus the phase plane plot might be called a *PARAMETRIC EQUATION PLOT*.

To obtain a phase plane plot on an X-Y recorder we connect the two variables desired into the two inputs of the recorder and, as they vary with time, the plot is traced out. Although each variable is a function of time, the time variable does not appear on the plot. As you observe the plotter during the construction of a phase plane plot you can see that the plotter moves slowly when the variables are changing slowly, and quickly when the variables are changing quickly, thus supplying a 'feel' for the speed of the system. By adding small timing marks on the plot, the speed of the system can be included. Phase plane plots for second order systems are usually plots of the dependent variable versus its derivative.

If the system equations have constant coefficients then the phase trajectories depend only on the initial state of the system and not on the history of the system. Thus if we have a phase trajectory corresponding to a solution of the system equations, we have the solution for all initial conditions lying on that phase trajectory, since if we start off at this initial condition the solution just traces out the remainder of the original trajectory.

To obtain a phase plane plot for the van der Pol equation we can set up the analog computer diagram from Figure 2.29 and plot $\dot{y}$ versus y on an X-Y plotter. We observed in Example 2.14 that the solution to this equation builds up to a steady state oscillation if the initial conditions are small, and if the initial conditions are large the magnitude of the response decreases to the steady state oscillation. Systems that have this type of cyclic response are said to have a *LIMIT CYCLE.*

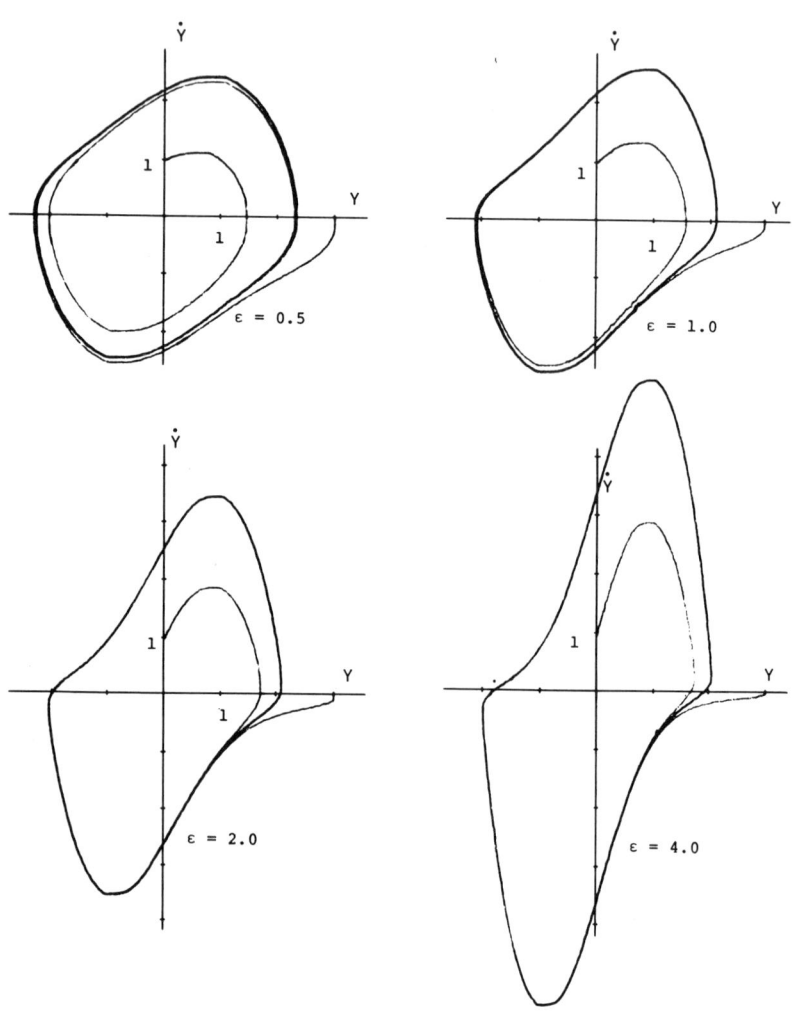

FIGURE 2.30 EXAMPLE 2.15 PHASE PLANE TRAJECTORIES FOR VAN DER POL EQUATION.

EXAMPLE 2.15

Obtain phase plane plots of the solution of the van der Pol equation for initial conditions inside the limit cycle and outside the limit cycle.

SOLUTION:

Figure 2.30 gives the phase plane plots for the initial conditions

(a) $y(0) = 0, \dot{y}(0) = 1$ and (b) $y(0) = 3.0$, $\dot{y}(0) = 0$ for various values of ε. ▼

The phase plane plots of a nonlinear system display the behavior of the system in terms of an exact response from some initial conditions, and give an insight into the type of transient response exhibited by the system. Furthermore, regions of system stability are determined in terms of the conditions for which the solution converges, diverges, or goes into a limit cycle.

A subroutine for plotting XY plots using the printer is presented in the appendix. XYPLT is called similarly to PLOT as demonstrated in the example below.

EXAMPLE 2.16

Give a digital phase plane plot of the van der Pol equation corresponding to the solution in the time series plots of Example 2.10.

SOLUTION:

The following program shows the minor variation from Example 2.10 required to obtain XYPLT instead of PLOT. Of course all of the plots for Example 2.10 and this example can be obtained from a single program. ▼

```
      REAL I1,I2,INTGRL,M1,M2
      DIMENSION TIMEO(201),PLOTI1(201),PLOTI2(201)
      COMMON TAU
C
C     INITIALIZATION STATEMENTS.
C
      TIME=0.0
      FINTIM=10.0
      TAU=0.01
      NUMOUT=200
      NSTEPS=FINTIM/TAU
      OUTDEL=FINTIM/NUMOUT
      TIMOUT=OUTDEL-0.5*TAU
      EPSILN=2.0
C
C     SET POTS, INITIAL CONDITIONS AND PLOT ARRAYS.
C
      POT1=EPSILN
      I1=-1.0
      I2=0.0
      NN=1
      TIMEO(1)=TIME
      PLOTI1(1)=-I1
      PLOTI2(1)=I2
C
C     SIMULATION LOOP...TO STATEMENT # 100.
C
      DO 100 K=1,NSTEPS
      M1=I2*I2
      S2=-(M1-1.)
      M2=I1*S2
      P1=POT1*M2
      S1=-(P1+I2)
      TIME=K*TAU
      I2=INTGRL(I2,I1)
      I1=INTGRL(I1,S1)
C
C     IS IT TIME TO STORE OUTPUT VALUES?
C
      IF(TIME.LT.TIMOUT) GO TO 100
      NN=NN+1
      TIMEO(NN)=TIME
      PLOTI1(NN)=-I1
      PLOTI2(NN)=I2
      TIMOUT=NN*OUTDEL-0.5*TAU
      IF(NN.GE.(NUMOUT+1)) GO TO 999
  100 CONTINUE
  999 CALL XYPLT(PLOTI2,PLOTI1,NN)
      STOP
      END
```

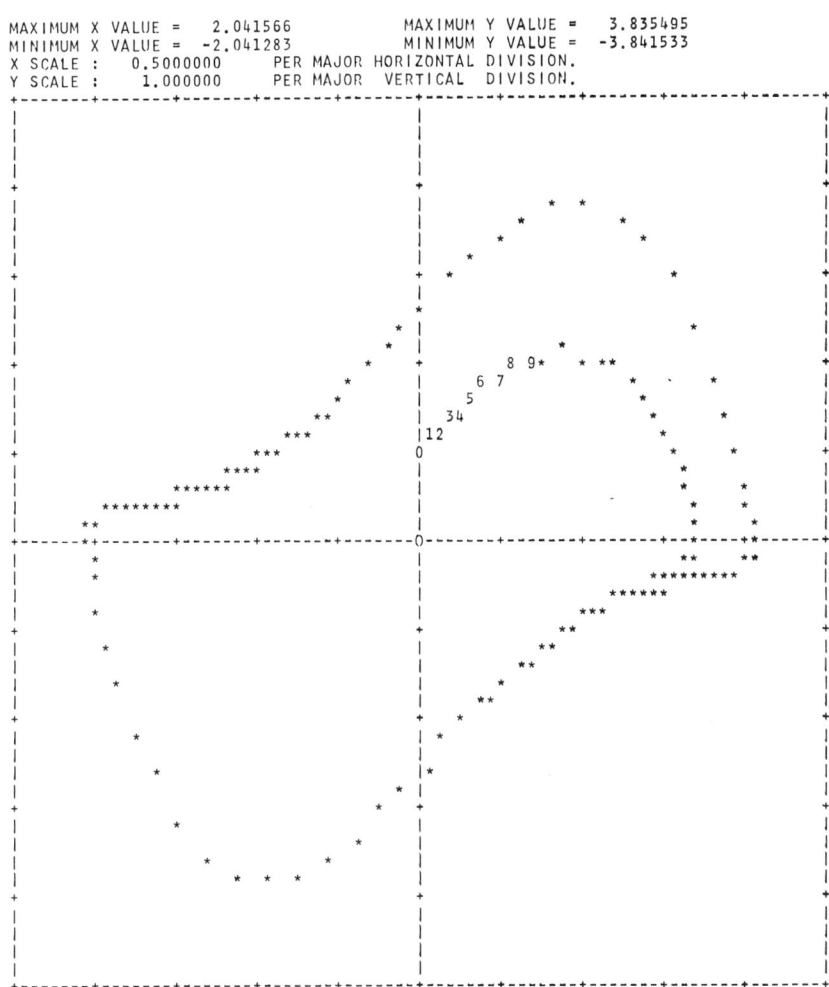

MAXIMUM X VALUE =   2.041566   MAXIMUM Y VALUE =   3.835495
MINIMUM X VALUE = -2.041283   MINIMUM Y VALUE = -3.841533
X SCALE :   0.5000000   PER MAJOR HORIZONTAL DIVISION.
Y SCALE :   1.000000   PER MAJOR VERTICAL DIVISION.

PROBLEMS

2.1 Give an analog computer block diagram to obtain
     the solution to the differential equation

$$\ddot{x} + 0.5\dot{x} + x = 10u(t) \qquad x(0) = 1.0$$

$$\dot{x}(0) = -0.5$$

$$u(t) = \begin{cases} 0 & \text{for } t < 0 \\ 1 & \text{for } t \geq 0 \end{cases}$$

The unit step u(t) is defined

(assume REF = 1)

2.2 Give an analog computer block diagram to simulate systems that are described by the differential equations below. Assume REF = 1 and x(t) are available.

(a) $\ddot{y} + 2\dot{y} + y = x(t)$, $y(0) = 1.0$, $\dot{y}(0) = -1.0$

(b) $\ddot{y} + 4\dot{y} = x(t)$, $y(0) = 0$, $\dot{y}(0) = 0.5$

(c) $y^{(3)} + 7\ddot{y} + y = 0$, $y(0) = 1.0$, $\dot{y}(0) = 0$,
    $\ddot{y}(0) = 0.5$

2.3 Go play around with the analog computer. Locate the POTs, the patchboard, and leaf through the instruction manual.

2.4 Wire an analog computer to correspond to the block diagram shown in Figure 2.1 and obtain the response for different values of the initial condition. How does changing the POT setting alter the solution?

2.5 Patch up a sinewave generator on an analog computer and obtain the output for various initial conditions and settings of the frequency ω.

2.6 On an analog computer set up the differential equation $\ddot{y} + 2\zeta\omega_n\dot{y} + \omega_n^2 y = 0$ with the initial condition $y(0) = 1$, $\dot{y}(0) = 0$ (assuming REF = 1) and obtain the natural response y(t) for $\omega_n = 1$ and $\zeta = 0$, 0.1, 0.2, 0.5, and 1.0. (Put all curves on a single set of axes.)

2.7 Give an analog computer block diagram to simulate the system $\ddot{y} + 4y = x(t)$ for $x(t) = u(t)$ where

$$u(t) = \begin{cases} 0, & t < 0 \\ 1, & t \geq 0 \end{cases}$$

Take $y(0) = \dot{y}(0) = 0$. Write a digital simulation program like that in Example 2.8 from the block diagram and obtain the step response for various values of τ.

2.8 The natural frequency of the system in Problem 2.7 is $\omega_n = 2$. Revise the program obtained in Problem 2.7 to obtain the response to $x(t) = \cos 50t$. Guess a reasonable value for the independent variable increment to obtain results correct to 2 decimal places over 5 periods of the excitation.

2.9 It would be nice to have the simulation program automatically adjust τ for a response to the desired accuracy. Adapt your program from Problem 2.8 so it will compare the results of the calculations with $\tau_0 = 0.01$ at the time $t = 0.1$, and the result $\tau = \tau_0/2$. Have the program continue halving τ until the calculated value of $y(0.1)$ does not change by more than 10^{-4}. Plot the relationship between τ and calculated $y(0.1)$ on a log-log scale. Of course, the TIME - TIMOUT logic will allow printout with a specified OUTDEL regardless of TAU.

2.10 Add the functions SUM(A) and POT(A,B) to the program given in Example 2.7 to replace their equivalent FORTRAN statements.

2.11 Set initial conditions on the variables I1 and I2 in the program in Example 2.8 and complete the calculations by hand one time through the last representation statement. Now repeat this procedure for the alternative representation given after Example 2.8 to show the equivalence.

2.12 From the instruction manual for the analog computer used for this course, read the description of each control on the front panel of the computer.

(a) In what control mode should the computer be when you change a patchboard?

(b) What is the reference (REF) voltage of your machine?

(c) In what mode do you set the POT?

(d) What is the minimum time to RESET your machine?

2.13 Set up an integrator with initial condition x = 0.1 and constant input of 0.1. Obtain a plot of the output x(t) while changing the operating mode alternately between RESET and COMPUTE.

2.14 (a) Sketch the response of the circuit shown in Figure 2.31 for the mode control shown.

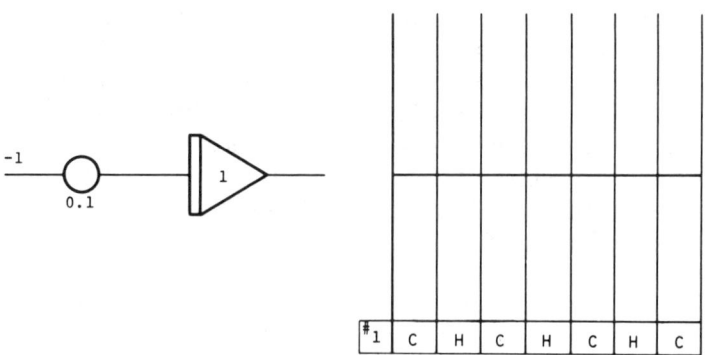

FIGURE 2.31   PROBLEM 2.14(A).

(b) Sketch the response of the circuit shown in Figure 2.32 for the mode control shown. (Sketch x(t) and −y(t).)

2.15 (a) Set up the circuit discussed in Problem 2.14 on an analog computer and obtain plots of the functions x(t), −y(t) and z(t) as a function of time while manually switching modes.

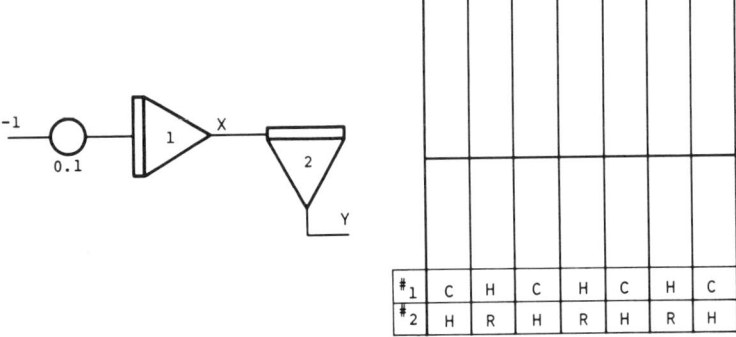

FIGURE 2.32   PROBLEM 2.14(B).

(b) Generate a time base on the computer for the horizontal deflection of the recorder in HOLD while Integrator 2 is in RESET, and in COMPUTE when Integrator 2 is in HOLD. Obtain a plot of $x(t)$ and $z(t)$ on the same axes.

(c) Repeat (a) and (b) with the computer in REPOP.

2.16 Sketch the signals $x(t)$, $-y(t)$ and $z(t)$ for the $x(t)$ and the circuit shown in Figure 2.33.

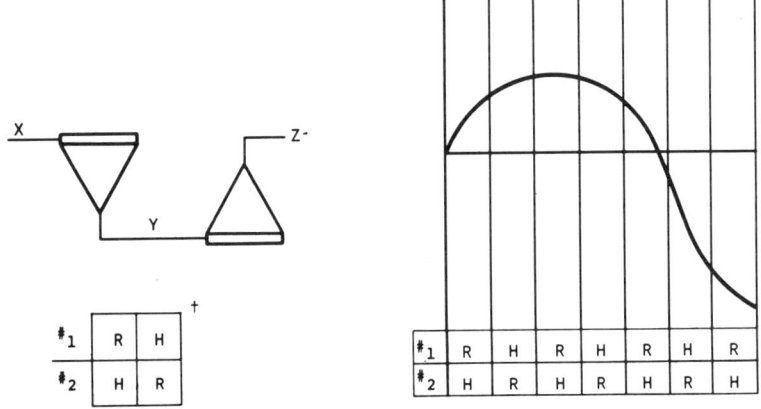

FIGURE 2.33   PROBLEM 2.16.

---

† This alternative notation indicates that integrator 1 switches between RESET and HOLD while Integrator 2 switches between HOLD and RESET.

2.17 Sketch the signals x(t), −y(t), and z(t) for the circuit shown in Figure 2.34 where y(0) = z(0) = 0.

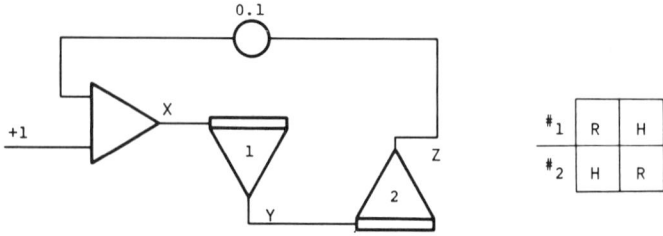

FIGURE 2.34 PROBLEM 2.17.

2.18 Give a block diagram including mode control which will generate the solution to the recursion relation

$$y_n = (1-a)y_{n-1} + ax_{n-1}$$

(Hint: This is similar to Problem 2.17 in principle.)

2.19 Give an analog computer block diagram to generate the time base needed to plot each of the functions shown in Figure 2.15.

2.20 (a) On the analog computer set up the program shown in Figure 2.20 with $\omega = 2$ and obtain the plots shown in that figure using time scale factors of 2 and 0.5.

(b) Repeat part (a) with the time base scaled along with the system and obtain output plots for k = 0.5, 1, and 2.

2.21 An estimate of the ratio of $\dot{y}_{max}$ to y_{max} for a second order system is the larger of ω_n and 2α where

$$\ddot{y} + 2\alpha\dot{y} + \omega_n y = 0$$

(a) Show that if $\alpha \gg \omega_n$, $\dot{y}_{max} \approx 2\alpha y_{max}$ is a good estimate.

(b) For what relation between α and ω_n are these estimates probably in greatest error?   What is the error, and are these still reasonable estimates if $\alpha \approx \omega_n$?

2.22 Give a scaled analog computer diagram in machine units (REF = 1) for the following equations. Give any time scale factors. (If the initial conditions or terms of the equation are negligible, indicate your reasoning.)

(a) $\ddot{x} + \dot{x} + 4x = 2$, $\dot{x}_{max} = 10$

(b) $\ddot{y} + 2x10^4\dot{y} + 10^8 y = 0$, $y_{max} = 10^3$

(c) $\ddot{y} + 2x10^4\dot{y} + 10^8 y = 0$, $y(0) = 100$, $\dot{y}(0) = 50$

(d) $\ddot{y} + 0.02\dot{y} + 100y = 50$, $y(0) = 0$, $\dot{y}(0) = 0$

2.23 How do you write representation statements in proper order for the analog computer block diagram for the equation $\ddot{y} - y = .0$? (Hint: Use a POT of value 1.) Why was this problem not encoutered in Example 2.5?

2.24 On an analog computer set up a track and hold circuit as shown in Figure 2.33. Use the circuit to sample a sinusoidal waveform (generate this) at various sample rates. Integrate the results to obtain the rectangular integration of the sine wave. Subtract this result from the integral of the sine wave to obtain a plot of the integration (truncation) error as a function of time. Finally, make a plot of the maximum truncation error as a function of the ratio ω_s/ω_0 where ω_0 is the sinusoidal frequency and $\omega_s = 2\pi/\omega$ is the sampling frequency.

2.25 Program the system shown in Example 2.12 by your digital analog simulation program and plot the response for various values of integration gain K by taking for the integrator function K*INTGRL(A,B).

2.26 A professor is standing on an elevated tee preparing to drive his golf ball toward the green. The tee is h feet above the green and a horizontal distance of D yards from the green. Give a program including logic and mode control that will solve the equations of motion $(d^2y/dt^2 = -g,\ d^2x/dt^2 = 0)$ with initial conditions $\dot{x}(0) = V_0\cos\theta$ and $\dot{y}(0) = V_0\sin\theta$. Have the program RESET when the ball hits the ground (that is, $y = 0$). (We use this solution in Chapter 6 for adjusting the initial conditions as an application of analog computers to the solution of boundary value problems.)

2.27 (a) Give an analog computer block diagram to solve Duffing's equation,

$$\ddot{x} + \omega_0^2 x + hx^3 = G\cos\omega_1 t$$

which represents a mass on a nonlinear spring.

(b) Obtain the solution to Duffing's equation for the initial conditions $x(0) = 0$ and $\dot{x}(0) = 0$. (Use $\omega_0 = 1$, $\omega_1 = 2$, and $G = 1$, $h = 0.1$.)

2.28 Repeat Problem 2.27 for the Bernoulli equation

$$\ddot{y} + aty - e^{\frac{at^2}{2}} y^2 = 0 \qquad \text{(take a=1)}$$

2.29 Repeat Problem 2.27 for the Mathieu equation

$$\ddot{y} + (a-2q\cos 2t)y + 0 \qquad \text{(take a=q=1)}$$

2.30 Estimate y_{max} and $\dot{y}_{max}$ for the following equations assuming zero initial conditions.

(a) van der Pol equation

(b) Mathieu equation

(c) Duffing equation

(d) Bernoulli equation

2.31 On an analog computer set up the block diagram given in Figure 2.29 for the solution to the van der Pol equation. Give phase plane plots of the response for values of the initial conditions inside and outside the limit cycle. Take $\varepsilon = 0.1, 0.5, 1,$ and 5.

2.32 Scale the van der Pol equation simulation given in Figure 2.29 for $y_{max} = 2$ and $\dot{y}_{max} = 5$ to be valid for $0.1 \leq \varepsilon \leq 10$.

2.33 Scale the analog block diagrams obtained in Problem 2.27 to machine units. Use the maximum variable values obtained in Problem 2.30.

2.34 Repeat Problem 2.33 for

(a) The Bernoulli equation block diagram from Problem 2.28 and

(b) the Mathieu equation block diagram from Problem 2.29.

2.35 *CRITICAL POINTS* in phase space are those points where the derivatives are zero. Thus the equations

$$\dot{x}_1 = -x_2$$
$$\dot{x}_2 = x_1 - 0.2x_1^2$$

have critical points where $x_1 = x_2 = 0$. Find these points and obtain a phase plane (x_2 versus x_1) plot of sufficient phase trajectories to estimate the trajectories throughout the plane. In particular obtain some trajectories near the critical points. Put arrows on the trajectories to indicate the direction of motion.

2.36 Patch up the following circuits on an analog computer and obtain plots of input and output as a function of time for a sinusoidal input signal. If you make a plot of output versus input on an xy recorder you obtain the given characteristic function shown.

(a) *bang bang*

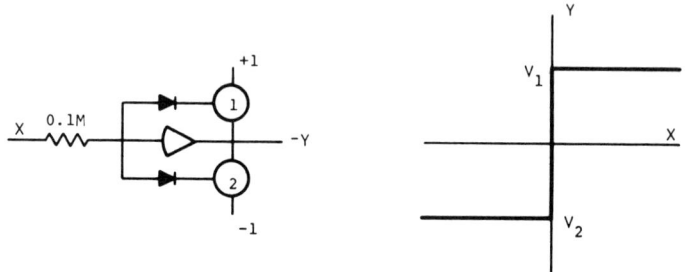

POTS 1 AND 2 ADJUST V_1 AND V_2

FIGURE 2.35   PROBLEM 2.36(A) BANG-BANG.

(b) *Dead zone*

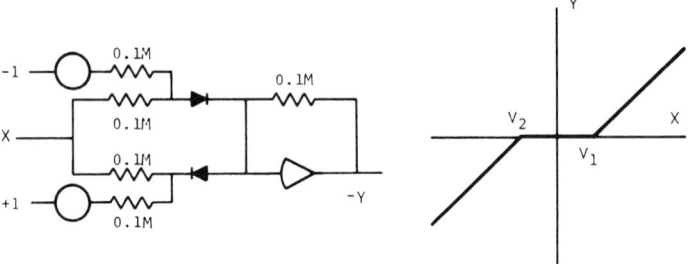

POTS 1 AND 2 ADJUST V_1 AND V_2

FIGURE 2.36   PROBLEM 2.36(B) DEAD-ZONE.

(c) *Half wave rectifier*

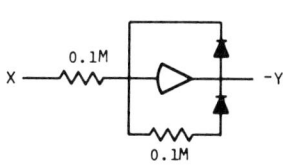 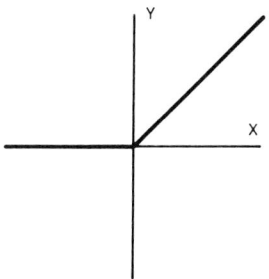

FIGURE 2.37   PROBLEM 2.36(c) HALF WAVE RECTIFIER.

(d) *Absolute value or full wave rectifier*

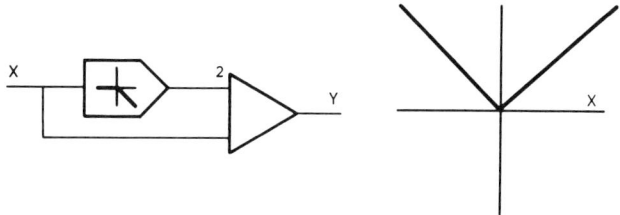

FIGURE 2.38   PROBLEM 2.36(D) ABSOLUTE VALUE.

2.37 Obtain a phase portrait for the systems given below. Choose sufficient sets of phase trajectories to display the character of the solution.

(a) $\ddot{y} + |\dot{y}|y + y = 0$

(b) $\ddot{y} + y^2 + y = 0$

(c) $\ddot{y} + y\dot{y} + y = 0$

(d) $\ddot{y} + |\dot{y}|\dot{y} + |y|y = 0$

(e) $\ddot{y} + |\dot{y}|\dot{y} + y = 0$

(f) $\ddot{y} + \dot{y}|1/\dot{y}| + y = 0$

(g) $\dot{x}_1 = (x_1^2/4) - x_2^2/4) - 1$

$\dot{x}_2 = (x_1^2/9) - (x_2^2/16) - 1$

(h) $\ddot{y} + 0.2\dot{y} + y = 1$

2.38 A simulation can be run with time run backward by inverting the input to each integration, that is, by using a time scale with a factor of −1. Demonstrate this by plotting a phase trajectory for Problem 2.35 in reverse time.

2.39 Devise a scheme to put timing marks on phase plane and time plots using REPOP where calculations are in OPERATE.

simulation of systems
described by
differential equations

Chapters 1 and 2 contain some basic analog and
digital computational techniques to provide a general
familiarity with analog and digital programming. In
Chapter 2 we obtain the analog computer solution to
linear and nonlinear ordinary differential equations,
discuss time and amplitude scaling for analog
computation, and mode control of analog computers.
This material is the foundation for all analog
simulation and the solution of ordinary differential
equations on the analog computer. The digital-analog
simulation techniques discussed in the last chapter
and used for the examples and problems comprise one
technique for obtaining digital simulations of physical systems. In this chapter we discuss several
techniques of digital simulation, starting with an
introduction to z-transforms, which greatly facilitate
the discussion of digital processes.

To conduct the simulation study of a dynamic system
we must first describe the model of the system to the
computer. This is accomplished on the analog computer

through the interconnection of analog computing elements on the block diagram, and on the patchboard through the interconnection using patchcords. On the digital computer the equations of the system simulation and other statements are usually presented to the computer by punched cards. The control of the independent variable is accomplished on the analog computer simulation through mode control and time scaling. On the digital computer, control of the independent variable is determined by the integration algorithm, the step size and the error criteria or desired accuracy.

Control of the simulation with regard to termination logic, logic for repeated runs or repetitive operation, and the handling of input/output data is usually accomplished on the analog computer by the operator and some programmed control logic such as mode control. In digital simulation the control of the simulation input/output is almost exclusively handled by programmed control logic and the programmer is not usually the computer operator as is the case in analog simulation. Finally, the documentation of the program, the input data, the initial conditions of the simulation, and the results complete the simulation.

3.1 DIGITAL COMPUTER SIMULATION

There are essentially two types of systems that one might simulate – the event oriented systems and the continuous or analog type of systems. Event oriented systems can be characterized by the occurrence of a sequence of events, some of which may be in parallel. In continuous systems, the variables are continuous functions of an independent variable, such as the systems described by differential equations.

An example of an event oriented system is a queue (line up) of patrons to buy tickets for a sporting event. The patrons arrive at the ticket windows and get on the shortest line for tickets. Each transaction takes a certain amount of time depending on the number of tickets required, whether or not the clerk must make change, and the efficiency of the ticket selling procedure.

A simulation of this system requires the generation of random numbers, to represent the timing of the arrival of patrons that is consistent with information about the character and distribution of such arrivals in real life (the system). This random number representation for the arrival times is then a model of that portion of the system. Furthermore, the time for a single transaction is a random quantity which can be modeled by a random number generator based on the measured characteristics for that type of transaction.

A typical event oriented simulation of this system might be used to evaluate the number of ticket windows required to maintain the longest line less than a prescribed length. Or it may be desirable to evaluate the expected maximum waiting time for a customer based upon the number of ticket clerks. There are several simulation languages for the simulation of event oriented systems. Probably the most popular are SIMSCRIPT and GPSS (General Purpose Simulation System). These are block oriented languages that have statements to represent operations such as storages, transactions, queues, and tables.

There are also simulation languages, like DYNAMO that are designed to solve difference equations while being incremented in an independent variable. These programs are oriented toward the solution of problems in industrial dynamics such as inventory control, and sales and income relationships where the system has a natural independent variable, usually time, and is described by difference equations.

Event or difference equation oriented simulations are very conveniently oriented to digital computers and are at best very cumbersome by analog computation. A study of the event oriented simulation languages to sufficient depth to be of real value would require a separate book. In this chapter the emphasis is on the digital computer simulation of systems described by differential equations. A digital computer simulation of a continuous process is a discrete system that behaves as nearly as possible like the continuous process being modeled. The digital-analog simulation program discussed in the last chapter results in a digital computer program similar to an analog computer description of the system being modeled, where every element can be identified with a component of the physical system. This block diagram oriented language for digital simulation allows the application of analog block diagram notation and techniques to a digital computer language and eliminates the scaling problems in analog computation.

A major problem in the digital simulation of continuous systems described by differential equations is the inability of discrete integrators to represent with precision the integral of continuous functions except for very special inputs. For example, rectangular integration is exact when the integrand is a constant. As you know, the operations must be performed in sequence with all inputs to a block updated before the output is updated.

The first published discussion of a digital-analog simulation program was presented by R. G. Selfridge at the 1955 Western Joint Computer Conference.† Since then at least 25 new digital analog simulation programs have been developed. Among them are COBLOC, DAS, DSL/90, DYSAC, MIDAS, and CSMP/360. (You have added one yourself while studying Chapter 2.) In fact there have been so many that they have been surveyed and cataloged.††

In these programs the increment size and integration schemes are variable and in many the statements of the program are automatically ordered by the program. Probably every added feature you can imagine has been incorporated into some of these programs. In fact, many of the programs are simply

---

† R. G. Selfridge, "Coding a General-Purpose Digital Computer to Operate as Differential Analyzer," *Proceedings of the Western Joint Computer Conference,* The Institute of Radio Engineers, New York, N. Y., 82-84 (1955).
†† R. N. Linebarger and R. D. Brennan, "A Survey of Digital Simulation: Digital-Analog Simulator Programs," *Simulation 3,* No. 6, 22-36 (Dec., 1964). J. J. Clancey and M. S. Finebert, "Digital Simulation Languages: A Critique and A Guide," *AFIPS Conference Proceedings,* Fall Joint Comp. Conf. Part I, 22-36 (Nov. 1965).

combinations of what someone considered to be the best features of programs existing at that time.

To change our digital-analog simulation program so that it automatically adjusts the increment size at the beginning of a run, we can apply an excitation to the system at the highest frequency expected using some initial estimate for the increment size and allow the simulation to run for 1/4 of a period at the highest excitation frequency. At the completion of this short run the increment size is decreased, the simulation rerun, and the results compared with those previously obtained. If the difference between the two results is less than the specified error criterion, it is assumed that the increment size is sufficiently small. Otherwise, the increment size is again reduced and the results compared. Starting at a reasonable value for τ it is convenient to sequence through the values $\tau = 1$, 0.5, 0.2, 0.1, 0.05, 0.02, and so on. It is also desirable to include a lower bound for τ so that if τ does not become sufficiently small to meet the error criterion the simulation will be run with this minimum and a warning printed stating that the error criterion was not met. The increment size must be small to obtain the desired accuracy with regard to truncation error. However, if the increment size is too small the computation time is excessive and round off errors become dominant.

Each integration scheme has associated with it an error estimate which can be used to maintain error control. After a discussion of several classical numerical integration schemes in terms of z-transforms and their error analyses we incorporate these methods into our program. Additionally, we obtain from the estimates another method for automatically adjusting the increment size to meet an error criterion. In fact, having an error estimate for each calculation allows the increment size to be changed while the program is in progress. Thus the program can take large steps when the inputs to the integrators vary slowly and small steps while the inputs vary rapidly.

3.2 Z-TRANSFORMS AND DIGITAL SIMULATION BY Z-TRANSFORM INTEGRATING OPERATORS

The z-transform is particularly applicable to the analysis and design of discrete systems and to the solution of difference equations or recursion relations as they arise in digital simulation. The z-transform is an operation that converts a sequence of numbers, such as a dependent variable sequence

$$\{y_n\} = [y_0, y_1, y_2, \ldots]$$

into a function of the complex variable z.

The z-transform was originally used to analyze sampled data control systems, but it has become very popular in the last few years as a method for representing discrete processes, recursive systems and, most recently, digital filters. The application of the techniques of this chapter to the design of digital filters is discussed in Section 3.5.

In digital computation and simulation, calculations are performed on sample values of the dependent variable. The z-transform provides convenient techniques for performing digital operations on sequences that represent system variables. The z-transform of a function is defined as the sum of the negative powers of the variable z by the equation

$$Z[y(t)] = Y(z) = y_0 + y_1 z^{-1} + y_2 z^{-2} + \ldots$$

$$= \sum_{k=0}^{\infty} y_k z^{-k} \tag{3.1}$$

Thus the *Z-TRANSFORM* is merely the series of decreasing powers of z having coefficients equal to the sequential values of $\{y_n\}$. In some cases it is possible to express the z-transform in closed form as shown in the following example.

EXAMPLE 3.1

Give the z-transform of the unit step function u(t).

SOLUTION:

$$u(t) = \begin{cases} 0, & t < 0 \\ 1, & t \geq 0 \end{cases}$$

Therefore the sequence representing u(t) has the values $u_n = 1$, (n=0,1,2, ...). The z-transform of u(t) is

$$Z[u(t)] = Z[1] = 1 + z^{-1} + z^{-2} + \ldots \quad U(z)$$

$$= \sum_{n=0}^{\infty} = \frac{1}{1-z^{-1}} \quad †$$

If two sequences have identical z-transform then their sequential values are equal since the coefficients of the z-transform series are equal. Also multiplication of a z-transform Y(z) by z^{-1} results in the z-transform of the sequence $\{y_{n-1}\}$. That is, we obtain the sequence $\{y_n\}$ delayed by one sample point. Figure 3.1(b) shows the z-transform $W(z) = z^{-1}Y(z)$ and the corresponding sequential values identical to those in Figure 3.1(a) delayed by one sample interval.

Since when written out in series form the z-transform has for its coefficients the sequential values of the function, it is clear that the z-transform can be used for keeping track of the times at which the sequential values occur.

---

† You can easily show by long division that

$$\frac{1}{1 - z^{-1}} = 1 + z^{-1} + z^{-2} + \ldots$$

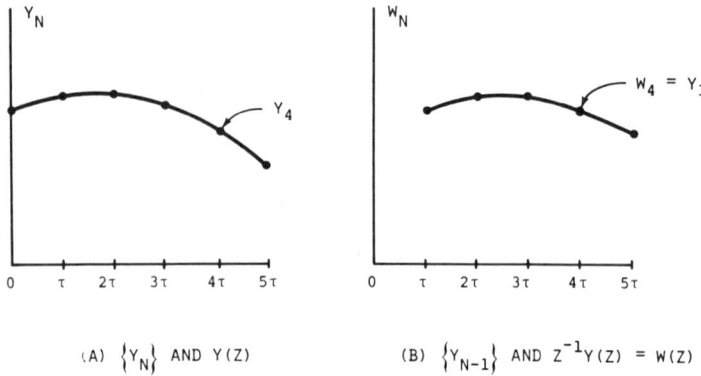

(A) $\{Y_N\}$ AND $Y(Z)$ (B) $\{Y_{N-1}\}$ AND $Z^{-1}Y(Z) = W(Z)$

FIGURE 3.1 z^{-1} AS A DELAY OPERATOR.

The linearity of the z-transformation is easily demonstrated by taking

$$Z[x(t) + y(t)] = \sum_{k=0}^{\infty} (x_k + y_k) z^{-k}$$

$$= \sum_{k=0}^{\infty} x_k z^{-k} + \sum_{k=0}^{\infty} y_k z^{-k}$$

$$= Z[x(t)] + Z[y(t)]$$

The linearity of the z-transform operation is very useful in writing z-transform equations from recursion relations.

Consider the recusion relationship representing the rectangular integration formula

$$y_{n+1} = y_n + \tau x_n \qquad (3.2)$$

which can be represented as shown in Figure 3.2. If we assume that we have an initial value for y_n and the input value x_n, we can obtain y_{n+1} as shown. Now if y_{n+1} is passed into a delay element we have the assumed y_n. (This is analogous to drawing an analog block diagram to represent a differential equation.)

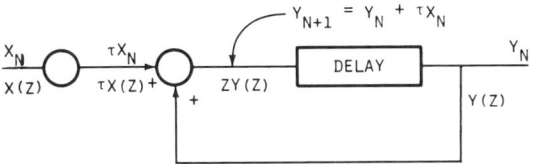

FIGURE 3.2 BLOCK DIAGRAM FOR THE DIFFERENCE EQUATION $y_{N+1} = y_N + \tau x_N$.

Equation (3.2) is a recursion relation that gives the relationship between the individual values of the sequences $\{x_n\}$ and $\{y_n\}$ We refer to the z-transform of $\{y_n\}$ and $\{x_n\}$ as, $Y(z)$ and $X(z)$ respectively. Then rewriting Equation (3.2) as

$$y_n = y_{n-1} + \tau x_{n-1} \qquad (3.3)$$

and taking the z-transform of both sides we have

$$Z[\{y_n\}] = Z[\{y_{n-1} + \tau x_{n-1}\}]$$

and by linearity of the z-transform

$$Y(z) = z^{-1}Y(z) + \tau z^{-1}X(z)$$

or

$$zY(z) = Y(z) + \tau X(z)$$

which can be written

$$Y(z) = \frac{\tau}{z - 1} X(z)$$

Thus to represent rectangular integration in terms of the z-transforms of the input sequence $X(z)$ and the output sequence $Y(z)$ we need only multiply $X(z)$ by the factor $\tau/(z-1)$. This function of z, which is the ratio of the z-transform of the output of a linear system to the z-transform of its input, is called the

PULSE TRANSFER FUNCTION. Since multiplication of the z-transform by z^{-1} corresponds to a delay of one sample interval τ we can obtain a block diagram representation of this equation by writing it as the equality of negative polynomials in z or,

$$(z-1)Y(z) = \tau X(z)$$

$$(1-z^{-1})Y(z) = \tau z^{-1}X(z)$$

$$Y(z) - z^{-1}Y(z) = \tau z^{-1}X(z)$$

$$Y(z) = z^{-1}Y(z) + \tau z^{-1}X(z)$$

Thus $Y(z)$ is the output of the delay element whose input is $Y(z) + \tau X(z)$. Clearly this is the block diagram shown in Figure 3.2.

It turns out that obtaining a recursion relation from a pulse transfer function is quite straightforward. If we have a pulse transfer function relating the z-transform $X(z)$ and $Y(z)$, then

$$Y(z) = H(z)X(z)$$

where $H(z)$ is the ratio of polynomials in z, for example

$$Y(z) = \frac{b_0 z^2 + b_1 z + b_2}{a_0 z^2 + a_1 z + a_2} X(z)$$

Dividing numerator and denominator by the highest power of z, in this case z^2, we have

$$Y(z) = \frac{b_0 + b_1 z^{-1} + b_2 z^{-2}}{a_0 + a_1 z^{-1} + a_2 z^{-2}} X(z) \qquad (3.4)$$

Multiplying both sides by the denominator of the right hand side of this equation gives

$$\left(a_0 + a_1 z^{-1} + a_2 z^{-2}\right)Y(z) = \left(b_0 + b_1 z^{-1} + b_2 z^{-2}\right)X(z)$$

$$(3.5)$$

We have on the left side of Equation (3.5) the sum of a_0 times the sequence $\{y_n\}$, a_1 times $\{y_n\}$ delayed by one sample interval, and a_2 times $\{y_n\}$ delayed by two intervals. On the right side of the equation we have the sum of b_0 times the sequence $\{x_n\}$, b_1 times the sequence $\{x_n\}$ delayed by one interval and b_2 times $\{x_n\}$ delayed by two intervals. If we write out the z-transforms of the left side and right side of Equation (3.5) and equate the coefficients of z^{-n} we obtain the recursion relation

$$a_0 y_n + a_1 y_{n-1} + a_2 y_{n-2} = b_0 x_n + b_1 x_{n-1} + b_2 x_{n-2}$$

$$(3.6)$$

The relationship between Equations (3.5) and (3.6) regarding the z-transforms and the recursion relation is apparent from the observation below.

$$\left[a_0 z^0 + a_1 z^{-1} + a_2 z^{-2}\right] Y(z) = \left[b_0 z^0 + b_1 z^{-1} + b_2 z^{-2}\right] X(z)$$

$$a_0 y_n + a_1 y_{n-1} + a_2 y_{n-2} = b_0 x_n + b_1 x_{n-1} + b_2 x_{n-2}$$

Thus by inspection we can write the correspondence between the z-transforms and the recursion relations. This equivalence between Equations (3.5) and (3.6) is the same as the relationship found in the discussion of Figure 3.2.

At this point it is convenient to introduce notation for the representation of differentiation and integration of dependent variables. Hereafter we can use $Dy = \dot{y}$, $D^2 y = \ddot{y}$, and so on. To represent integration of the dependent variables we can write

$$Iy = \int_0^t y(t)\,dt, \qquad I^2 y = \int_0^t \int_0^\alpha y(\beta)\,d\beta\,d\alpha, \quad \ldots$$

With the appropriate selection for initial conditions on the value of the integral, I is the reciprocal operation of D. Also note that I and D have no sign reversal.

EXAMPLE 3.2

Write the differential equation $\ddot{y} + a\dot{y} + by = x(t)$ in the operator notation given above.

SOLUTION:

$$\ddot{y} + a\dot{y} + by = x$$
$$D^2 y + aDy + by = x$$
$$(D^2 + aD + b)y = x$$

$$\left(1 + a\,\frac{1}{D} + b\,\frac{1}{D^2}\right)y = \frac{1}{D^2}\,x$$
$$(1 + aI + bI^2)y = I^2 x$$

▼

One technique for obtaining a recursion relationship to represent a differential equation is to take the z-transform of y and x by substituting $Y(z)$ for y, $X(z)$ for x, and the pulse transfer function $\tau/(z-1)$ for the integrating operator I. This technique is demonstrated in the following example.

EXAMPLE 3.3

Give a recursion relation to solve the following differential equation using rectangular integration

$$\ddot{y} + 6\dot{y} + 5y = x$$

SOLUTION:

(This is the recipe)

1. Write the equation in I operator notation.

$$D^2 y + 6Dy + 5y = x$$
$$(D^2 + 6D + 5)y = x$$
$$(1 + 6I + 5I^2)y = I^2 x$$

2. Replace y and x by $Y(z)$ and $X(z)$, and replace I by $\tau/(z-1)$.

$$\left[1+6\left(\frac{\tau}{z-1}\right) + 5\left(\frac{\tau}{z-1}\right)^2\right]Y(z) = \left(\frac{\tau}{z-1}\right)^2 X(z)$$

3. Write the equation in polynomials in z times $Y(z)$ and $X(z)$ (multiply through the equation by $(z-1)^2$).

$$\left[(z-1)^2 + 6\tau(z-1) + 5\tau^2\right]Y(z) = \tau^2 X(z)$$

or

$$\left[z^2 + (6\tau-2)z + (1-6\tau+5\tau^2)\right]Y(z) = \tau^2 X(z)$$

4. Divide the equation by the highest power of z (z^2 for this example).

$$\left[1 + (6\tau-2)z^{-1} + (1-6\tau+5\tau^2)z^{-2}\right]Y(z) = \tau^2 z^{-2} X(z)$$

5. Write the recursion relation by inspection as shown above.

$$y_n + (6\tau-2)y_{n-1} + (1-6\tau+5\tau^2)y_{n-2} = \tau^2 x_{n-2}$$

or

$$y_n = (2-6\tau)y_{n-1} - (1-6\tau+5\tau^2)y_{n-2} + \tau^2 x_{n-2}$$

This single recursion relation is the digital simulation of the given system. ▼

Substitution of $(\tau/z-1)^2$ for I^2 in the above example is equivalent to the integration of the variable twice by rectangular integration. That is, the function being integrated is approximated by a piecewise constant function and then integrated. This result is in turn integrated by rectangular integration to obtain the final result. Figure 3.3 shows two rectangular integration pulse transfer functions $(\tau/z-1)$ cascaded to obtain the integral of a function $x(t)$ along with the equivalent time function approximations and their integrals.

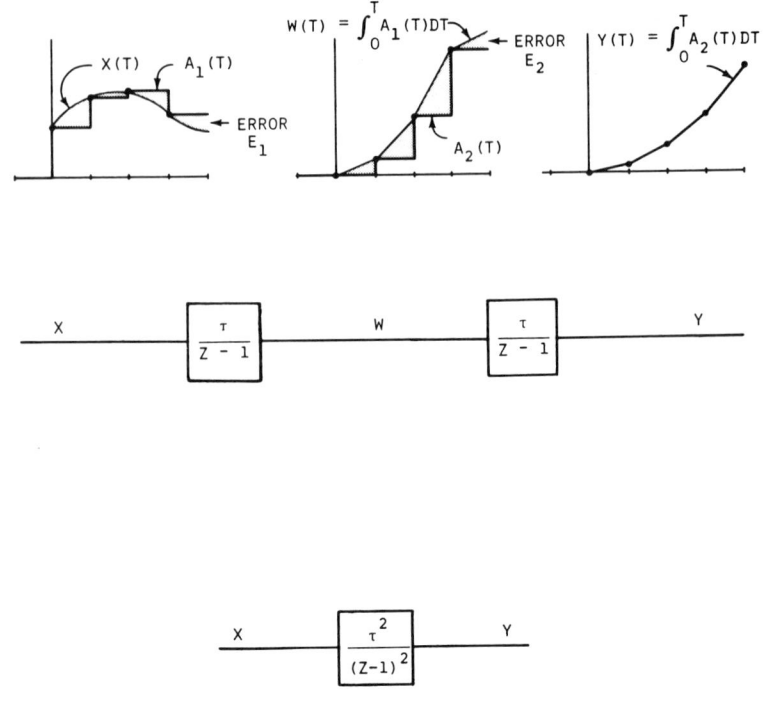

FIGURE 3.3 APPROXIMATION AND ERRORS IN CASCADED RECTANGULAR INTEGRATORS.

The first rectangular integrator in the figure exactly integrates $a_1(t)$, the piecewise constant approximation of the input $x(t)$, giving the result $\int_0^t a_1(t)\,dt$ which is piecewise linear. Now if this is integrated by rectangular integration it is approximated by the piecewise constant function $a_2(t)$ which is exactly integrated by a rectangular integrator. Thus using the operator $(\tau/z-1)$ for I^2 introduces an error in two distinct approximations.

It seems more appropriate to perform the second integration by assuming that the function to be integrated is piecewise linear. Figure 3.4 shows a typical integral of a function and its piecewise linear approximation. The integral of the approximation is the area of the trapezoid

$(\tau/2)(x_{n-1}+x_n)$ resulting in the so-called trapezoidal rule for integration,

$$y(t) = \int_0^t x(t)\,dt \iff y_n = y_{n-1} + \frac{\tau}{2}(x_{n-1}+x_n)$$

(3.7)

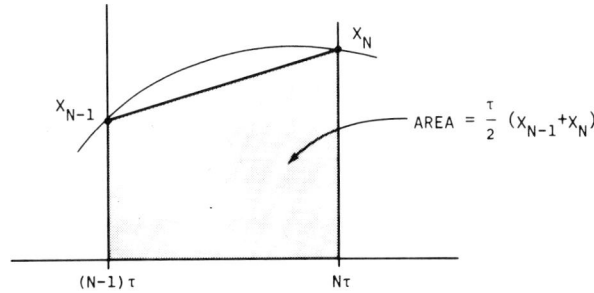

FIGURE 3.4 TRAPEZOIDAL INTEGRATION: $y_N = y_{N-1} + \frac{\tau}{2}(x_{N-1}+x_N)$.

The pulse transfer function corresponding to trapezoidal integration is obtained from the recursion relation following the procedures given above.

$$y_n = y_{n-1} + \frac{\tau}{2}(x_{n-1}+x_n)$$

taking the z-transform of both sides we have

$$Y(z) = z^{-1}Y(z) + \frac{\tau}{2} z^{-1}X(z) + X(z)$$

and solving for $Y(z)$ gives

$$Y(z) = \frac{\tau}{2} \frac{(z+1)}{(z-1)} X(z)$$

Thus the pulse transfer function for trapezoidal integration is

$$I_{T1} = \frac{\tau}{2} \frac{(z+1)}{(z-1)}$$

(3.8)

Now if we desire to integrate a function twice assuming it is piecewise constant we can perform the first integration by rectangular integration as in Figure 3.3. However, instead of approximating w(t) as piecewise constant we can integrate it exactly by trapezoidal integration by passing it through the pulse transfer function I_{T1} obtained above as shown in Figure 3.5.

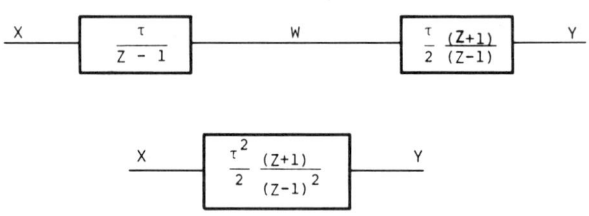

FIGURE 3.5 EXACT SECOND INTEGRAL OF A PIECEWISE CONSTANT APPROXIMATION.

Thus the pulse transfer function which exactly gives the second integral of a piecewise constant function is

$$I_{R1}I_{T1} = I_{R2} = \frac{\tau^2}{2} \frac{(z+1)}{(z-1)^2} \qquad (3.9)$$

Since the output of this pulse transfer function is the second integral of a piecewise constant function it is itself a piecewise quadratic function. It can be demonstrated that the piecewise quadratic function is exactly integrated by Simpson's one-third rule,

$$y_n = y_{n-2} + \frac{\tau}{3}\left(x_n + 4x_{n-1} + x_{n-2}\right) \qquad (3.10)$$

which has the pulse transfer function I_{Q1} (Q for quadratic).

$$I_{Q1} = \frac{\tau}{3} \frac{(z^2+4z+1)}{(z^2-1)} \tag{3.11}$$

To integrate a piecewise constant function three times exactly we can use the pulse transfer function $I_{R1}I_{T1}I_{Q1} = I_{R2}I_{Q1} = I_{R3}$. The I_{R3} notation means we use a rectangular approximation and integrate three times. Similarly to integrate a piecewise linear function two times we can use $I_{T1}I_{Q1} = I_{T2}$. The details of the derivation of these types of integrating operators are contained in Cuénod and Durling[†] and are summarized in Table 3.1.

TABLE 3.1 INTEGRATION OPERATORS FOR I, I^2, I^3, AND I^4 WHICH ASSUME THE FUNCTIONS INTEGRATED ARE PIECEWISE CONSTANT OR PIECEWISE LINEAR.

		EXACT FOR PIECEWISE CONSTANT FUNCTION		EXACT FOR PIECEWISE LINEAR FUNCTION
I	I_{R1}	$\dfrac{\tau}{(Z-1)}$	I_{T1}	$\dfrac{\tau}{2}\dfrac{(Z+1)}{(Z-1)}$
I^2	I_{R2}	$\dfrac{\tau^2}{2}\dfrac{(Z+1)}{(Z-1)^2}$	I_{T2}	$\dfrac{\tau^2}{6}\dfrac{(Z^2+4Z+1)}{(Z-1)^2}$
I^3	I_{R3}	$\dfrac{\tau^3}{6}\dfrac{(Z^2+4Z+1)}{(Z-1)^3}$	I_{T3}	$\dfrac{\tau^3}{24}\dfrac{(Z^3+11Z^2+11Z+1)}{(Z-1)^3}$
I^4	I_{R4}	$\dfrac{\tau^4}{24}\dfrac{(Z^3+11Z^2+11Z+1)}{(Z-1)^4}$	I_{T4}	$\dfrac{\tau^4}{120}\dfrac{(Z^4+26Z^3+66Z^2+26Z+1)}{(Z-1)^4}$

[†] M. Cuénod and A. Durling, *A Discrete Time Approach for System Analysis*, Academic Press, Inc., 1969.

EXAMPLE 3.4

Given an input sequence $\{x_n\}$ obtain a recursion relation to simulate a system described by the differential equation

$$8\dot{y} + y = x(t), \quad y(0) = y_0$$

SOLUTION:

Follow the recipe

$$8Dy + y = x$$

$$8y + Iy = Ix$$

1. Assuming both y and x can be integrated satisfactorily by rectangular integration, we have

$$8Y(z) + I_{R1}Y(z) = I_{R1}X(z)$$

$$\left[8 + \frac{\tau}{(z-1)}\right]Y(z) = \left[\frac{\tau}{(z-1)}\right]X(z)$$

which corresponds to the recursion relation

$$y_n = \frac{1}{8}\left[(8-\tau)y_{n-1} + \tau x_{n-1}\right]$$

(Verify this.)

2. Assuming both y and x can be satisfactorily integrated by trapezoidal integration, we have

$$8Y(z) + I_{T1}Y(z) = I_{T1}X(z)$$

or

$$y_n = \frac{1}{(8 + \frac{\tau}{2})}\left[(8 - \frac{\tau}{2})y_{n-1} + \frac{\tau}{2}(x_{n-1}+x_n)\right]$$

(Verify this.)

3. We know that y(t) is continuous for finite inputs (the step response approaches the final value exponentially with no discontinuity). So for a discontinuous input the best approximation is probably to assume that x is piecewise constant and that y is piecewise linear. Doing this we have

$$8Y(z) + I_{T1}Y(z) = I_{R1}X(z)$$

or

$$y_n = \frac{1}{(8 + \frac{\tau}{2})}\left[(8 - \frac{\tau}{2})y_{n-1} + \tau x_{n-1}\right] \quad \blacktriangledown$$

If the excitation to a differential equation is finite, the solution or output is continuous. Therefore, a piecewise linear approximation yields more accurate results than a piecewise constant approximation. However, if the excitation is discontinuous, like a step function, a piecewise linear approximation introduces more error than a rectangular integration. (Verify this by sketching a piecewise constant and piecewise linear interpolation between the sample points of a unit step. Also compare the integral of a unit step by using I_{R1} and I_{T1}.)

EXAMPLE 3.5

Give a recursion relation to obtain the solution to the differential equation

$$\ddot{y} + \dot{y} = x, \quad x = 1 \qquad y(0) = 0$$

$$\dot{y}(0) = 0$$

(take x = 0 for t < 0)

SOLUTION:

$$D^2 y + y = x$$

$$y + I^2 y = I^2 x$$

Assuming y is piecewise linear and x is piecewise constant, which it is, we have

$$Y(z) + I_{T2}Y(z) = I_{R2}X(z)$$

$$\left[1 + \frac{\tau^2}{6}\frac{(z^2+4z+1)}{(z-1)^2}\right]Y(z) = \left[\frac{\tau^2}{2}\frac{(z+1)}{(z-1)^2}\right]X(z)$$

$$\left[(z-1)^2 + \frac{\tau^2}{6}(z^2+4z+1)\right]Y(z) = \left[\frac{\tau^2}{2}(z+1)\right]X(z)$$

$$\left[(1 + \frac{\tau^2}{6})z^2 + (\frac{2}{3}\tau^2-z)z + (1 + \frac{\tau^2}{6})\right]Y(z)$$

$$= \left[\frac{\tau^2}{2}(z+1)\right]X(z)$$

or

$$y_n = \frac{1}{(1 + \frac{\tau^2}{6})}\left[-(\frac{2}{3}\tau^2-z)y_{n-1} - (1 + \frac{\tau^2}{6})y_{n-2}\right.$$

$$\left. + \frac{\tau^2}{2}(x_{n-1}+x_{n-2})\right]$$

where x_n = 1 for all n $\geq$ 0. ▼

   The program corresponding to the recursion relation in Example 3.5 is straightforward. We just replace y_n, y_{n-1}, and y_{n-2} by Y(N), Y(N-1) and Y(N-2) respectively and we have a digital simulation of a system satisfying the given differential equation. Therefore the z-transform substitution technique results in a single recursion relation representing the total system input-output relationship.
   To start the recursion relation for the calculation of y_1 we must have the values of y_0 and y_{-1} called starting values. The generation of starting values is a problem in all integration schemes that use

information other than the immediate past value y_{n-1}.
Clearly we can choose $y_0 = y(0)$, but how do we pick
y_{-1}, y_{-2}, and so on?

An nth order differential equation must have n
initial (or boundary) conditions to have a unique
solution. These initial conditions must be converted
to starting values for the recursion relation. It
turns out that using the z-transform substitution we
need all of the initial conditions to obtain the
required starting values.

For a second order system where $y(0)$ and $y(0)$ are
given we can choose $y_0 = y(0)$ and y_{-1} such that the
slope of the linear segment between y_{-1} and y_0 is $\dot{y}(0)$
as shown in Figure 3.6.

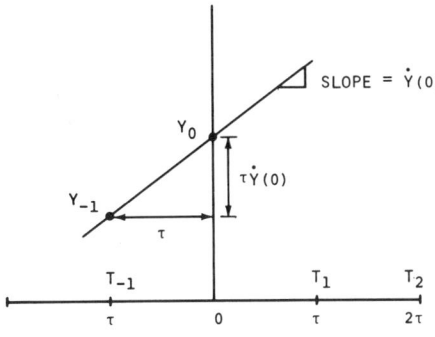

FIGURE 3.6 STARTING VALUES FOR A SECOND ORDER SYSTEM DIFFERENCE EQUATION.

Thus we can take y_{-1} to satisfy

$$y_0 = y_{-1} + \tau \dot{y}(0)$$

or

$$y_{-1} = y_0 - \tau \dot{y}(0)$$

By definition of $X(z)$, and since we start off at
the correct values for the response by the procedure
shown in Figure 3.6, we take the starting value of the

input $\{x_n\}$ as $x_k = 0$ for $k < 0$.

The assumption that the slope $\dot{y}(0)$ is constant in the interval from $-\tau$ to 0 is equivalent to integrating the differential equation backward in time by rectangular integration. The ease of the generation of starting values is a factor in the selection of an integration algorithm and is discussed further after the introduction of some additional integration techniques.

The recursion relation in Example 3.5 can be written in FORTRAN as

```
      DO 100 N=2,NSTEPS
100  Y(N)=B1*Y(N-1)+B2*Y(N-2)+A1*X(N-1)+A2*X(N-2)
```

with appropriate statements for setting initial values and obtaining output values. However since each calculated value is not used as an output value it is not necessary nor desirable to dimension Y to take on NSTEPS values. We need only store the number of values to be printed or plotted. This can be accomplished by the following strategy.

If $x(t) = \sin 2t$ we can program the equation as

```
      DO 100 N=1,NSTEPS
      Y=B1*YNM1+B2*YNM2+A1*SIN((N-1)*TAU)+A2*SIN((N-2)*TAU)
      YNM2=YNM1
100  YNM1=Y
```

where YNM2 is an acronym for y_{n-2}. (Verify that this program accomplishes the desired result with an undimensioned variable Y.)

3.3 NUMERICAL INTEGRATION OF DIFFERENTIAL EQUATIONS

Thus far we have obtained solutions to differential equations, or the simulation of systems described by differential equations, by arranging the equations into a form where the variables can be obtained by integrating the input functions. In general we can write a differential equation in the form

$$\dot{y} = f(y, x, t)^{\dagger}$$ (3.12)

where t is the independent variable, $x(t)$ the excitation, and $y(t)$ the solution.

Equations of higher than first order can be written as a set of first order equations. An analog computer diagram is just such a representation. If each of the integrator outputs is designated as a system variable, then the equation describing each integrator is a first order equation of the form (3.12) where the derivative of the output variable is a function of that variable, the other variables and the excitation. The next section generalizes the representation of systems in vector notation by the application of state variable analysis. So, we can consider the integration of first order equations of the form (3.12) and apply the results to each of the first order equations in parallel. The notation of Equation (3.12) is demonstrated in the example below.

EXAMPLE 3.6

Integrate the equation

$$\dot{y} = y^2 + 3$$

from the initial value $y_0 = 2$ for two steps by rectangular integration with $\tau = 0.01$.

SOLUTION:

By rectangular integration we have

$$y_n = y_{n-1} + \tau f(y_{n-1})$$

$$= y_{n-1} + \tau(y_{n-1}^2 + 3)$$

---

† For simplicity in notation we use $\dot{y} = f(y)$ keeping in mind that when f is a function of x and t also these variables should be included as in Equation (3.12).

For $\tau = 0.01$ we have

$$y_n = y_{n-1} + 0.01y_{n-1}^2 + 0.03$$

starting from $y_0 = 2$ we have $(y(0.01) = y_1)$

$$y_1 = y_0 + \tau f(y_0)$$

$$y_1 = y_0 + 0.01y_0^2 + 0.03$$

$$= 2 + 0.01(4) + 0.03$$

$$= 2.07$$

then

$$y_2 = y_1 + \tau f(y_1)$$

$$= y_1 + 0.01y_1^2 + 0.03$$

$$= 2.07 + 0.01(4.285) + 0.03$$

$$= 2.14285 \quad \blacktriangledown$$

Integrating the equation $\dot{y} = f(y)$ by trapezoidal integration yields the recursion relation

$$y_n = y_{n-1} + \frac{\tau}{2}\left(f(y_{n-1}) + f(y_n)\right) \qquad (3.13)$$

which cannot be directly solved for y_n. Chapter 4 discusses several methods for the iterative solution of equations like Equation (3.13) but they are time consuming and generally not suitable for solving at every step of the simulation unless it is unavoidable. Fortunately a more straightforward procedure exists in the so-called PREDICTOR-CORRECTOR methods.

PREDICTOR-CORRECTOR METHODS

The difficulty mentioned above in the evaluation of y_n where the right hand side of the integration equation like (3.13) contains a nonlinear function of y_n can be circumvented by predicting the value of y_n on the right hand side by a formula which does not use the value y_n, called an *OPEN FORMULA*. For example, we can *PREDICT* using rectangular integration to obtain the approximate value for y_n

$$p_n = y_{n-1} + \tau f(y_{n-1}) \qquad (3.14a)$$

and then *CORRECT* this value by using p_n on the right side of the *CLOSED FORMULA*

$$y_n = y_{n-1} + \frac{\tau}{2}\left(f(y_{n-1}) + f(p_n)\right) \qquad (3.14b)$$

Equations (3.14a) and (3.14b) make up a *PREDICTOR-CORRECTOR* pair.

EXAMPLE 3.7

Integrate the equation

$$\dot{y} = y^2 + 3$$

from the initial value $y_0 = 2$ one step using a rectangular predictor and trapezoidal corrector with an increment $\tau = 0.01$.

SOLUTION

Predict

$$p_1 = y_0 + \tau f(y_0)$$

$$= 2 + 0.01(2^2+3)$$

$$= 2.07$$

Correct

$$y_1 = y_0 + \frac{\tau}{2}\Big(f(y_0) + f(p_1)\Big)$$

$$= 2 + 0.005(2^2+3+2.07^2+3)$$

$$= 2.071425 \ \blacktriangledown$$

Predictor-corrector methods can be applied to system simulation by use of the digital-analog simulation program as demonstrated in the following example.

EXAMPLE 3.8

Use a rectangular predictor and trapezoidal corrector to write a simulation program to represent the system described by the van der Pol equation.

$$\ddot{y} - \varepsilon(1-y^2)\dot{y} + y = 0$$

SOLUTION:

Figure 2.18 shows a block diagram and variable names for the van der Pol equation. The predictor program using rectangular integration is identical to Example 2.10 with PRE1 and PRE2 representing the predicted outputs of integrators 1 and 2 respectively. The trapezoidal integration can be used for the corrector formula. (Write the predictor-corrector program **below** using two function subprograms; one for rectangular integration INTGRL, and the other for the trapezoidal integration INTGTR.)

```
            DO 153 K=1,N
            TIME=K*TAU
C
C   PREDICT.
C
            M1=I2*I2
            S2=-(M1-1.)
            M2=S2*I1
            P1=POT1*M2
            S1=-(P1+I2)
            PRE2=I2-TAU*I1
            PRE1=I1-TAU*S1
C
C   CORRECT.
C
            A1=S1
            M1=PRE2*PRE2
            S2=-(M1-1.)
            M2=S2*PRE1
            P1=POT1*M2
            S1=-(P1+PRE2)
            I2=I2-TAU*(I1+PRE1)/2.
            I1=I1-TAU*(A1+S1)/2.
C
C   INSERT OUTPUT STATEMENTS.
C
        153 CONTINUE            ▼
```

In the example above, the corrector formula using trapezoidal integration needs both the past value and the predicted value of the integrator inputs. Thus before S1 is changed in the corrector portion it must be saved, say as A1, to be used in the integration for the corrected value of I1. In general, all the integrator inputs must be saved to be used in the corrector portion of the program. The state variable representation of systems presented in the next section provides an organized technique for keeping track of the integrator inputs and outputs in a vector or array that greatly simplifies the numerical solution of high order differential equations.

Rectangular integration requires a single calculation of the slope $\dot{y} = f(y)$ for each step of the simulation, while the trapezoidal rule requires the evaluation of the slope at two points to determine the new value of the solution. Section 3.2 and Table 3.1 present integration schemes that make use of several past values of the solution to obtain the new values. These formulas, which use multiple past values, are called *MULTISTEP METHODS*.

The rectangular predictor-trapezoidal corrector procedure used in the previous examples are very useful. However, it is frequently desirable to use higher order formulas for the predictor-corrector pair. There are many predictor-corrector methods in general use. Perhaps the most popular are the *FOURTH-ORDER MILNE METHOD*.

PREDICTOR

$$P_n = y_{n-4} + \frac{4\tau}{3}\left[2f(y_{n-1}) - f(y_{n-2}) + 2f(y_{n-3})\right]$$

(3.15a)

CORRECTOR

$$y_n = y_{n-2} + \frac{\tau}{3}\left[f(p_n) + 4f(y_{n-1}) + f(y_{n-2})\right]$$

(3.15b)

and the *ADAMS MOULTON METHOD*

PREDICTOR

$$P_n = y_{n-1} + \frac{\tau}{24}\left[55f(y_{n-1}) - 59f(y_{n-2})\right.$$

$$\left. + 37f(y_{n-3}) - 9f(y_{n-4})\right]$$

(3.16a)

CORRECTOR

$$y_n = y_{n-1} + \frac{\tau}{24}\left[9f(p_n) + 19f(y_{n-1})\right.$$

$$\left. - 5f(y_{n-2}) + f(y_{n-3})\right]$$

(3.16b)

The error in the fourth order Milne and Adams-Moulton methods are of the order τ^5. (Fourth order means either that the method is correct to 4th order or the error is of 5th order.) Many other multistep predictor-corrector methods are discussed at length in any standard text on numerical analysis. (See for example reference [4].) These standard discussions also include derivations of the methods and the associated error estimates.

The predictor-corrector methods provide a convenient estimate of the error in one step of the calculation. If the method uses the same order predictor as corrector, as do the Milne and Adams-Moulton methods, the difference between the predicted and corrected values give an indication of the magnitude of the error. Thus if the error estimate is too large the increment can be decreased. Occasionally the results of the corrector formula are used as predicted values and the corrector formula is interated until there is no change in corrected value or until successive outputs do not differ by some predetermined amount. However, if the errors in the predictor and corrector formulas are of the same order little accuracy is gained by these iterations. When the corrector is iterated the method is called a *MODIFIED ADAMS METHOD*.

Applying the Adams-Moulton formulas (3.16a) and (3.16b) to even a third order system by the techniques used thus far in the digital-analog simulation program indicate that we must save four past values of the inputs to each of the integrators during each step of the simulation. This should give us increased motivation for the orderly bookkeeping procedures provided by the state variable representation of the system which we discuss in Section 3.4.

Multistep methods such as those presented above have one distinct limitation - you must supply the simulation with starting values. One technique for supplying these values might be to integrate the equations using rectangular integration, which requires only initial values for each variable, using a small increment size until sufficient starting values are obtained for the higher order formula to be used with a larger increment. For example, you might choose $\tau = 10^{-5}$ and use rectangular integration for 400 steps to obtain 4 values to start the *ADAMS-MOULTON* method with $\tau = 10^{-4}$ (See problem 3.15). A more popular method for generating starting values which can also be used for the entire simulation is application of *RUNGE-KUTTA METHODS*.

RUNGE-KUTTA METHODS

The Runge-Kutta Methods are a class of one step methods based on evaluation of slopes $\dot{y} = f(y)$ within the interval of calculation. Because of the considerable amount of algebra involved in their development we present here the general idea as applied to the second order and several other formulas, for application to the simulation program.

For the differential equation $\dot{y} = f(y)$ Runge-Kutta formulas use a value for the slope as evaluated at several points within the interval of calculation. The second order Runge-Kutta formula uses two estimates for the slope to give

$$y_n = y_{n-1} + \frac{\tau}{2}(k_1 + k_2) \qquad (3.17)$$

where

$$k_1 = f(y_{n-1})$$

and

$$k_2 = f(y_{n-1} + \tau k_1)$$

This formula evaluates the slope k_1 at $(n-1)\tau$, and uses this slope to go across the interval to obtain $y_{n-1} + \tau k_1$ which is used to evaluate the second estimate for the slope $k_2 = f(y_{n-1} + \tau k_1)$ The average of these slope estimates is used to go across the interval to obtain y_n.

Standard numerical analysis books develop higher order Runge-Kutta formulas extensively, the most popular of which is the *FOURTH ORDER RUNGE-KUTTA FORMULA*

$$y_n = y_{n-1} + \frac{\tau}{6}(k_1 + 2k_2 + 2k_3 + k_4) \qquad (3.18)$$

where

$$k_1 = f(y_{n-1})$$

$$k_2 = f(y_{n-1} + \frac{\tau}{2} k_1)$$

$$k_3 = f(y_{n-1} + \frac{\tau}{2} k_2)$$

and

$$k_4 = f(y_{n-1} + \tau k_3)$$

This fourth order formula uses the slope k_1 to go halfway across the interval where k_2 is evaluated. Then k_2 is used to go over the same half interval where we evaluate k_3. k_3 is then used to go all the way across the interval where the slope estimate k_4 is evaluated. Finally y_n is evaluated by Equation (3.18) using the weighted average of the slopes to minimize the error.

EXAMPLE 3.9

Use the second order Runge-Kutta formula to take the first step in the solution of the equation.

$$\dot{y} = y^2 + 3$$

from the initial value $y_0 = 2$ with $\tau = 0.01$.

SOLUTION:

First,

$$
\begin{aligned}
k_1 &= f(y_0) \\
    &= (2)^2 + 3 \\
    &= 7
\end{aligned}
$$

then

$$k_2 = f(y_0 + \tau k_1)$$

$$= f(2.07)$$

$$= (2.07)^2 + 3$$

$$= 7.285$$

finally,

$$y_1 = y_0 + \frac{\tau}{2}(k_1 + k_2)$$

$$= 2 + 0.005(7+7.285)$$

$$= 2.0714 \quad \blacktriangledown$$

As with the predictor corrector methods, we clearly need the bookkeeping procedures of state variables if a high order formula like (3.18) is applied to a high order system.

A COMPARISON OF THE METHODS

We conclude this brief introduction to predictor-corrector and Runge-Kutta methods with some considerations for the selection of a method for a particular problem.

While the multistep predictor-corrector formulas require few evaluations of the function $f(y)$ for each calculation, since they operate using past stored values, starting values must be generated to begin the calculation. On the other hand, the Runge-Kutta formulas don't require starting values but do require the evaluation of the function $f(y)$ several times for each interval. Thus there is a tradeoff between the calculation time and the ease of programming for the two methods. An excellent and popular compromise is to use a fourth order Runge-Kutta formula to begin the solution until sufficient starting values have been generated to proceed with the Adams-Moulton method. A distinct advantage of switching over to a predictor-corrector formula lies in the simplicity of obtaining error estimates for these formulas, like the difference between the predicted and corrected values. It is generally difficult to obtain accurate running estimates of the error in Runge-Kutta methods.

Most differential equation solving and system simulation programs automatically adjust the increment τ during the calculations to maintain the local relative error below some specified value like 10^{-6}. The local error is obtained in predictor-corrector methods by standard error formulas and by comparing the predicted and corrected values, while the Runge-Kutta programs can compare the results obtained by computing with two half steps and one full step. Although the latter is not efficient computationally, it is easy to program and very effective for obtaining quick accurate results.

Automatically adjusting the increment for the Runge-Kutta methods is straightforward since it requires starting values other than the immediate past value which is known. Thus a Runge-Kutta solution can change the increment frequently without additional difficulty. On the other hand, alteration of the step size with a predictor-corrector method is quite difficult since usually not all of the past values corresponding to the new increment are available. Usually the increment is changed by a factor of 2 or 1/2 to allow some of the past values to be used. Then intermediate new past values can be interpolated or calculated by a single step method. Alternatively, each time the increment is changed, the solution can be restarted with a Runge-Kutta formula. However, if the increment is changed too frequently we lose the computational speed advantage of the predictor-corrector method through continual restarting.

As promised, the next section discusses *STATE-VARIABLE ANALYSIS* which allows the representation of an nth order equation as n coupled first order equations. These equations, written in vector notation, can be solved in the same way as the first order equations treated in this section.

3.4 STATE VARIABLE METHODS IN SYSTEM SIMULATION

It is clear from the preceeding section that a systematic method for reducing high order differential equations to a set of first order differential equations is useful for the application of the integration schemes for first order equations. The state-variable approach provides such facility plus convenient techniques for the bookkeeping involved in the numerical integration or digital simulation of complex systems.

A great amount of work has been done on the theory of linear time invariant systems in terms of the state variable representation of the systems. This section presents the state-variable representation of linear and nonlinear systems with emphasis on its application to system simulation. Let's start with the definition of the state of a system.

THE STATE OF A SYSTEM: A minimum set of numbers or variables that contains sufficient information about the past history of the system to completely specify the future of the system; provided the present condition of the system, present and future excitations, and the equations describing the system are given is called the *STATE OF THE SYSTEM*.

The *STATE VARIABLES* are usually chosen to have some physical significance, like the energy stored in each element of the system, the settings of digital circuits in a digital computer, or the position and velocity of a particle. The selection of state variables is not unique but different sets of state variables merely provide a different type of description for the same thing. A convenient selection of state variables for the systems discussed so far are the outputs of the integrators of the simulation. Note that when all integrator initial conditions (the state of the system at t = 0) and the excitation to the simulation are given the behavior of the simulation is uniquely determined. So the integrator outputs comprise a sufficient set of variables to describe the system, and, if one of the integrator initial values is unspecified, the response is not uniquely determined. Thus, all of those variables are required and this selection makes a minimal set of state-variables.

An nth order system has n independent state variables. A convenient way to write an nth order differential equation as n first order equations (in the n state variables) is to let $x_1 = y$, $x_2 = \dot{x}_1$, .. , and so on.

$$x_1 = y$$

$$x_2 = \dot{x}_1 = \dot{y}$$

$$x_3 = \dot{x}_2 = \ddot{y}$$

$$\begin{array}{ccc} \cdot & \cdot & \cdot \\ \cdot & \cdot & \cdot \\ \cdot & \cdot & \cdot \end{array}$$

$$x_n = \dot{x}_{n-1} = \frac{d^{n-1}y}{dt^{n-1}} \qquad (3.19a)$$

Solving the nth order equation for the higher derivative,

$$\frac{d^n y}{dt^n} = f\left(y, \dot{y}, \ldots, \frac{d^{n-1}y}{dt^n}, u(t)\right)$$

$$= f\left(x_1, x_2, \ldots, x_n, u(t)\right) \qquad (3.19b)$$

where $u(t)$ is the excitation. Since $d^n y/dt^n = x_n$, (3.19a) and (3.19b) can be written in the form

$$\dot{x}_1 = x_2$$

$$\dot{x}_2 = x_3$$

$$\begin{array}{cc} \cdot & \cdot \\ \cdot & \cdot \\ \cdot & \cdot \end{array}$$

$$\dot{x}_{n-1} = x_{n-2}$$

$$\dot{x}_n = f(x_1, x_2, \ldots, x_n, u) \qquad (3.20)$$

EXAMPLE 3.10

Write the van der Pol equation with $\varepsilon = 1$ in state variable form defining the state variables by Equation (3.19a).

SOLUTION:

In the van der Pol equation

$$\ddot{y} - (1-y^2)\dot{y} + y = 0$$

let $x_1 = y$ and $\dot{x}_2 = \dot{y}$, then

$$\dot{x}_2 = \ddot{y} = (1-y^2)\dot{y} - y$$

$$= (1-x_1^2)x_2 - x_1$$

which gives the two equations

$$\dot{x}_1 = x_2$$

$$\dot{x}_2 = (1-x_1^2)x_2 - x_1 \qquad \blacktriangledown \qquad (3.21)$$

Since the state variables are an array of numbers for each value of the independent variable, it is convenient to write the set of equations given in (3.20) in the vector notation

$$\dot{X} = F(X,u) \qquad\qquad (3.22)$$

where X is the state vector, U is the excitation and F is a vector valued function of X and U. Writing Equation (3.20) in the form of Equation (3.22) we have

$$\dot{X} = \begin{bmatrix} \dot{x}_1 \\ \dot{x}_2 \\ \cdot \\ \cdot \\ \cdot \\ \dot{x}_n \end{bmatrix} = \frac{d}{dt} \begin{bmatrix} x_1 \\ x_2 \\ \cdot \\ \cdot \\ \cdot \\ x_n \end{bmatrix} = \begin{bmatrix} x_2 \\ x_3 \\ \cdot \\ \cdot \\ x_{n-1} \\ f(x_1, x_2, \ldots x_n, u) \end{bmatrix} \qquad (3.23)$$

EXAMPLE 3.11

Write the equations given in (3.21) in the vector notation of Equation (3.22).

SOLUTION:

$$\dot{X} = \begin{bmatrix} \dot{x}_1 \\ \dot{x}_2 \end{bmatrix} = \begin{bmatrix} x_2 \\ (1-x_1^2)x_2 - x_1 \end{bmatrix}, \qquad (3.24)$$ ▼

Equations (3.22), (3.23) and (3.24) each give the rule for determining the numbers $\dot{X}$ as a function of the numbers X and U. Specifically in (3.24), if X is given, say as an initial condition X_0, then the two elements of X_0 are substituted into (3.24) to obtain the two elements of $\dot{X}$. The example demonstrates this evaluation and the simplicity of this state variable representation for digital simulation.

EXAMPLE 3.12

Evaluate the first step of the second order Runge-Kutta integration of Equation (3.24) from the initial conditions $x_1(0) = 3$ and $x_2(0) = 2$; $(y(0) = 3,$ $y(0) = 2$ with $\tau = .01$.

SOLUTION:

Note now that the slopes K_1 and K_2 are vectors.

$$\dot{X} = F(X), \qquad x(0) = \begin{bmatrix} 3 \\ 2 \end{bmatrix}$$

$$K_1 = F(X_0) = \begin{bmatrix} x_2(0) \\ (1-x_1^2(0))x_2(0) - x_1(0) \end{bmatrix}$$

$$= \begin{bmatrix} 2 \\ (1-3^2)2 - 3 \end{bmatrix}$$

$$= \begin{bmatrix} 2 \\ -19 \end{bmatrix}$$

$$K_2 = F(X_0 + \tau K_1)$$

$$= F(\begin{bmatrix} 3 \\ 2 \end{bmatrix} + 0.01 \begin{bmatrix} 2 \\ -19 \end{bmatrix})$$

$$= F(\begin{bmatrix} 3.02 \\ 1.81 \end{bmatrix})$$

$$= \begin{bmatrix} 1.81 \\ (1-3.02^2)(1.81) - 3.02 \end{bmatrix}$$

$$= \begin{bmatrix} 1.81 \\ -18.6 \end{bmatrix}$$

finally

$$X_1 = X_0 + \frac{\tau}{2}(K_1 + K_2)$$

$$= \begin{bmatrix} 3 \\ 2 \end{bmatrix} + 0.005 \left(\begin{bmatrix} 2 \\ -19 \end{bmatrix} + \begin{bmatrix} 1.81 \\ -18.6 \end{bmatrix} \right)$$

$$= \begin{bmatrix} 3 \\ 2 \end{bmatrix} + \begin{bmatrix} 0.019 \\ -0.188 \end{bmatrix}$$

$$X(0.01) = X_1 = \begin{bmatrix} 3.019 \\ 1.812 \end{bmatrix}$$

This completes one step in the numerical integration of the given system equations.▼

The numerical integration of state equations $\dot{X} = F(X)$ by Runge–Kutta methods requires repeated evaluations of the function f as encountered in the previous section. Accordingly it is convenient to write a subroutine describing the system which has for input numbers the elements of X and for output numbers the elements of f.

In Example 3.12 we would use the subroutine called SYSTEM for the system equations

```
SUBROUTINE SYSTEM(X1,X2,F1,F2)
F1=X2
F2=(1-X1*X1)*X2-X1
RETURN
END
```

Then the fourth order Runge–Kutta simulation of Equation (3.21) or any other system is given by Equation (3.19) (for one step) as †

```
CALL SYSTEM(X1,X2,K11,K12)
CALL SYSTEM(X1+T*K11/2.,X2+T*K12/2.,K21,K22)
CALL SYSTEM(X1+T*K21/2.,X2+T*K22/2.,K31,K32)
CALL SYSTEM(X1+T*K31,X2+T*K32,K41,K42)
X1=X1+(T/6)*(K11+2.*K21+2.*K31+K41)
X2=X2+(T/6)*(K12+2.*K22+2.*K32+K42)
```

---

† Don't forget REAL K11, K12, ...

Kll is the first estimate of the slope of Xl. In general KIJ is the Ith calculation of the slope of XJ during this interval. With appropriate initial conditions and output this is the entire program. Note that changing SYSTEM is all that is necessary to change the simulation. The method remains the same.

To generalize the program to simulate or integrate an Nth order system we need only define the arrays or vectors X(N) and F(N) for the system equations and K1(N), K2(N), K3(N), K4(N) are the successive slope estimates, giving the brief program

```
CALL SYSTEM(X,K1)
CALL SYSTEM(X+T*K1/2,K2)
CALL SYSTEM(X+T*K2/2,K3)
CALL SYSTEM(X+T*K3,K4)
X=X+(T/6)*(K1+2*K2+2*K3+K4)
```

where the matrix operations are performed by matrix subroutines.

LINEAR SYSTEMS

In the analysis of linear systems the state equations $\dot{X} = F(X,U)$ are of course linear differential equations. When F is a linear combination of X and U, we can write the state equations in the form

$$\dot{X} = AX + BU \qquad (3.25)$$

where X is the n dimensional state vector, U is a vector of inputs (there may be more than one input), and A and B are matrices of appropriate dimension so as to make the equation consistent. If X has n elements and U has r elements, then A has n rows and n columns while B has n rows and r columns. †

The linear differential equation

$$\frac{d^n y}{dt^n} + a_n \frac{d^{n-1} y}{dt^{n-1}} + \ldots + a_1 \frac{dy}{dt} + a_0 y = u(t)$$

$$(3.26)$$

---

† Some useful matrix subroutines are given in Example 4.8 of the next chapter.

can be written in the form of Equation (3.25) by the selection of state variables in Equation (3.19) to obtain

$$\dot{x}_1 = x_2$$

$$\dot{x}_2 = x_3$$

$$\vdots$$

$$\dot{x}_{n-1} = x_n$$

$$\dot{x}_n = -(a_0 x_1 + a_1 x_2 + \ldots a_{n-1} x_n) + u$$

$$(3.27)$$

Equations (3.27) represent the analog computer block diagram shown in Figure 3.7. For this section it is convenient to draw block diagrams with integrators, I, that do not have sign reversals as shown in this figure. These state equations of the system are said to be in *PHASE VARIABLE FORM* which is the form we used exclusively in Chapter 2.

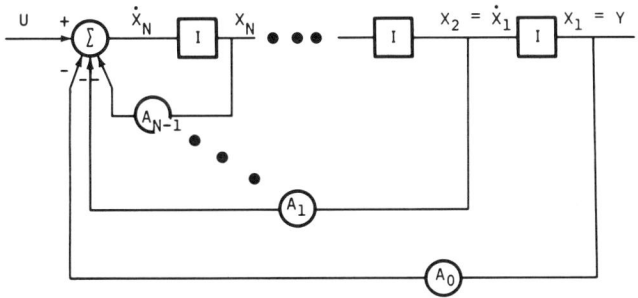

FIGURE 3.7 ANALOG DIAGRAM FOR THE DIFFERENTIAL EQUATION
$$Y^{(N)} + A_{N-1} Y^{(N-1)} + \ldots + A_1 \dot{Y} + A_0 Y = U(T)$$
IN PHASE VARIABLE FORM.

Writing the state equations using matrix and vector notation in the form of Equation (3.25) we have

$$\dot{X} = AX + BU$$

where X is the vector

$$X = \begin{bmatrix} x_1 \\ x_2 \\ \cdot \\ \cdot \\ \cdot \\ x_n \end{bmatrix}$$

U is the scalar u(t), and A and B are the matrices

$$A = \begin{bmatrix} 0 & 1 & 0 & \cdots & 0 \\ 0 & 0 & 1 & \cdots & 0 \\ \cdot & \cdot & \cdot & & \cdot \\ \cdot & \cdot & \cdot & & \cdot \\ \cdot & \cdot & \cdot & & \cdot \\ 0 & 0 & 0 & & 1 \\ -a_0 & -a_1 & -a_2 & \cdots & -a_{n-1} \end{bmatrix} \qquad B = \begin{bmatrix} 0 \\ 0 \\ \cdot \\ \cdot \\ \cdot \\ 0 \\ 1 \end{bmatrix}$$

In general if a system has r inputs, $u_1 \ldots u_r$, and n state variables, $x_1 \ldots x_n$, the state equation can be written

$$\dot{X} = F(X,U,t) \qquad\qquad (3.28a)$$

Note that the system is given as a function of time. Additionally, the output variables, $y_1 \ldots y_m$, are usually some function of X, U and t of the form

$$Y = G(X,U,t) \qquad\qquad (3.28b)$$

where

$$X = \begin{bmatrix} x_1 \\ x_2 \\ \vdots \\ x_n \end{bmatrix} \quad U = \begin{bmatrix} u_1 \\ u_2 \\ \vdots \\ u_r \end{bmatrix} \quad Y = \begin{bmatrix} y_1 \\ y_2 \\ \vdots \\ y_m \end{bmatrix}$$

$$F(X,U,t) = \begin{bmatrix} f_1(X,U,t) \\ f_2(X,U,t) \\ \vdots \\ f_n(X,U,t) \end{bmatrix} \quad g(X,U,t) = \begin{bmatrix} g_1(X,U,t) \\ g_2(X,U,t) \\ \vdots \\ g_m(X,U,t) \end{bmatrix}$$

If the system is linear then the vector functions $F(X,U,t)$ and $G(X,U,t)$ are linear and the elements of X and Y are linear combinations of the elements of X and U, so they can be written

$$\dot{x}_1 = a_{11}x_1 + a_{12}x_2 + \ldots + a_{1n}x_n + b_{11}u_1 + \ldots + b_{1r}u_r$$

$$\dot{x}_2 = a_{21}x_1 + a_{22}x_2 + \ldots + a_{2n}x_n + b_{21}u_1 + \ldots + b_{2r}u_r$$

$$\ldots \qquad\qquad \ldots \qquad\qquad \ldots$$

$$\dot{x}_n = a_{n1}x_1 + a_{n2}x_2 + \ldots + a_{nn}x_n + b_{n1}u_1 + \ldots + b_{nr}u_r$$

and

$$y_1 = c_{11}x_1 + \ldots + c_{1n}x_n + d_{11}u_1 + \ldots + d_{1r}u_r$$

$$\ldots \qquad\qquad \ldots \qquad\qquad \ldots$$

$$y_m = c_{m1}x_1 + \ldots + c_{mn}x_n + d_{m1}u_1 + \ldots + d_{mr}u_r$$

Rewriting in matrix notation as above, we have the linear state equations

$$\dot{X} = AX + BU$$

$$Y = CX + DU$$

(3.29)

where

$$A = \begin{bmatrix} a_{11} & a_{12} & \cdots & a_{1n} \\ a_{21} & a_{22} & \cdots & a_{2n} \\ \vdots & \vdots & & \vdots \\ a_{n1} & a_{n2} & \cdots & a_{nn} \end{bmatrix} \qquad B = \begin{bmatrix} b_{11} & b_{12} & \cdots & b_{1r} \\ b_{21} & b_{22} & \cdots & b_{2r} \\ \vdots & \vdots & & \vdots \\ b_{n1} & b_{n2} & \cdots & b_{nr} \end{bmatrix}$$

and

$$C = \begin{bmatrix} c_{11} & c_{12} & \cdots & c_{1n} \\ c_{21} & c_{22} & \cdots & c_{2n} \\ \vdots & \vdots & & \vdots \\ c_{m1} & c_{m2} & \cdots & c_{mn} \end{bmatrix} \qquad D = \begin{bmatrix} d_{11} & d_{12} & \cdots & d_{1r} \\ d_{21} & d_{22} & \cdots & d_{2r} \\ \vdots & \vdots & & \vdots \\ d_{m1} & d_{m2} & \cdots & d_{mr} \end{bmatrix}$$

The elements of the *SYSTEM MATRIX* A depend upon the mathematical description of the system, and the elements of the distribution matrix B show how the input variables affect the state variables. C and D show how the state variables and input variables directly affect the output variables. In general linear systems the elements of the A, B, C and D matrices are time varying, although they have been written as constants in the above equations.

EXAMPLE 3.13

(a) Write the following equations in phase variable form and (b) give a block diagram showing the state variables.

$$\ddot{y} + a_1\dot{y} + a_0y = u(t)$$

SOLUTION:

(a) Let $x_1 = y$ and $x_2 = \dot{y}$; then $\dot{x}_1 = x_2$ and $\dot{x}_2 = -a_1x_2 - a_0x_1 + u$

The state equations can be written in matrix form

$$\dot{X} = \frac{d}{dt}\begin{bmatrix} x_1 \\ x_2 \end{bmatrix} = \begin{bmatrix} 0 \\ -a_0 \end{bmatrix}\begin{bmatrix} 1 \\ -a_1 \end{bmatrix}\begin{bmatrix} x_1 \\ x_2 \end{bmatrix} + \begin{bmatrix} 0 \\ 1 \end{bmatrix}u$$

which is in the form of Equation (3.29).

(b) The block diagram is shown in Figure 3.8. As with the analog block diagrams, assume you have $\ddot{y}$ and integrate twice, then feed back the outputs of the integrators with appropriate gains and signs. With noninverting integrators sign reversals are incorporated into the POTs of the diagram or indicated at the summing junction.

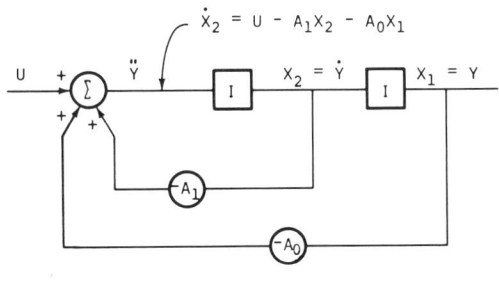

FIGURE 3.8 BLOCK DIAGRAM FOR STATE EQUATIONS IN EXAMPLE 3.13.

Defining the output of the last integrator x_1 and proceeding to the first integrator naming the output of the kth integrator from the end x_k gives a definition of the state variables in phase variable form. Clearly this selection corresponds to the equations obtained in part (a).▼

A general system differential equation involves not only the excitation a but its derivatives in the form of Equation (3.30).

$$(D^n + a_{n-1}D^{n-1} + \ldots + a_1 D + a_0) y = (b_n D^n + \ldots + b_1 D + b_0) u$$

$$(3.30)$$

y is the solution or response, u is the excitation and $D = d/dt$. The block diagram in Figure 3.9 is easily shown to be equivalent to Equation (3.30). (See Problem 3.14)

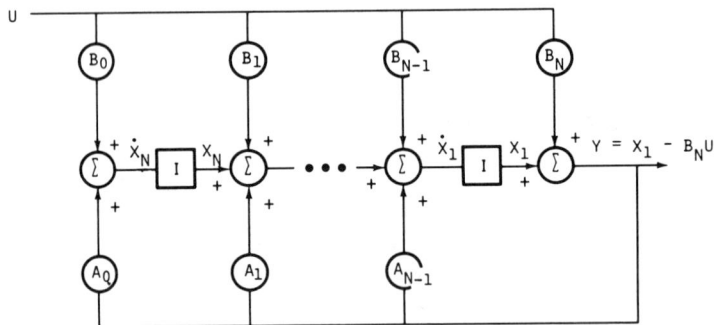

FIGURE 3.9 ·BLOCK DIAGRAM REPRESENTATION OF SYSTEM DESCRIBED BY EQUATION (3.30).

By defining the state variables to be the outputs of the integrators as shown in the figure, we have the state equations

$$\dot{x}_1 \quad = x_2 + b_{n-1}u - a_{n-1}y$$

$$\vdots$$

$$\dot{x}_{n-1} = x_n + b_1 u - a_1 y$$

$$\dot{x}_n \quad = b_0 u - a_0 y \qquad\qquad (3.31)$$

Substitution of the output equation

$$y \quad = x_1 + b_n u \qquad\qquad (3.32)$$

into Equation (3.11) gives

$$\dot{x}_1 \quad = -a_{n-1}x_1 + x_2 + (b_{n-1} - a_{n-1}b_n)u$$

$$\vdots$$

$$\dot{x}_{n-1} = -a_1 x_1 \quad + x_n + (b_1 - a_1 b_n)u$$

$$\dot{x}_n \quad = -a_0 x_1 \qquad\qquad + (b_0 - a_0 b_n)u \qquad (3.33)$$

Equation (3.32) and (3.33) are now in the form of Equation (3.29)

$$\dot{X} = AX + BU$$
$$\qquad\qquad\qquad\qquad\qquad (3.29)$$
$$Y = CX + DU$$

where y and u are scalars or one dimensional vectors and

$$A = \begin{bmatrix} -a_{n-1} & 1 & 0 & 0 & \cdots \\ -a_{n-2} & 0 & 1 & 0 & \cdots \\ \vdots & \vdots & \vdots & \vdots & \\ \vdots & \vdots & \vdots & \vdots & \\ -a_1 & & & 0 & 1 \\ -a_0 & & & 0 & 0 \end{bmatrix} \qquad B = \begin{bmatrix} b_{n-1} - a_{n-1}b_n \\ b_{n-2} - a_{n-2}b_n \\ \vdots \\ \vdots \\ b_1 - a_1 b_n \\ b_0 - a_0 b_n \end{bmatrix}$$

$$C = [1 \quad 0 \quad \cdots \quad 0]$$

$$D = [b_n]$$

EXAMPLE 3.14

Give a block diagram and state equation representation of a system described by the differential equation

$$(D^3 + 5D^2 + 2D)y = (2D^2 + D)u \qquad (3.34)$$

SOLUTION:

Comparing the given equation with Equation (3.30) we have ($n = 3$)

$$a_2 = 5, \quad a_1 = 2, \quad a_0 = 0, \quad b_3 = 0, \quad b_2 = 2,$$
$$b_1 = 1, \text{ and } b_0 = 0$$

Substitution into Equation (3.29) gives the state equations

$$\dot{X} = \frac{d}{dt}\begin{bmatrix} x_1 \\ x_2 \\ x_3 \end{bmatrix} = \begin{bmatrix} -5 & 1 & 0 \\ -2 & 0 & 1 \\ 0 & 0 & 0 \end{bmatrix}\begin{bmatrix} x_1 \\ x_2 \\ x_3 \end{bmatrix} + \begin{bmatrix} 2 \\ 1 \\ 0 \end{bmatrix}u$$

and the output equation

$$y = \begin{bmatrix} 1 & 0 & 0 \end{bmatrix}\begin{bmatrix} x_1 \\ x_2 \\ x_3 \end{bmatrix} + [0]u$$

The substitution of the coefficients into the block diagram in Figure 3.9 results in the block diagram shown in Figure 3.10. ▼

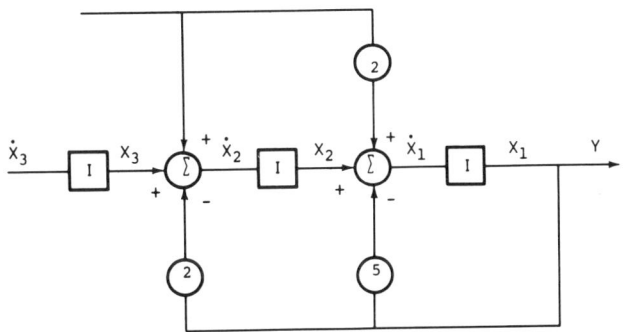

FIGURE 3.10 EXAMPLE 3.14 BLOCK DIAGRAM OF THE SYSTEM $(D^3+5D^2+2D)Y = (2D^2+D)U$.

Note that in the above example the first integrator has no input, so that x_3 never changes from its initial value. This is also clear from the system equations because the third row of the System Matrix has all zero elements. This situation is the result of the common factor in the right and left sides of Equations (3.19). Integration of the system differential equation once gives the equivalent system which is a second order system with two independent states. It is left for the reader to obtain a block diagram and system state equations for this equivalent equation and to observe its relationship to Figure 3.10. For other equivalent block diagram representations and their corresponding state equations see Problems 3.17 and 3.18.

What we now desire are discrete representations of the state equations. Since the output vector Y can be easily obtained from the state X and the input U by Equation (3.28b), we concentrate on digital solutions to the state Equation (3.28a)

$$\dot{X} = F(X, U, t) \tag{3.28a}$$

which is the same as the first order equation discussed in the last section in terms of scalar functions.

Integrating Equation (3.28a) over the interval $[(n-1)\tau, n\tau]$ gives

$$X(n\tau) - X((n-1)\tau) = \int_{t_{n-1}}^{t_n} F(X, U, t)\, dt$$

or

$$X_n = X_{n-1} + \int_{t_{n-1}}^{t_n} F(X, U, t)\, dt \tag{3.35}$$

where $t_n = n\tau$.[†]

The system equations can be numerically solved by approximation of the integral in Equation (3.35).

---

† The subscript on the state vector X_n corresponds to the independent variable increment $X_n = X(n\tau)$ while the subscript on the scalar x_k corresponds to the kth element of the state vector.

Evaluating the integral by Euler (rectangular) integration gives

$$X_n = X_{n-1} + \tau F(X_{n-1}, U_{n-1}, t_{n-1})$$

or

$$X_n = X_{n-1} + \tau F_{n-1} \qquad\qquad (3.36)$$

where by analogy to the scalar case we denote $F_n = F(X_n, U_n, t)$.

Rectangular integration of Equations (3.28) and (3.35) to obtain Equation (3.36) is identical to the rectangular integration used in the last chapter for the digital-analog simulation program. All of the state variables (integrator outputs) are calculated from the past values of the integrator inputs. The evaluation of F_{n-1} from X_{n-1}, U_{n-1} and t_n is the evaluation of all integrator inputs from the present integrator outputs (X_{n-1}) and the system inputs U_{n-1}. When all of the integrator inputs (F_{n-1}) are evaluated a new set of outputs is calculated by Equation (3.36).

The application of the predictor-corrector and Runge-Kutta methods to the solution of the system equations (3.28) is straightforward as demonstrated in Example 3.12. The relationship between using these integration schemes as subroutines in the digital-analog simulation program and directly integrating the state variable equations is explored in greater depth in the problems, and a general purpose program for the integration of differential equations by the 4th order Runge-Kutta formula is given in the appendix.

The direct application of the integration formulas to the state equations of linear systems provides some insight into recent methods for the simulation of linear systems using state variables. Consider the solution of the unforced linear system ($U = 0$)

$$\dot{X} = AX \qquad\qquad (3.37)$$

by Euler integration (a first order method),

$$X_n = X_{n-1} + \tau F(X_{n-1})$$

$$= X_{n-1} + \tau A X_{n-1}$$

or

$$X_n = (I + \tau A) X_{n-1} \qquad\qquad (3.38)$$

by modified Euler (rectangular predictor and trapezoidal corrector – a second order method),

$$X_n = X_{n-1} + \frac{\tau}{2}\Big(F(X_{n-1}) + F(X_{n-1} + F(X_{n-1}))\Big)$$

$$= X_{n-1} + \frac{\tau}{2}\Big(A X_{n-1} + A(X_{n-1} + \tau A X_{n-1})\Big)$$

or

$$X_n = \Big(I + \tau A + \frac{\tau^2}{2} A^2\Big) X_{n-1} \qquad\qquad (3.39)$$

and finally by the 4th order Runge Kutta formula

$$X_n = X_{n-1} + \frac{\tau}{6}\Big(K_1 + 2K_2 + 2K_3 + K_4\Big)$$

where

$$K_1 = A X_{n-1}$$

$$K_2 = A\Big(X_{n-1} + \frac{\tau}{2} X_1\Big)$$

$$K_3 = A\Big(X_{n-1} + \frac{\tau}{2} K_2\Big)$$

and

$$K_4 = A\Big(X_{n-1} + \tau K_3\Big)$$

which can be simplified to obtain

$$X_n = \left(I + \tau A + \frac{\tau^2}{2} A^2 + \frac{\tau^3}{6} A^3 + \frac{\tau^4}{24} A^4 \right) X_{n-1}$$

(3.40)

Comparison of Equations (3.38), (3.39), and (3.40) indicates that higher order formulas appear to be the equivalent of taking more terms of a Taylor series expansion of the solution, which is indeed the case. In fact, any 4th order method applied to Equation (3.37) yields the same first 5 terms of the series given in Equation (3.40). In the following discussion we show that the exact solution of Equation (3.37) is just Equation (3.40) with the sum replaced by the infinite series

$$\Phi(\tau) = \sum_{k=0}^{\infty} \frac{\tau^k A^k}{k!}$$

(3.41)

It is reasonable to expect that the first order linear matrix differential equation has a solution of exponential form since it is merely a set of scalar linear equations which have an exponential type of solution. The introduction of the concept of a matrix exponential leads to several useful and accurate methods for digital simulation.

The *MATRIX EXPONENTIAL* e^{At}, which is a function of the matrix A, may be defined by the infinite series of matrices

$$e^{At} = I + At + \frac{A^2 t^2}{2!} + \frac{A^3 t^3}{3!} + \cdots$$

$$e^{At} = \sum_{k=0}^{\infty} \frac{A^k t^k}{k!}$$

(3.42)

where $A^0 = I$.

It is easy to show from Equation (3.42) that the matrix exponential has the following properties. (See Problem 3.2.)

$$Ae^{At} = e^{At}A \qquad (3.43a)$$

$$e^{At}e^{Bt} = e^{(A+B)t} \qquad (3.43b)$$

$$e^{A0} = I \qquad (3.43c)$$

and

$$\frac{d(e^{At})}{dt} = Ae^{At} \qquad (3.43d)$$

It is also easy to show that the solution to the homogeneous equation $\dot{X} = AX$, is

$$X(t) = e^{At}X(0) \qquad (3.44)$$

Clearly Equation (3.44) satisfied the initial condition $X(0)$ since $e^{A0} = I$. Also substituting $e^{At}x(0)$ into Equation (3.37) gives

$$\dot{X} = AX$$

$$\frac{d(e^{At}X(0))}{dt} = A(e^{At}X(0)) \qquad (3.37)$$

and using (3.43d) gives

$$Ae^{At}X(0) = Ae^{At}X(0)$$

demonstrating that (3.44) is indeed a solution.

For the linear system

$$\dot{X} = AX + BU \qquad (3.29)$$

We can obtain the solution by rewriting the equation

$$\dot{X} - AX = BU$$

Multiplying on the left of each term by e^{-At} gives

$$e^{-At}\dot{X} - e^{-At}AX = e^{-At}BU$$

The left side is the derivative of $e^{-At}X$ so we have

$$e^{-At}\dot{X} - e^{-At}AX = \frac{d}{dt}(e^{-At}X) = e^{-At}BU$$

Integrating both sides from 0 to t and using the fact that $e^{A0} = I$ gives

$$e^{-At}X = X(0) + \int_0^t e^{-A\alpha}BU(\alpha)\,d\alpha$$

Finally multiplying on the left of each term by e^{At} gives the exact solution

$$X(t) = e^{At}X(0) + e^{At}\int_0^t e^{-A\alpha}BU(\alpha)\,d\alpha \qquad (3.45)$$

Clearly if $U = 0$ Equation (3.45) reduces to the solution to the homogenous system given in Equation (3.44).

The first term on the right of Equation (3.48) is called the *FORCE FREE RESPONSE*, as it is the system response in the absence of a forcing function or excitation U. The second term is called the *ZERO*

STATE RESPONSE (that is, all state variables are initially zero.) The matrix exponential e^{At} is called the fundamental matrix of the system and is denoted

$$\Phi(t) = e^{At} = \sum_{k=0}^{\infty} \frac{A^k t^k}{k!} \qquad (3.46)$$

The force free response is, of course, the solution to the homogeneous equation $\dot{X} = AX$. The following example demonstrates the application of Equation (3.45) to a first order scalar equation.

EXAMPLE 3.15

Evaluate and sketch the zero state and force free response of the differential equation

$$\dot{y} + y = 2$$

with the initial value $y(0) = -1$.

SOLUTION:

Rewriting the equation, we get

$$\dot{y} = -y + 2$$

so $A = -1$, $y(0) = -1$, and $u = 2$.

From Equation (3.45) we have

$$y(t) = e^{-t} y(0) + e^{-t} \int_0^t e^{-\alpha} 2d\alpha$$

$$= -e^{-t} + 2(1 - e^{-t})$$

$$= y_{ff} + y_{zs}$$

Figure 3.11 shows the force free response y_{ff}, zero state response y_{zs}, and total response $y(t)$. ▼

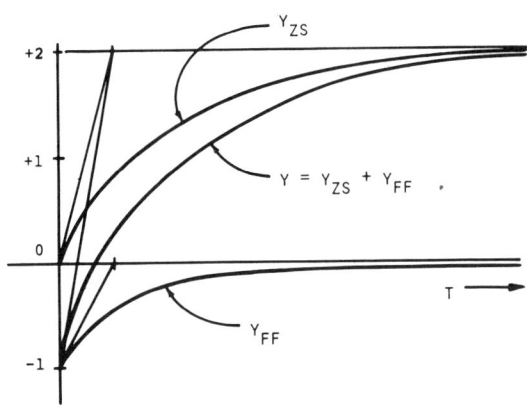

FIGURE 3.11 EXAMPLE 3.15 ZERO STATE, FORCE FREE, AND TOTAL RESPONSE
FOR THE SYSTEM $\dot{y} + y = 2$, $y(0) = -1$.

In the above example the fundamental matrix is merely the scalar exponential $e^{At} = e^{-t}$. Some additional insight into the meaning of the fundamental matrix is obtained by considering the force free response in general. Writing

$$X(t) = e^{At}X(0)$$

$$= \Phi(t)X(0)$$

or

$$X(t) = \begin{bmatrix} x_1(t) \\ \cdot \\ \cdot \\ \cdot \\ x_n(t) \end{bmatrix} = \begin{bmatrix} \phi_{11}(t) & \phi_{12}(t) & \cdots & \\ \phi_{21}(t) & \phi_{22}(t) & \cdots & \\ \cdot & \cdot & & \\ \cdot & \cdot & & \\ \cdot & \cdot & & \phi_{nn}(t) \end{bmatrix} \begin{bmatrix} x_1(0) \\ \cdot \\ \cdot \\ \cdot \\ x_n(0) \end{bmatrix}$$

$$(3.47)$$

Now if each of the state variables is initially zero except the ith which is 1 we have

$$X(t) = \begin{bmatrix} x_1(t) \\ \cdot \\ \cdot \\ \cdot \\ x_n(t) \end{bmatrix} = \begin{bmatrix} \phi_{11}(t) & \phi_{12}(t) & \cdots \\ \phi_{21}(t) & \phi_{22}(t) & \cdots \\ \cdot & & \\ \cdot & \cdot & \\ \cdot & \cdot & \phi_{nn}(t) \end{bmatrix} \begin{bmatrix} 0 \\ \cdot \\ \cdot \\ 0 \\ 1 \\ 0 \\ \cdot \\ \cdot \\ 0 \end{bmatrix}$$

or

$$\begin{bmatrix} x_1(t) \\ \cdot \\ \cdot \\ \cdot \\ x_n(t) \end{bmatrix} = \begin{bmatrix} \phi_{1i}(t) \\ \cdot \\ \cdot \\ \cdot \\ \phi_{ni}(t) \end{bmatrix}$$

So the prescribed initial condition gives the response $X(t)$ which is the ith column of the fundamental matrix, or the ith column of the fundamental matrix is the state vector response of the system if all the initial conditions are set equal to zero except that on the ith state variable (x_i), which is set at 1. For a general initial state $X(0)$ the force free response is the superposition of the weighted initial conditions on each element of $X(0)$.

To obtain the elements of the fundamental matrix as a function of time we can set up an analog computer simulation similar to Figures 3.7 or 3.9 with no excitation or forcing function. Then the element $\phi_{ij}(t)$ of the fundamental matrix is the response of the jth state variable $x_j(t)$ to the initial conditions

$$x_k(0) = \begin{cases} 0 & \text{for } k \neq i \\ 1 & \text{for } k = i \end{cases}$$

EXAMPLE 3.16

Give an analog computer block diagram to generate each element of the fundamental matrix corresponding to the system $\dot{X} = AX$ where

$$A = \begin{bmatrix} 0 & 1 \\ -6 & -5 \end{bmatrix}$$

SOLUTION: Since $\dot{X} = AX$ we have

$$\begin{bmatrix} \dot{x}_1 \\ \dot{x}_2 \end{bmatrix} = \begin{bmatrix} 0 & 1 \\ -6 & -5 \end{bmatrix} \begin{bmatrix} x_1 \\ \dot{x}_2 \end{bmatrix}$$

or

$$\dot{x}_1 = x_2$$
$$\dot{x}_2 = -6x_1 - 5x_2$$

Taking the state variables x_1 and $-x_2$ as integrator outputs we have the block diagram shown in Figure 3.12. (Verify that the block diagram corresponds to the system equations and compare this with Figure 3.7.)

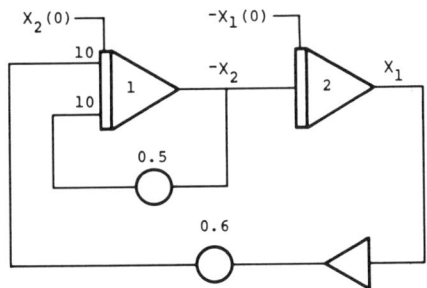

FIGURE 3.12 BLOCK DIAGRAM FOR EXAMPLE 3.16 IN STATE VARIABLE FORM.

The solution $X(t) = e^{At}X(0) = \Phi(t)x(0)$ can be written in the form of Equation (3.47)

$$X(t) = \begin{bmatrix} x_1(t) \\ x_2(t) \end{bmatrix} = \begin{bmatrix} \phi_{11}(t) & \phi_{12}(t) \\ \phi_{21}(t) & \phi_{22}(t) \end{bmatrix} \begin{bmatrix} x_1(0) \\ x_2(0) \end{bmatrix}$$

The application of the initial conditions

$$X(0) = \begin{bmatrix} x_1(0) \\ x_2(0) \end{bmatrix} = \begin{bmatrix} 1 \\ 0 \end{bmatrix}$$

gives the response

$$X(t) = \begin{bmatrix} x_1(t) \\ x_2(t) \end{bmatrix} = \begin{bmatrix} \phi_{11}(t) \\ \phi_{21}(t) \end{bmatrix}$$

That is, if the initial conditions on integrators 1 and 2 are 0 and 1 respectively, then $x_1(t)$ traces out $\phi_{11}(t)$ and $x_2(t)$ traces out $\phi_{21}(t)$. When this is implemented we obtain the plots of ϕ_{11} and ϕ_{21} shown in Figure 3.13a. Similarly, the initial condition

$$X(0) = \begin{bmatrix} x_1(0) \\ x_2(0) \end{bmatrix} = \begin{bmatrix} 0 \\ 1 \end{bmatrix}$$

yields the response

$$X(t) = \begin{bmatrix} x_1(t) \\ x_2(t) \end{bmatrix} = \begin{bmatrix} \phi_{21}(t) \\ \phi_{22}(t) \end{bmatrix}$$

which is shown in Figure 3.13b. ▼

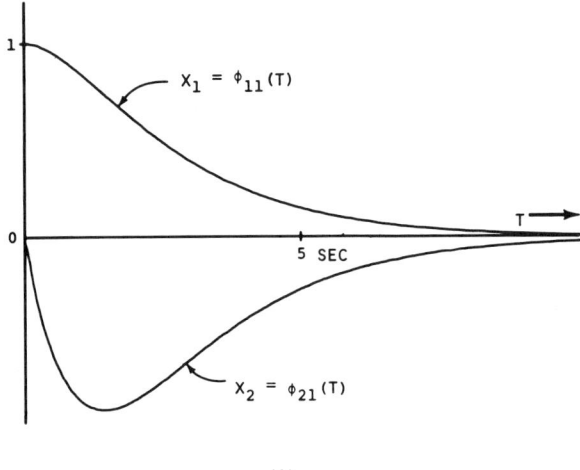

(A)

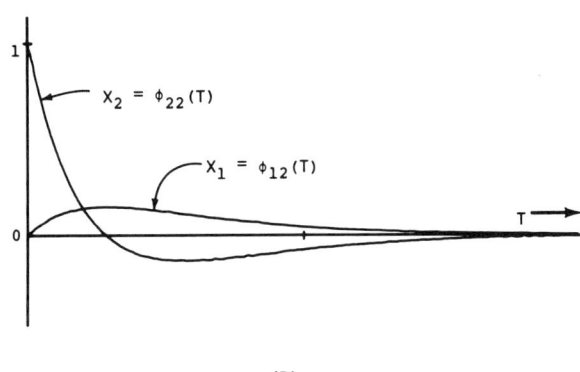

(B)

FIGURE 3.13 ELEMENTS OF THE FUNDAMENTAL MATRIX FOR EXAMPLE 3.16.

In Figure 3.13, ϕ_{21} is the derivative of ϕ_{11} and ϕ_{22} is the derivative of ϕ_{12} since $x_2 = \dot{x}_1$.
To obtain the force-free response to arbitrary initial conditions, such as

$$X(0) = \begin{bmatrix} x_1(0) \\ x_2(0) \end{bmatrix} = \begin{bmatrix} \alpha \\ \beta \end{bmatrix}$$

we merely take the superposition of the individual responses weighted by the magnitude of the initial conditions, α and β.

To calculate the force free response on the digital computer we must evaluate

$$X(t) = e^{At}X(0)$$

at discrete increments of time $t_n = n\tau$, to obtain

$$X_n = X(n\tau) = e^{An\tau}X(0) = e^{An\tau}X_0$$

But

$$X_{n-1} = e^{A(n-1)\tau}X_0$$

so

$$X_n = e^{A\tau}\left[e^{A(n-1)\tau}X_0\right]$$

$$X_n = e^{A\tau}X_{n-1} \qquad\qquad (3.48)$$

Thus the matrix $e^{A\tau}$ takes discrete steps through the force free response of the system. For this reason the matrix $e^{A\tau}$ for a fixed number τ is called the *DISCRETE STATE TRANSITION MATRIX*. For the digital calculation of Equation (3.48), $e^{A\tau}$ must be evaluated.

It is impractical to evaluate $e^{A\tau}$ on the analog computer as discussed in Example 3.15. We did that example only to gain insight into the fundamental matrix $e^{A\tau}$. To evaluate the discrete state transition matrix on the digital computer in a practical way we merely take a sufficient number of terms in the defining series so that each element of the matrix is correct to the desired accuracy.

EXAMPLE 3.17

Evaluate the discrete state transition matrix $e^{A\tau}$ for the system

$$\ddot{y} + 3\dot{y} + 2y = 0$$

where $\tau = 0.1$.

SOLUTION:

Writing the equation in phase variable form $\dot{X} = AX$

$$\begin{bmatrix} \dot{x}_1 \\ \dot{x}_2 \end{bmatrix} = \begin{bmatrix} 0 & 1 \\ -2 & -3 \end{bmatrix} \begin{bmatrix} x_1 \\ x_2 \end{bmatrix}$$

so

$$e^{A\tau} = \sum_{k=0}^{\infty} \frac{A^k \tau^k}{k!} = \sum_{k=0}^{\infty} B_k$$

where $B_k = A^k \tau^k / k!$, $\tau = 0.1$, and,

$$A = \begin{bmatrix} 0 & 1 \\ -2 & -3 \end{bmatrix}$$

Evaluating the first few terms by hand gives

$$e^{A\tau} = I + \tau A + \frac{\tau^2}{2!} A^2 + \frac{\tau^3}{3!} A^3 + \cdots$$

$$\approx \begin{bmatrix} 1 & 0 \\ 0 & 1 \end{bmatrix} + 0.1 \begin{bmatrix} 0 & 1 \\ -2 & -3 \end{bmatrix} + 0.005 \begin{bmatrix} 0 & 1 \\ -2 & -3 \end{bmatrix} \begin{bmatrix} 0 & 1 \\ -2 & -3 \end{bmatrix}$$

$$\approx \begin{bmatrix} 1 & 0 \\ 0 & 1 \end{bmatrix} + \begin{bmatrix} 0 & 0.1 \\ -0.2 & -0.3 \end{bmatrix} + \begin{bmatrix} -0.01 & -0.015 \\ 0.03 & 0.035 \end{bmatrix}$$

So

$$e^{A\tau} \approx \begin{bmatrix} 0.99 & 0.085 \\ -0.17 & 0.735 \end{bmatrix} \blacktriangledown$$

Evaluation of $e^{A\tau}$ on the digital computer is efficiently performed by recognizing that the general term of the sum defining $e^{A\tau}$

$$B_k = \frac{A^k \tau^k}{k!}$$

can be written

$$B_k = \frac{\tau A}{k} \left(\frac{A^{k-1} \tau^{k-1}}{(k-1)!} \right)$$

or

$$B_k = \frac{\tau A^k}{k} B_{k-1}$$

(Also $B_0 = I$)

So, the kth term of the series Equation (3.49) can be obtained from the (k-1)th term by a single matrix multiplication of the last term by A and then multiplication by the scalar τ/k. (Subroutines to perform matrix manipulations are developed in Problem 3.24 at the end of the chapter.)

A program for the evaluation of the discrete state transition matrix from a given system matrix A, for a specified increment τ and correct to a desired accuracy in each element of the matrix can be written using the subroutines mentioned above. (See Problem 3.28). The valuable experience gained by writing this program is not spoiled by the listing of a program in this text. However, evaluation of the discrete state transition matrix for Example 3.16 using such a program correct to 4 decimal places gives the results

$$e^{A\tau} = \begin{bmatrix} 0.99094 & 0.08611 \\ -0.17221 & 0.73262 \end{bmatrix}$$

It is wise to always evaluate a few terms by hand to insure that the results are reasonable. Compare these results with those obtained in Example 3.16.

To obtain plots of the fundamental matrix e^{At} at discrete values of t we can take the starting values

$$E^{A0} = I$$

and calculate the next value by the matrix multiplication

$$e^{An\tau} = e^{A\tau} e^{A(n-1)\tau}$$

(See Problem 3.29)

The *FORCED RESPONSE* or the response to an excitation is obtained by including a forcing function $U \neq 0$.

The state equation (3.29)

$$\dot{X} = AX + BU \qquad (3.29)$$

can be integrated from $t = (n-1)\tau$ to $t = n\tau$ to obtain

$$X_n = X_{n-1} + \int_{(n-1)\tau}^{n\tau} AX(\alpha) + BU(\alpha) d \qquad (3.49)$$

and the integral on the right can be evaluated by any of the classical integration formulas.

A more accurate technique for the digital simulation of the system in Equation (3.29) is to make use of the discrete state transition matrix that arises from the exact solution. As observed above we can multiply Equation (3.29) by e^{-At} to obtain

$$e^{-At}\dot{X} - e^{At}AX = e^{-At}BU$$

or

$$\frac{d}{dt}(e^{-At}X) = e^{-At}BU \qquad (3.50)$$

Integrating (3.50) from $(n-1)\tau$ to $n\tau$ gives

$$e^{-An\tau}X_n - e^{-A(n-1)\tau}X_{n-1} = \int_{(n-1)\tau}^{n\tau} e^{-A\alpha}BU(\alpha)\,d\alpha$$

or

$$X_n = e^{A\tau}X_{n-1} + e^{An\tau}\int_{(n-1)\tau}^{n\tau} e^{-A\alpha}BU(\alpha)\,d\alpha \qquad (3.51)$$

which is a recursion relation giving X_n in terms of the past state X_{n-1}, the discrete state transition matrix $e^{A\tau}$, and the excitation $U(t)$ in the interval from $(n-1)\tau$ to $n\tau$.

Now, approximate evaluation of Equation (3.51) by a numerical integration scheme is more accurate than a similar approximation in Equation (3.49), since a portion of the solution, the force free response, is exact. An even more accurate simulation is obtained if only the excitation $U(t)$ is approximated in the interval of integration. In particular, if $U(t)$ is assumed to be a polynomial in the interval, then the integral in Equation (3.51) can be integrated exactly. For example, suppose U is the piecewise constant vector excitation function

$$U(t) = U_\ell \quad \text{for} \quad \ell\tau \le t < (\ell+1)\tau \qquad (3.52)$$

Then Equation (3.51) becomes

$$X_n = e^{A\tau}X_{n-1} + e^{An\tau}\int_{(n-1)\tau}^{n\tau} e^{-A\alpha}\,d\alpha\, BU_{n-1}$$

$$= e^{A\tau} X_{n-1} + \int_{(n-1)\tau}^{n\tau} e^{A(n\tau-\alpha)} d\alpha B U_{n-1}$$

Substituting the defining series for $e^{A(n\tau-\alpha)}$ gives

$$X_n = e^{A\tau} X_{n-1} + \int_{(n-1)\tau}^{n\tau} \sum_{k=0}^{\infty} \frac{A^k (n\tau-\alpha)^k}{k!} d\alpha \quad B U_{n-1}$$

Interchanging the summation and integration gives

$$X_n = e^{A\tau} X_{n-1} + \sum_{k=0}^{\infty} \frac{A^k}{k!} \int_{(n-1)\tau}^{n\tau} (n\tau-\alpha)^k d\alpha B U_{n-1}$$

$$= e^{A\tau} X_{n-1} + \sum_{k=0}^{\infty} \frac{A^k}{k!} \left[- \frac{(n\tau-\alpha)^{k+1}}{(k+1)} \right]_{(n-1)\tau}^{n\tau} B U_{n-1}$$

or

$$X_n = e^{A} X_{n-1} + \sum_{k=0}^{\infty} \frac{A^k \tau^{k+1}}{(k+1)!} B U_{n-1}$$

which can be written in the recursion relation form

$$X_n = F X_{n-1} + G U_{n-1} \tag{3.53}$$

where

$$F = \sum_{k=0}^{\infty} \frac{A^k \tau^k}{k!} \quad \text{and} \quad G = \left(\sum_{k=0}^{\infty} \frac{A^k \tau^{k+1}}{(k+1)!} \right) B$$

The F and G matrices can be evaluated once for a given value of τ and used throughout the calculation or simulation. In the evaluation of F as described above the general term $B_k = A^k \tau^k / k!$ is multiplied by A and by the scalar τ/k to obtain B_{k+1}.   If B_k is first multiplied by τ/k we have $A^k \tau^{k+1}/(k+1)!$ which is the general term of the series defining G.   Figure 3.14 outlines an efficient algorithm for calculating both F and G with essentially the same computation as that required for F alone.   The general term starts off at the identity which is accumulated into the F matrix; the term is then multiplied by $\tau/1$, which is accumlated in the G' matrix (G = G' B).   Now the term $I\tau/1$ is multiplied by the matrix A to get the next term for F and so on.   Sufficient accuracy is obtained when each element of the matrix term to be added is less than the prescribed error, at which time the summation is terminated and G' multiplied by B to obtain G.

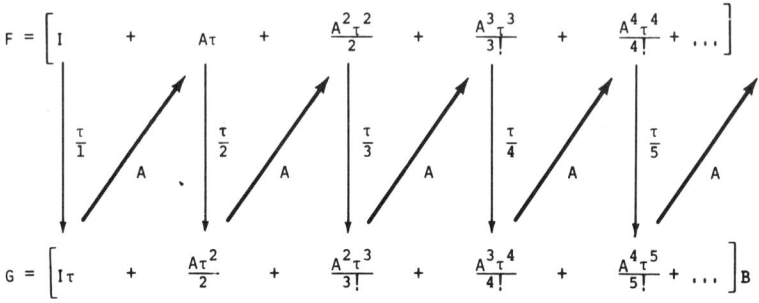

FIGURE 3.14   EVALUATION OF F AND G MATRICES FOR $x_N = Fx_{N-1} + Gu_{N-1}$.

EXAMPLE 3.18

Evaluate the F and G matrices in Equation (3.53) for the system $\ddot{y} + 3\dot{y} + 2y = w(t)$. with $\tau = 0.1$. Obtain the step response of the system.

SOLUTION:

Writing the equations in phase variable form we have

$$\begin{bmatrix} \dot{x}_1 \\ \dot{x}_2 \end{bmatrix} = \begin{bmatrix} 0 & 1 \\ -2 & -3 \end{bmatrix} \begin{bmatrix} x_1 \\ x_2 \end{bmatrix} + \begin{bmatrix} 0 \\ 1 \end{bmatrix} u$$

where $x_1 = y$, $x_2 = \dot{x}_1$ and $u = w$.

Clearly

$$A = \begin{bmatrix} 0 & 1 \\ -2 & -3 \end{bmatrix} \quad \text{and} \quad B = \begin{bmatrix} 0 \\ 1 \end{bmatrix}$$

A straightforward extension of the calculations given in Example 3.16 and the above discussion yields

$$G = \begin{bmatrix} -0.0997 & 0.0045 \\ -0.0090 & 0.0861 \end{bmatrix} \begin{bmatrix} 0 \\ 1 \end{bmatrix}$$

or

$$G = \begin{bmatrix} 0.0045 \\ 0.0861 \end{bmatrix}$$

(Verify this! See Problem 3.31)

The following program uses the previously obtained F matrix together with this G matrix to give the step response of the system for $X(0) = 0$ and $u = 1$ for $t \geq 0$. The zero state step response $X(nt)$ is plotted.▼

```
      DIMENSION F(2,2),G(2,1),X(2,1),FX(2,1)
      DIMENSION PLOTX1(51),PLOTX2(51),TIMEO(51)
      DATA F/.99094,-.17221,.08611,.73262/,G/.0045,.0861/
      DATA X/2*0.0/,U/1./,TIMEO,PLOTX1,PLOTX2/153*0.0/
      DO 100 K=2,51
C
C     X =FX    +GU
C     N    N-1   N-1
C
      CALL MULT(F,X,FX,2,2,1)
      CALL MTX(1.,FX,U,G,X,2,1)
      TIMEO(K)=(K-1)*0.1
      PLOTX1(K)=X(1,1)
      PLOTX2(K)=X(2,1)
  100 CONTINUE
      CALL PLOT(TIMEO,PLOTX1,51)
      CALL PLOT(TIMEO,PLOTX2,51)
      STOP
      END
```

```
SUBROUTINE MTX(CA,A,CB,B,C,N,M)
DIMENSION A(N,M),B(N,M),C(N,M)
DO 1 I=1,N
DO 1 J=1,M
1 C(I,J)=CA*A(I,J)+CB*B(I,J)
RETURN
END
```

```
SUBROUTINE MULT(A,B,C,N,M,K)
DIMENSION A(N,M),B(M,K),C(N,K)
DO 1 I=1,N
DO 1 J=1,K
C(I,J)=0.
DO 1 L=1,M
1 C(I,J)=C(I,J)+A(I,L)*B(L,J)
RETURN
END
```
†

If $U(t)$ is assumed to be piecewise linear in the interval of integration it can be written

$$U(t) = U_{nc} + U_{n\ell}t \quad \text{for} \quad n\tau \leq t < (n+1)\tau$$

$$(3.54)$$

And the integral in Equation (3.51) can be integrated to obtain a matrix recursion relation for the solution of the system equations. Higher order methods can be derived by other methods of approximating the integral.

The procedures discussed above for the solution of linear time variant differential equations are quite useful and popular. However, there are some computational difficulties that limit their applicability.

First, although the series representation of the discrete state transition matrix converges uniformly for any finite τ, the larger τ is the more terms must be taken in the series, thus increasing the number of calculations and the round off error. As the number of terms becomes very large the round off error invalidates the resultant matrix. For example, in evaluating e^{100} the 100th term of the series is $100^{100}/100!$ which is greater than the 99th term. Only at the 101th term do the terms start to decrease, and then very slowly. The 200th term is only 1/2 the previous term since it is multiplied by 100/200.

---

† The printer plots resulting from this program appear on the following two pages.

```
MAXIMUM =    0.4928005
MINIMUM =    0.0

                                          (OUTPUT SCALED BY: 1.0E-01)
                        0 + =====>
                        0.0   0.5   1.0   1.5   2.0   2.5   3.0   3.5   4.0   4.5   5.0
     TIME     OUTPUT    +-----+-----+-----+-----+-----+-----+-----+-----+-----+-----+
     0.0      0.0       +     |     |     |     |     |     |     |     |     |     |
     0.100    0.045     I+    |     |     |     |     |     |     |     |     |     |
     0.200    0.164     I*+   |     |     |     |     |     |     |     |     |     |
     0.300    0.335     I***+ |     |     |     |     |     |     |     |     |     |
     0.400    0.542     I******+    |     |     |     |     |     |     |     |     |
     0.500    0.773     I********+  |     |     |     |     |     |     |     |     |
     0.600    1.016     I***********+     |     |     |     |     |     |     |     |
     0.700    1.265     I**************+  |     |     |     |     |     |     |     |
     0.800    1.514     I*****************+     |     |     |     |     |     |     |
     0.900    1.759     I*********************+ |     |     |     |     |     |     |
     1.00     1.995     I************************+    |     |     |     |     |     |
     1.10     2.223     I**************************+  |     |     |     |     |     |
     1.20     2.439     I***************************+|    |     |     |     |     |
     1.30     2.643     I*****************************+   |     |     |     |     |
     1.40     2.835     I*******************************+ |     |     |     |     |
     1.50     3.014     I*********************************+     |     |     |     |
     1.60     3.181     I**********************************+    |     |     |     |
     1.70     3.336     I************************************+  |     |     |     |
     1.80     3.480     I**************************************+ |     |     |     |
     1.90     3.612     I****************************************+|    |     |     |
     2.00     3.734     I*****************************************+     |     |     |
     2.10     3.846     I*******************************************+   |     |     |
     2.20     3.949     I********************************************+  |     |     |
     2.30     4.044     I**********************************************+ |     |     |
     2.40     4.130     I***********************************************+|    |     |
     2.50     4.209     I************************************************+     |     |
     2.60     4.281     I*************************************************+    |     |
     2.70     4.346     I**************************************************+   |     |
     2.80     4.406     I***************************************************+| |     |
     2.90     4.460     I****************************************************+ |     |
     3.00     4.510     I*****************************************************+|     |
     3.10     4.555     I******************************************************+     |
     3.20     4.596     I*******************************************************+    |
     3.30     4.633     I*******************************************************+    |
     3.40     4.667     I********************************************************+   |
     3.50     4.698     I*********************************************************+   |
     3.60     4.726     I****-*****************************************************+  |
     3.70     4.751     I**********************************************************+  |
     3.80     4.774     I***********************************************************+ |
     3.90     4.795     I***********************************************************+ |
     4.00     4.814     I************************************************************+|
     4.10     4.831     I*************************************.***********************+|
     4.20     4.846     I************************************************************+|
     4.30     4.860     I************************************************************+|
     4.40     4.873     I*************************************************************+|
     4.50     4.885     I*************************************************************+|
     4.60     4.895     I*************************************************************+|
     4.70     4.905     I*************************************************************+|
     4.80     4.913     I*************************************************************+|
     4.90     4.921     I*************************************************************+|
     5.00     4.928     I*************************************************************+|
                        |_____|_____|_____|_____|_____|_____|_____|_____|_____|_____|
```

```
MAXIMUM =   0.2500287
MINIMUM =   0.0
                                          (OUTPUT SCALED BY: 1.0E-01)
                      0 +  =====>
                      0.0   0.5   1.0   1.5   2.0   2.5   3.0   3.5   4.0   4.5   5.0
    TIME    OUTPUT  +-----+-----+-----+-----+-----+-----+-----+-----+-----+-----+
    0.0     0.0     +     |     |     |     |     |     |     |     |     |     |
    0.100   0.861   I*********+    |     |     |     |     |     |     |     |     |
    0.200   1.484   I*****************+   |     |     |     |     |     |     |     |
    0.300   1.920   I*********************+   |     |     |     |     |     |     |
    0.400   2.210   I*****************************+   |     |     |     |     |     |
    0.500   2.387   I******************************+|    |     |     |     |     |
    0.600   2.476   I********************************+    |     |     |     |     |
    0.700   2.500   I********************************+    |     |     |     |     |
    0.800   2.475   I********************************+    |     |     |     |     |
    0.900   2.413   I*******************************+|    |     |     |     |     |
    1.00    2.326   I******************************+  |    |     |     |     |     |
    1.10    2.222   I*****************************+   |    |     |     |     |     |
    1.20    2.106   I***************************+    |     |     |     |     |     |
    1.30    1.984   I**************************+    |     |     |     |     |     |
    1.40    1.859   I************************+    |     |     |     |     |     |
    1.50    1.735   I**********************+   |     |     |     |     |     |     |
    1.60    1.613   I*******************+    |     |     |     |     |     |     |
    1.70    1.495   I******************+    |     |     |     |     |     |     |
    1.80    1.382   I****************+|    |     |     |     |     |     |     |
    1.90    1.274   I**************+  |     |     |     |     |     |     |     |
    2.00    1.172   I*************+    |     |     |     |     |     |     |     |
    2.10    1.077   I************+    |     |     |     |     |     |     |     |
    2.20    0.987   I***********+    |     |     |     |     |     |     |     |
    2.30    0.904   I**********+|    |     |     |     |     |     |     |     |
    2.40    0.827   I*********+  |     |     |     |     |     |     |     |     |
    2.50    0.756   I********+    |     |     |     |     |     |     |     |     |
    2.60    0.690   I*******+   |     |     |     |     |     |     |     |     |
    2.70    0.629   I*******+    |     |     |     |     |     |     |     |     |
    2.80    0.574   I******+    |     |     |     |     |     |     |     |     |
    2.90    0.522   I*****+    |     |     |     |     |     |     |     |     |
    3.00    0.476   I*****+    |     |     |     |     |     |     |     |     |
    3.10    0.433   I****+|    |     |     |     |     |     |     |     |     |
    3.20    0.394   I****+|    |     |     |     |     |     |     |     |     |
    3.30    0.358   I****+    |     |     |     |     |     |     |     |     |
    3.40    0.325   I****+    |     |     |     |     |     |     |     |     |
    3.50    0.296   I***+    |     |     |     |     |     |     |     |     |
    3.60    0.269   I***+    |     |     |     |     |     |     |     |     |
    3.70    0.244   I**+    |     |     |     |     |     |     |     |     |
    3.80    0.221   I**+    |     |     |     |     |     |     |     |     |
    3.90    0.201   I*+    |     |     |     |     |     |     |     |     |
    4.00    0.183   I*+    |     |     |     |     |     |     |     |     |
    4.10    0.166   I*+    |     |     |     |     |     |     |     |     |
    4.20    0.151   I*+    |     |     |     |     |     |     |     |     |
    4.30    0.137   I*+    |     |     |     |     |     |     |     |     |
    4.40    0.124   I+    |     |     |     |     |     |     |     |     |
    4.50    0.113   I+    |     |     |     |     |     |     |     |     |
    4.60    0.102   I+    |     |     |     |     |     |     |     |     |
    4.70    0.093   I+    |     |     |     |     |     |     |     |     |
    4.80    0.085   I+    |     |     |     |     |     |     |     |     |
    4.90    0.077   I+    |     |     |     |     |     |     |     |     |
    5.00    0.070   I+    |     |     |     |     |     |     |     |     |
                    I___|___|___|___|___|___|___|___|___|___|
```

It appears that the accuracy of a simulation response at the mth step $X(m\tau)$ can be checked for the accumulation and buildup of truncation and round off errors by a single calculation which uses a new discrete state transition matrix with a new $\tau' = m\tau$ and evaluates $X(m\tau)$ in a single step. However, the round off errors discussed above make this almost useless for any significant increase in τ.

A particular advantage of the state space approach is that for some special classes of inputs the recursion relations give the 'exact' response within the errors discussed above. Also the matrix representation of the system equations allows working with a first order equation regardless of the order of the system and the number of system variables, allowing the programmer to consider large scale systems and high order equations without undue complexity. Finally, the phase variable form of the equations is identical to the representation of the system on an analog computer (with due consideration for sign reversals). The theoretical advantage of this analysis is that the increment τ is selected independently of the natural frequencies and time constants of the system. τ depends only on the frequency of the excitation and the desired output resolution. Computationally it is not quite so simple as observed above.

3.5 A COMPARISON OF THE METHODS OF DIGITAL SIMULATION

Digital-analog simulation languages have had great appeal among engineers who want to conduct their own simulation studies. These languages allow a digital simulation through a block oriented program without extensive programming of the individual steps of the simulation. A particular advantage of programs like MIMIC, DSL/90, and CSMP (the IBM Continuous System Modeling Program) is that of an automatic sort of the statements. The programmer can arrange the statements in essentially any order and the processor rearranges them into an order so that all of the inputs to a block are calculated before other operations which use those values are attempted.

The MIMIC language uses a 4th order variable-step Runge-Kutta method. The increment size is automatically adjusted to keep the local relative error below a specified value, where the error is estimated by comparing the results obtained in each step, with the computed results using two calculations across the interval with half the step size.

More recent languages allow the selection of the integration method, which adds flexibility to the program but requires the programmer to have a better understanding of the advantages and disadvantages of each method. The System/360 CSMP allows selection among five fixed-step methods (Runge-Kutta, Simpson, trapezoidal, rectangular, and second order Adams) and two variable-step methods (fifth-order Milne predictor-corrector, and fourth order Runge-Kutta). The variable-step methods vary the step size during execution by comparing an error estimate with a user specified bound. The objective is to choose the method that performs the simulation in the minimum time, with the required accuracy. As the complexity of the method increases, the accuracy and computation time increase. However, some sophisticated methods have good stability and accuracy properties which allow large increment size. Typical questions to ask about all methods concern which method is more accurate, integration by trapezoidal rule with an increment τ or rectangular integration with an increment $\tau/2$, which one takes longer, and which one is more stable.

ACCURACY

A rough idea of the accuracy of the various methods is obtained by integrating an undamped second order linear system and comparing the magnitude or frequency of the response to the known response. To keep the period of the solution correct within 0.01 percent Euler (rectangular) integration Equation (1.6) requires about 500 steps per period, while the Modified Euler (rectangular predictor-trapezoidal corrector), Equations (3.19 and (3.18), needs about 250 steps per period. The open formula (predictor only) of Simpson's rule, Equation (3.10) or (3.15b), needs about 40 steps per period while the Milne predictor-corrector (3.15a and b), and 4th order Runge-Kutta integration by Equation (3.18) need only about 20 steps per period. (See Problem 3.41.) Additional problems discuss a similar comparison for correct magnitude response at a particular frequency and the generation of plots of the error in magnitude and period as a function of the step size.

Another consideration of complex methods, like Runge-Kutta is that they may lead to large errors due to the round off errors in the many evaluations of the system function. In general, the Runge-Kutta methods are good as starting methods for the Milne or Adams-Moulton predictor-corrector used to integrate the entire solution, as in MIMIC or CSMP if no method is specified, poor error estimates are available. So although it is easy to vary the step size with Runge-Kutta methods it is difficult to determine when to change the step size. The usual procedure when using Runge-Kutta integration is to use a pessimistic error estimate and integrate with a small fixed interval. The ease of programming the Runge-Kutta method makes it suitable for a single run of a simulation with a small increment size but not particularly appropriate for a general differential equation solver program to be used repeatedly. The system complexity and system function may be such that, contrary to the usual situation, the Runge-Kutta method is faster than a predictor-corrector

In general, open formulas of the Adams-Bashforth type which use past values of the slopes f in $\dot{y} = f(y,x,t)$ and ignore past values of the solution y are very successful when a fast and accurate solution is desired. The Adams-Moulton predictor-corrector methods which use past slopes and a predictor for an estimate of the present slope give an indication of the error in each step by comparing the predicted value with the corrected value. However, this value is merely the error made during the corrector calculation. Repeated correction to get the successive values identical is generally unsuccessful in increasing accuracy. The additional calculations of f at each step to iterate the corrector can usually be more efficiently utilized by decreasing the increment size and correcting only once throughout the solution.

Several methods have been developed for modifying the Runge–Kutta and predictor–corrector methods to render them more suitable for variable step operation. These modifications usually require additional calculations in the Runge–Kutta algorithm to provide accurate error estimates, or more elaborate multistep methods with nonuniform step lengths to render the multistep methods self starting when the step size is changed. Benyon's excellent review article gives several references for these techniques.†

STABILITY OF NUMERICAL INTEGRATION METHODS

For large values of the increment size integration schemes become unstable. The rectangular integration of the equation $\dot{y} + y = 0$ approximates the system for small increment sizes by the recursion relation

$$y_n = (1-\tau)y_{n-1} \qquad (3.55)$$

which is equivalent to taking the first two terms of the exact recursion relation

$$y_n = e^{-\tau}y_{n-1} \qquad (3.56)$$

The solution to difference Equations (3.58) and (3.59) are

$$y_n = (1-\tau)^n y_0 \qquad (3.57)$$

and

$$y_n = (e^{-\tau})^n y_0 \qquad (3.58)$$

respectively.

---

† P. R. Benyon, "A Review of Numerical Methods for Digital Simulation," *Simulation*, 219–238 (November, 1968).

The known system solution is an exponential function decaying to zero. The approximate expression in Equation (3.57) also decays to zero approximately exponentially for small τ. However, if τ is large, Equation (3.57) does not so accurately approximate Equation (3.58). In fact for $\tau \geq 2$, the y_n values in Equation (3.57) increase in magnitude with successive n. Thus we have a discrete representation of a stable system which is itself unstable for $\tau > 2$. The first order equation $\dot{y} + ay = 0$, when integrated by rectangular integration, gives an unstable recursion relation for step sizes larger than 2/a. Second order and higher order stable systems also yield unstable recursion relations by using rectangular integration for large values of τ. (See Problem 3.43.) Integration of linear systems by the trapezoidal rule is stable for all values of the increment size. (See Problem 3.48.)

Z-transform methods are useful for analyzing the stability of linear discrete processes. If the poles† of a pulse transfer function lie inside the unit circle then the corresponding recursion relation is stable. Consider the following example.

EXAMPLE 3.19

Determine the location of the poles of the pulse transfer function

$$H(z) = \frac{1}{1-rz^{-1}} \qquad (3.59)$$

and discuss the stability of the corresponding recursion relation $(Y(z) = H(z)X(z))$

$$y_n = ry_{n-1} + x_n \qquad (3.60)$$

---

† H(z) is said to have a pole at $z = z_1$ if

    1) $\lim_{z \to z_1} H(z) = \infty$

and

    2) $\lim_{z \to z_1} H(z)(z-z_1)^k$ exists for some k.

SOLUTION:

H(z) has poles where the denominator is zero ($1-rz^{-1} = 0$). Thus H(z) has one pole located at $z = r$, and H(z) represents a stable recursion relation for $|r| < 1$ (the pole located inside the unit circle), and unstable for $|r| > 1$.

The response Y(z) to the excitation X(z) = 1 is

$$Y(z) = H(z)X(z) = H(z) = \frac{1}{1-rz^{-1}} = \sum_{k=0}^{\infty} r^k z^{-k}$$

$$(3.61)$$

Thus $y_n = r^n$, which indeed diverges for $|r| > 1$ and converges for $|r| < 1$. ▼

In general the system

$$\dot{X} = AX + BU$$

when integrated by rectangular integration is stable if the roots of the equation

$$\det[zI - (I+A\tau)] = 0 \qquad (3.62)$$

lie inside the unit circle. (See Problem 3.43)

For complex systems and integration methods a stability analysis is not trivial nor always straightforward, particularly since the evaluation of the roots of an nth degree ploynomial with coefficients in the variable τ is required.

The trapezoidal rule is stable for all step sizes when integrating linear differential equations. However, usually closed formulas which require present output values along with the past values cannot be used directly and an open formula using only past values must be used for predicting these values. Generally the expressions for the stability of the Euler method can be considered worst case bounds in the increment size τ and the trapezoidal rule bounds as best case. The predictor-corrector and Runge-Kutta methods usually have stability ranges between these extremes. However, it is shown in Problem 3.49 that a closed corrector formula may not improve the stability properties of a method over those of the predictor alone. Of course, these are merely guides for estimating stability in nonlinear systems. Procedures for stability analysis of·nonlinear systems are not known so it is extremely important to insure that simulated solutions are the desired response and not simply due to the method of implementation. Generally, if the simulation is run with different increments sizes and slightly differing initial conditions and similar results are obtained, it is assumed that the simulation indeed represents the system under study.

3.6 THE DESIGN OF DIGITAL FILTERS

A digital filter is a discrete system or recursion relation which approximates the input output characteristics of a continuous system.[†]

There are essentially three methods for the design of digital filters. Although there are presentations of digital filter design based on frequency characteristics, time response, interpolation, data reconstruction, and classical numerical integration, they can all be shown to be equivalent to an approximation to the integrals in Equations (3.49) or (3.51).

Consider the approximation of the integral in Equation (3.49)

$$X_n = X_{n-1} + \int_{(n-1)\tau}^{n\tau} AX(\alpha) + BU(\alpha)\,d\alpha \qquad (3.49)$$

[†] More generally it is any recursion relation, pulse transfer function $H(z)$, or system which has a sequence of numbers for its input and another sequence of numbers for its output.

by taking the integrand to be piecewise constant, corresponding to rectangular integration. This gives the recursion relation or digital filter

$$X_n = X_{n-1} + \tau[AX_{n-1} + BU_{n-1}]$$

$$= [I + \tau A]X_{n-1} + \tau BU_{n-1} \qquad (3.63)$$

(Look familiar?) Similarily any numerical scheme gives an approximate digital system.

The approximation of Equation (3.51) is similar but the integral relating the X has been evaluated.

$$X_n = e^{A\tau}X_{n-1} + e^{An\tau}\int_{(n-1)\tau}^{n\tau} e^{-A\alpha}BU(\alpha)d\alpha \qquad (3.51)$$

Evaluating the integral by rectangular integration (the entire integrand is piecewise constant) gives

$$X_n = e^{A\tau}X_{n-1} + e^{A\tau}BU_{n-1}\tau \qquad (3.64)$$

If we assume that the integrand is constant with the value at $t_n = n\tau$ (the right side of the interval) we get

$$X_n = e^{A\tau}X_{n-1} + BU_n\tau \qquad (3.65)$$

which sequences through the same sample values

$$X_n = e^{An\tau}X_0$$

for $U(t) = 0$ as the sampled values of the continuous system

$$X(t) = e^{At}X(0)$$

Since these two systems have similar responses to initial conditions or unit pulse inputs, Equation (3.65) is called the *IMPULSE INVARIANT DIGITAL FILTER*. A complete discussion of the frequency characteristics and implementation of the filter is not within the scope of the current text. However, it is emphasized that digital filter design is just digital simulation as presented in this chapter. Finally, approximation of the integral in Equation (3.51) by making assumptions about the character of the input $U(t)$ similar to Equation (3.53) leads to another set of recursion relations as obtained in the last section. Digital filters and several other topics in digital systems are presented in Kuo and Kaiser.[8]

PROBLEMS

3.1 Write a subprogram for the digital-analog simulation program which will integrate the solution for 1/4 period of the highest frequency of the excitation (assume $\omega_{max} = 5$) by taking successively smaller values of τ until the result changes by less than 4 significant figures.

3.2 Give a recursion relation for the digital simulation of the differential equation

$$\ddot{y} + 0.2\dot{y} + y = x(t), \quad \begin{array}{l} y(0) = 0 \\ \dot{y}(0) = 0 \end{array}$$

which assumes

(a) $y(t)$ is piecewise constant and $x(t)$ is piecewise constant, and

(b) $y(t)$ is piecewise linear and $x(t)$ is piecewise constant.

3.3 Implement the recursion relations obtained in Problem 3.2 to obtain the step response of the differential equation given. Choose an increment size to give two decimal places of accuracy.

3.4 (a) Give a recursion relation to simulate the equation

$$\ddot{y} + 3\dot{y} + y = \dot{x} + x$$

which assumes both x and y are piecewise linear. Obtain a plot of the response $y(t)$ to the input $x = tu(t)$.

(b) In part (a) above $\dot{x} + x = (t+1)u(t) = tu(t) + u(t)$. Obtain a plot of the response $y(t)$ using a simulation which integrated $\dot{x} + x$ exactly. That is, it assumes x is piecewise linear and $\dot{x}$ is piecewise constant. Compare the results with part (a).

3.5 (a) Using rectangular integration give a recursion relation suitable for simulating the Bernoulli equation

$$\ddot{y} - aty - e^{-t^2/2}y^2 = 0$$

(b) Using the trapezoidal rule, attempt to find a difference equation suitable for simulating the Bernoulli equation. Discuss any difficulty.

(c) Use rectangular integration for prediction and trapezoidal integration for correction to develop a set of difference equations suitable for numerically integrating the Bernoulli equation.

3.6 (a) z-transform integrating operators can frequently be applied to nonlinear systems. In the simulation of the van der Pol equation

$$\ddot{y} - \varepsilon(1-y^2)\dot{y} + y = 0$$

consider $w_n = y^2$ and obtain a recursion relation where past values of w_n are obtained from y_n directly and the present value of w_n if needed is calculated by the extrapolation

$w_n = 2w_{n-1} - w_{n-2}$.   (Justify this extrapolation.)

(b) Plot the response of (a) for $\varepsilon = 5$ and compare your results with Example 2.10.

3.7 Integrating the van der Pol equation once gives

$$Dy + \varepsilon y - \frac{\varepsilon}{3} y^3 + Iy = 0$$

(a) Give a recursion relation corresponding to this equation where $w_n = y^3$.

(b) Plot the response $y(t)$ for $\varepsilon = 5$ and compare the results with Example 2.10.

3.8 Implement the program given in Example 3.8 for the solution of the van der Pol equation.

3.9 Alter the program in Problem 3.8 to have the corrector formula iterate until the maximum difference between successive corrected values is less than 10^{-3} for $\tau = 0.01$.

3.10 Obtain a solution (phase plane) to the van der Pol equation given in Problem 3.6 using the 3rd order Runge-Kutta formula for integration with $\tau = .01$.

$$y_n = y_{n-1} + \frac{\tau}{6}(k_1 + 4k_2 + k_3)$$

$$k_1 = f(y_{n-1})$$

$$k_2 = f(y_{n-1} + \frac{\tau}{2} k_1)$$ $$X_0 = \begin{bmatrix} 1 \\ 0 \end{bmatrix}$$

$$k_3 = f(y_{n-1} + 2 k_2 - \tau k_1)$$

3.11 Change the program in Problem 3.9 to double when the relative change in y or $\dot{y}$ is less than 0.1 percent and halve τ when either y or $\dot{y}$ has a relative change greater than 0.1 percent. That is, make the program decrease τ when the variables are changing rapidly and increase τ when the variables are relatively constant. (If the change in the variables is sufficiently large to merit decreasing τ do not use the last calculated values. Have the program go back and recompute for the new τ. If τ is increased the last calculation can be retained.)

3.12 Compare the predictor and corrector equations

$$p_n = y_{n-1} + \frac{\tau}{2}(3f(y_{n-1}) - f(y_{n-2}))$$

$$c_n = y_{n-1} + \frac{\tau}{2}(f(y_{n-1}) + f(p_n))$$

with the Taylor series for the exact value of y_n

$$y_n = y_{n-1} + \tau \dot{f}(y_{n-1}) + \frac{\tau^2}{2!} \ddot{f}(y_{n-1})$$

$$+ \frac{\tau^3}{3!} \dddot{f}(y_{n-1}) + \ldots$$

to obtain $[(y_n)_{exact} - p_n]$ and $[(y_n)_{exact} - c_n]$ and show that the error is highest order for

$$(y_n)_{exact} \approx c_n - \frac{1}{6}(c_n - p_n)$$

Therefore the predictor-corrector pair can be augmented with the final equation

$$y_n = \frac{1}{6}(5c_n + p_n)$$

3.13 Substitute the back difference $\dot{y}_n = (y_n - y_{n-1})/\tau$ in the truncated Taylor series

$$y_{n+1} = y_n + \tau \dot{y}_n + \frac{\tau^2}{2} \ddot{y}_n$$

to obtain the second order Adams-Bashforth formula

$$y_{n+1} = y_n + \frac{\tau}{2}[3\dot{y}_n - \dot{y}_{n-1}]$$

3.14 Repeat Problem 3.9 using the 3rd order Runge-Kutta formula to obtain starting values. Then continue integrating with Milne's formula (3.15).

3.15 Use rectangular integration to generate starting values for the solution of the van der Pol equation given in Problem 3.6 and continue with the Adams-Moulton formula (3.16).

   (a) Use $\tau = 0.01$ for starting and the total solution.

   (b) Use $\tau = 0.001$ for the initial calculations until sufficient values at $\tau = 0.01$ have been obtained to continue with the Adams-Moulton method.

3.16 Show that the block diagram given in Figure 3.10 satisfied Equation (3.34).

3.17 Systems that can be described by differential equations of the form

$$(D^n + a_1 D^{n-1} + \ldots + a_n) y = (b_0 D^k + \ldots + b_k) x$$

are called *PROPER SYSTEMS* if $k < n$. That is, systems in which the input does not appear in higher derivatives than the response are proper systems. Proper systems can be simulated on analog computers without the use of differentiators by the methods discussed in Chapter 2.

Figure 3.15(a) shows a block diagram for the system

$$D^2 y = -a_1 Dy - a_2 y + b_0 D^2 x + b_1 Dx + b_2 x$$

where the blocks containing D and I are differentiators and integrators respectively. If an integrator is placed in series with b_0 and b_1 and the signal added in after the first integrator, the system equations are preserved. Then the I and D in series cancel, resulting in the diagram shown in Figure 3.15(b).
   Continue using similar block diagram

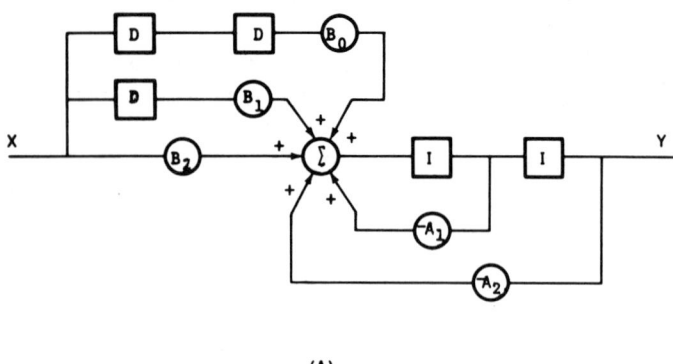

(A)

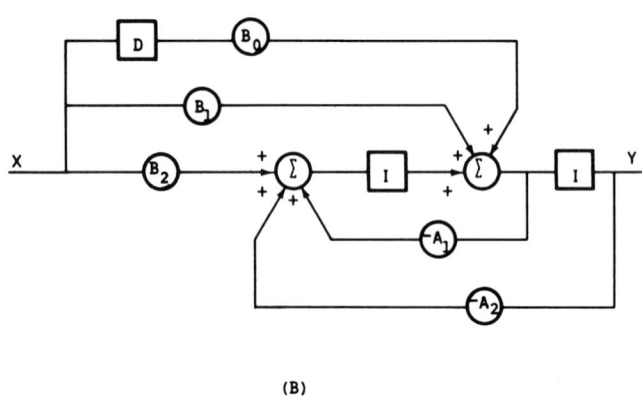

(B)

FIGURE 3.15 PROBLEM 3.17.

manipulations to obtain a block diagram
containing no differentiation. Define the state
variables as the integrator outputs and write the
resultant state equations.

3.18 Show that the block diagram in Figure 3.16
satisfies the equation

$$(D^k + a_1 D^{k-1} + \ldots + a_{k-1} D + a_k) y = (b_0 D^k + \ldots + b_k) x$$

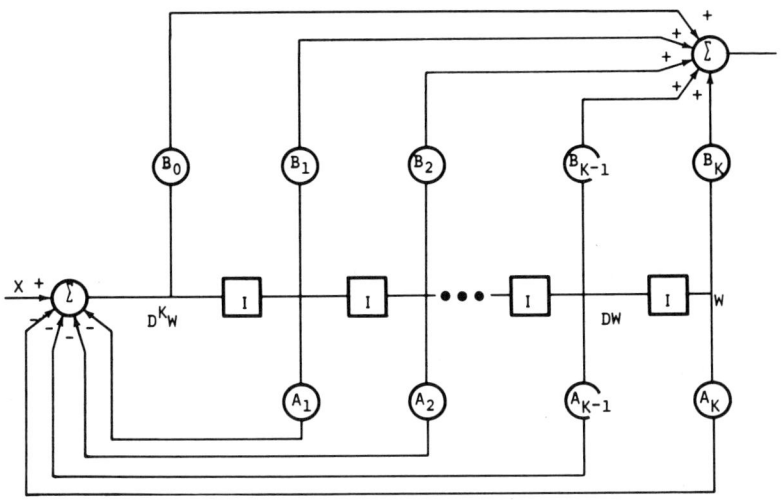

FIGURE 3.16 PROBLEM 3.18.

3.19 Draw a unit step $u(t)$ and mark sample points at values $t = n\tau$ for some τ. Assume that $x_n = x(n\tau^+)$† so that $x_0 = 1$.

(a) Now indicate the area under $u(t)$ as computed using a piecewise constant interpolation (that is, rectangular integration).

(b) Indicate the area under $u(t)$ as computed using piecewise linear interpolation (that is, trapezoidal integration).

(c) Finally evaluate the sample values of $I_{R1}U(z)$ and $I_{T1}U(z)$ where $U(z)$ is the z-transform of $u(t)$, and compare these with the results of (a) and (b).

3.20 Draw a time axis with the point $(n-1)\tau$ and $n\tau$ marked about two inches apart. Above the point

---

† $x(n\tau^+) = \lim_{\substack{\varepsilon \to 0 \\ |\varepsilon| > 0}} x(n\tau + \varepsilon)$

$(n-1)\tau$ mark an arbitrary point y_{n-1}. Now assume a slope $f(y_{n-1})$ and extend a line to the value obtained using a rectangular predictor $P_n = y_{n-1} + \tau f(y_{n-1})$. Indicate a slope through P_n to be $f(p_n)$ and finally use the trapezoidal corrector

$$y_n = y_{n-1} + \frac{\tau}{2}(f(y_{n-1}) + f(p_n))$$

to obtain y_n and show this point on the diagram.

3.21 Draw and label an axis as in Problem 3.20. Indicate the following values and slopes from the second order Runge-Kutta formula (3.17) on the diagram: y_{n-1}, k_1, $y_{n-1} + \tau k_1$, k_2, y_n.

3.22 Draw a figure similar to those described in Problems 3.20 and 3.22 to justify use of the open trapezoidal formula

$$y_n = y_{n-1} + \frac{\tau}{2}(3f(y_{n-1}) - f(y_{n-2}))$$

3.23 Look up the error estimates for the Milne integration formulas in a numerical analysis book and adapt your digital-analog simulation program to automatically adjust τ as a function of the per step error.

3.24 Write subroutines to perform the following matrix operations where A, B and C are matrices of arbitrary order and α is a scalar.

(a) SUBROUTINE MADD   (   ) --- [C=A+B]

(b) SUBROUTINE AEQB   (   ) --- [A=B]

(c) SUBROUTINE MTC   (   ) --- [B=αA]

(d) SUBROUTINE MSUB   (   ) --- [C=A-B]

(e) SUBROUTINE MMULT   (   ) --- [C=AB]

(f) SUBROUTINE MTX   (   ) --- [C=αA+βB]

(g) SUBROUTINE MMX   (   ) ---- [x=Fx]

3.25 Write the following equations in state variable form.

(a) $\ddot{y} + 4y = \cos 3t$

(b) $\ddot{y} + 2y^2\dot{y} + y = u(t)$

(c) Duffing equation

$$\ddot{y} + \omega_0^2 y + hy^3 = G \cos \omega_1 t$$

(d) The Bernoulli Equation

$$\ddot{y} + aty - e^{\frac{at^2}{2}} y^2 = 0$$

(e) The Mathieu equation

$$\ddot{y} + (a-2q \cos 2t)y = 0$$

3.26 Give recursion relations in vector form which integrate the equations obtained in Problem 3.25 by rectangular integration. Implement the vector recursion relations on the computer and obtain plots of the response for some selection of constants and initial conditions.

3.27 Use the power series definition of e^{At} to show that for (n×n) matrices A and B;

(a) $Ae^{At} = e^{At}A$

(b) $e^{At}e^{Bt} = e^{(A+B)t}$

[If A and B commute, i.e. AB = BA]

(c) $e^{At}e^{-At} = 1$

(d) $e^{A0} = 1$

(e) $\dfrac{d(e^{At})}{dt} = Ae^{At}$

3.28 Evaluate e^{AT} to 4 decimal places accuracy for the following A Matrices, for the τ shown. Take sufficient terms of the series so the last term added to the series changes no element of the truncated sum by more than 0.0001. How many terms were necessary for this accuracy? Calculate the first few terms by hand and write a program and evaluate the matrix to the desired accuracy.

(a) $\tau = 0.1$

$$\begin{bmatrix} 0 & 1 \\ -6 & -5 \end{bmatrix}$$

(b) $\tau = 0.1$

$$\begin{bmatrix} 0 & 5 \\ -5 & -1 \end{bmatrix}$$

(c) $\tau = 1$

$$\begin{bmatrix} 0.01 & 0 \\ 0 & 100 \end{bmatrix}$$

(d) $\tau = 0.01$

$$\begin{bmatrix} 0 & 2 \\ 3 & 0 \end{bmatrix}$$

3.29 (a) For $e^{AT}(=0.1)$ obtained in Problem 3.28(a), plot the column vectors of $\Phi(t) = e^{At}$, (ϕ_{11} and ϕ_{21} on a single sheet and ϕ_{12} and ϕ_{22} on another by taking

$$\Phi(n\tau) = e^{An\tau}$$

$$= e^{A\tau}e^{A(n-1)\tau}$$

$$= e^{A\tau}\Phi((n-1)\tau)$$

(b) Simulate the differential equation having the system matrix given in Problem 3.28(a) on an analog computer and plot the state variable response to the initial conditions which give the elements of (t). Compare these results with the results of part (a) above and Example 3.16.

3.30 Repeat Problem 3.29 for the systems given in

(a) Problem 3.28(b)

(b) Problem 3.28(c)

(c) Problem 3.28(d)

3.31 Write a program to evaluate the F and G matrices for Equation (3.53). If you have done Problem 3.28 just add a few statements to get G as a useful by-product from that program. Verify the results given in Example 3.17 with your program. Use the recursion relation, Equation (3.53), to obtain the step response of the system given in Example 3.17.

3.32 Use the program from Problem 3.31 to evaluate the F and G matrices, and obtain the step response of the following systems. Assume all initial conditions are zero.

(a) $\ddot{y} + 4\dot{y} + y = u(t)$

(b) $\ddot{y} + 4y = \cos 10t$

(c) $(D^5 + 3D^3 + 2D^2 + D + 1)y = u(t)$

(d) $(D^3 + 5D^2 + 25)y = \cos 10t$

3.33 Determine infinite series representations of F and G in Equation (3.53) if the excitation u(t) is assumed piecewise linear in the interval by Equation (3.54).

3.34 Obtain the response of the systems given in Problem 3.32 assuming that the input is piecewise linear. Use the program for Problem 3.33 to evaluate the F and G matrices.

3.35 Give a recursion relation for a 4th order Runge Kutta solution to the equation $\dot{X} = AX + BU$ where $U = 0$. Where possible relate the recursion relation to the matrix exponential.

3.36 Give a recursion relation which results from integrating the equation $\dot{X} = AX + BU$ by Simpson's rule.

3.37 Sketch the locus of the poles of the pulse transfer matrix which results from integrating the differential equation $\ddot{y} + 3\dot{y} + 2y = 0$ by rectangular integration. For what values of τ is the recursion relation stable?

3.38 Repeat Problem 3.37 for the differential equation $\ddot{y} + y = u$ for (a) Euler integration, (b) Trapezoidal rule.

3.39 Give an analog computer block diagram (including sign reversals in the integrators) for the equation $\ddot{y} + 6\dot{y} + 5y = u$. Give the system matrix A if the integrator outputs are defined as the state variables.

3.40 Problem 2.7 involved the integration of the equation $\ddot{y} + 4y = x(t)$ by rectangular integration. For what values of τ is the integration algorithm stable?

3.41 Obtain a digital simulation of the equation $\ddot{y} + y = 0$ using Euler Equation (1.6), Modified Euler (3.9), Simpson's rule (3.15b), Milne (3.15) and 4th order Runge Kutta (3.18).

   (a) Determine the step size τ, in terms of the period of the system response, to obtain the period correct to 0.01 percent by each method.

   (b) Plot the error in the period for each method as a function of the increment size τ.
   (This is a rather extensive problem.)

3.42 Repeat Problem 3.41 to obtain 0.01 percent error in magnitude of the value of $y(2\pi)$ if $y(0) = 1$ and $\dot{y}(0) = 0$.

3.43 Integrating the system equation $\dot{X} = AX + BU$ by rectangular integration gives the vector recursion relation

$$X_n = [I+A\ ]X_{n-1} + BU_{n-1}$$

Show that this recursion relation is stable if the roots of the equation

$$\det[zI-(I+A\ )] = 0$$

lie inside the unit circle. That is, the roots z_i have the property that $|z_i| < 1$.

3.44 Determine the range of step size τ for which the following systems are stable if integrated by rectangular integration. (See Problem 3.43.)

$$\dot{X} = AX + BU$$

(a) $A = \begin{bmatrix} 0 & 1 \\ -2 & -3 \end{bmatrix}$ (b) $A = \begin{bmatrix} 0 & 1 \\ 0 & -2 \end{bmatrix}$

(c) $A = \begin{bmatrix} 1 & 0 \\ 0 & 10^3 \end{bmatrix}$ (d) $A = \begin{bmatrix} 0 & 1 & 0 \\ 0 & 0 & 1 \\ -1 & -1 & -1 \end{bmatrix}$

3.45 Prove the following conditions which frequently give ranges of τ for which a discrete system is unstable. The polynomial

$$D(z) = (z-p_1)(z-p_2) \ \cdots \ (z-p_n) = 0$$

has the roots $p_1, p_2 \cdots p_n$.

(a) If $D(1) < 0$ then $|P_i| > 1$ for some i

(b) If $D(-1) < 0$ for n odd then $|p_i| > 1$ for some i (or $D(-1) > 0$ for n even)

3.46 (a) Show that the second order system $\ddot{y} + 2\zeta\omega_n\dot{y} + \omega_n^2 y = 0$ is stable when integrated by rectangular integration for all positive $\tau < 2\zeta/\omega_n$ and give the location poles of the pulse transfer function in terms of τ, ζ and ω_n.

(b) Use the results of (a) to show that the critically damped system ($\zeta=1$) is stable for τ equal to twice the system time constant consistent with the previously obtained results.

3.47 (a) Show that integration of the equation $\dot{X} = AX + BU$ by the trapezoidal rule is stable if the roots of

$$\det\left[z\left(I - \frac{\tau}{2} A\right) - \left(I + \frac{\tau}{2} A\right)\right] = 0$$

are inside the unit circle.

(b) In (a) above consider the scalar equation $\dot{y} + ay = u$, $(A=a)$ and show that the pole is located at the point

$$p = \frac{1 - \frac{\tau}{2} a}{1 + \frac{\tau}{2} a}$$

which is inside the unit circle for all $\tau > 0$ if $a > 0$.

(c) How about the case $a < 0$ in (b) above?

3.48 Show that the integration of linear systems by the trapezoidal rule is stable as shown in Problem 3.47 and is valid if $A = -a$ is complex (that is, $a = \alpha + j\beta$).   Thus   the   stability   is extended   to   second   order   systems   that can be considered as cascaded first order   systems   with perhaps   complex   coefficients.   Similarly   the stability is extended to any order linear   system as the cascade of first order systems.

3.49 Show   that   the   modified   Euler   (rectangular-predictor, trapezoidal-corrector) integration   of the   equation $\dot{X} = AX + BU$ is   stable   for $0 < \tau < a/2$ in   the   scalar   case   where $A = -a$. Thus   the   predictor-corrector   has the   stability properties   of   the   predictor,   and   using   the trapezoidal   rule to correct does not improve the stability properties.

3.50 To   evaluate   the   effect   of   choice   of   state variables in simulation of a system   with   widely separated   time   constants,   consider   the differential equation

$$\ddot{y} + 1001\dot{y} + 1000y = 0$$

(a) Let $x_1 = y$, $x_2 = \dot{y}$ and set up the state equations. Call this system α).

(b) Let $x_1 = y$, $x_2 = \dot{y} + y$ and set up the state equations. Call this system β).

(c) Either   by   hand   or   using   the   computer, evaluate $e^{A\tau}$ for each system, $\tau = 0.0001$ and $\tau = 1.0$.

(d) Compare   your   numerical   results.   Is $(e^{0.0001A})^{10000} = e^A$? Why?

(e) Simulate   the   system for $y(0) = 1$, $\dot{y}(0) = 0$. Using the state equations for   system α and system β.   Plot the error (by computing the analytic solution) as a function of time   for a   sufficient   duration   to   see the solution reach steady state.

(f) Repeat part (e) with $y(0) = 1.0$, $\dot{y}(0) = -1.0$.

3.51 Consider a system described by the differential equation

$$\dot{X} = AX + BU$$

where

$$A = \begin{bmatrix} 0 & 1 \\ 0 & -1 \end{bmatrix} \quad \text{and} \quad B = \begin{bmatrix} 0 \\ 1 \end{bmatrix}$$

which has solution

$$X(t) = e^{At}X(0) + e^{At}\int_0^t e^{-A\alpha}BU(\alpha)\,d\alpha.$$

(a) For $\tau = .01$ evaluate $e^{A\tau}$ (first 3 terms are sufficient)

(b) Give a recursion relation that will generate $\Phi(t) = e^{At}$ for discrete values of $t = n\tau$.

(c) Draw an analog block diagram representing this system.

(d) Sketch the functions ϕ_{11}, ϕ_{12}, ϕ_{21}, ϕ_{22} where

$$\Phi(t) = e^{At} = \begin{bmatrix} \phi_{11} & \phi_{12} \\ \phi_{21} & \phi_{22} \end{bmatrix}$$

(e) Give an expression for integrating the state equation by rectangular integration.

(f) Suppose (e) is used as a predictor in a predictor-corrector scheme and give a recursion relation for a corrector formula using trapezoidal integration.

3.52 Consider the boundary value problem

$$\ddot{y} + 0.2\dot{y} + y = 0 \qquad \begin{aligned} y(0) &= 1 \\ y(0.2) &= 0 \end{aligned}$$

(a) Write the equation in state space notation [that is, $\dot{X} = AX + BU$, give A, X, B and U].

(b) Give an exact expression for the solution $X(t)$ for arbitrary initial conditions $X(0)$.

(c) Sketch the elements of the matrix $\Phi(t) = e^{At}$.

(d) Evaluate the exponential matrix e^{At} for some value of t.

(e) Determine the value of $\dot{y}(0)$ which together with $y(0) = 1$ satisfies the condition $y(0.2) = 0$.

3.53 (a) Write the differential equation

$$\ddot{y} + 4y = f(t)$$

as a state equation

(b) Evaluate the exponential matrix $e^{A\tau}$ for $\tau = 0.1$ correct to 3 decimal places.

(c) Sketch the elements of $e^{At} = \Phi(t)$. (Give scales.)

3.54 Suppose $\dot{X} = AX + BU$ where X and U are (nx1) and (rx1) vector respectively. Give a recursion relation for the exact solution if $U(t)$ is the sequence of impulses

$$U(t) = \sum_{k=0}^{\infty} U_k \delta(t - k\tau - \frac{\tau}{2})$$

$(\delta(t) = 0$ for $t \neq 0$ and $\int_{-\infty}^{\infty} \delta(t)\,dt = 1)$

3.55 (a) Write the equation

$$\dddot{y} + 3\ddot{y} + y = u$$

in state variable form

(b) Evaluate $e^{A\tau}$ correct to 4 decimal places for $\tau = 0.01$ (You do not have to use the computer, but you may.)

If $y(0) = 0$, $\dot{y}(0) = 1$, $\ddot{y}(0) = 0$ and $u(t) = 0$, find $y(0.01)$ and $y(0.03)$.

3.56 Compare the number of scalar and matrix multiplications required to evaluate the truncated series for $e^{A\tau}$ by the following methods.

(a) $e^{A\tau} = \sum_{k=0}^{\infty} \frac{(A\tau)^k}{k!}$

$$= I + A\tau(I + \frac{A\tau}{2}\{I + \frac{A\tau}{3}[I + \cdots$$

(b) $e^{A\tau} = \sum_{k=0}^{\infty} \frac{(A\tau)^k}{k!}$

$$= I + A\tau + \frac{(A\tau)^2}{2} + \frac{(A\tau)^3}{3!} + \cdots$$

(c) The procedure given in Figure 3.14.

3.57 Compare the accuracy of Runge-Kutta and predictor-corrector methods for discontinuous inputs. Discuss and give some experimental results.

3.58 (a) Define a subroutine SYSTEM that takes all of the integrator outputs from your digital-analog simulation program and calculates the corresponding integrator inputs.

(b) Use this subroutine to perform the integration by 4th order Runge-Kutta integration.

(c) Obtain the solution to the van der Pol equation using this program.

(d) Make the program automatically decrease τ when any integrator input changes by more than 0.01 percent, and increase τ when the changes are all less than 0.001 percent.

3.59 Compare the computation time to calculate sin θ for $0 \le \theta \le \pi/2$ to the same accuracy by the two following methods.

(a) The "canned" function in the computer.

(b) The solution of the differential equation by numerical integration.

3.60 Generate a table of sin θ using the discrete state transition matrix.

3.61 An analog computer integrator can be used to obtain a piecewise constant representation of a waveform.

(a) Set up the circuit shown in Figure 3.17 to obtain a sinusoidal waveform and its piecewise constant approximation. (See Problem 2.24.)

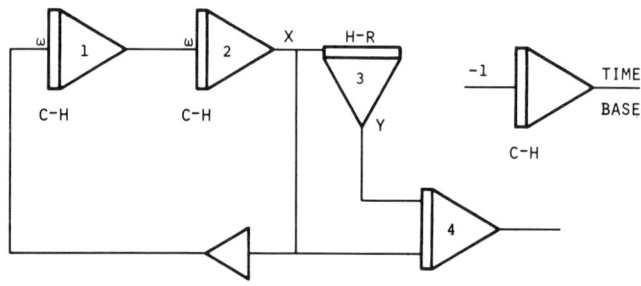

FIGURE 3.17 PROBLEM 3.61.

(b) To integrate x and y to obtain the error in rectangular integration what should the mode control be on integrator 4?

(c) Plot the integration error e(t) as a function of time for several values of ω.

(d) Plot the maximum error e_{max} as a function of ω.

(e) Obtain an analytical expression for $e_{max}(\omega)$ if possible and compare it with the results of (d) above.

3.62 Get the manual for the digital-analog simulation program (CSMP, DSL/90 and so on) that resides in your computer and use it to simulate a system which you have already simulated by another method.

3.63 Show that if A^{-1} exists, then in Equation (3.53)

$$G = A^{-1}(e^{A\tau}-I)$$

3.64 Use the z-transform integrating operators to simulate systems described by the equations

(a) $\ddot{y} + 5\dot{y} + 6y = 1$
$y(0) = 0, \dot{y}(0) = 0$

(b) $\ddot{y} + 100y = \cos 10t$
$y(0) = 0, \dot{y}(0) = 0$

4 iterative methods and successive approximations

In an attempt to integrate the nonlinear differential equation

$$\dot{y} = f(y,x,t)$$

by the trapezoidal rule in the last chapter we encounter the difficulty that the new value of the dependent variable y_n appears on both sides of the equation. For linear equations this presents no problem but in the nonlinear case we have equations of the form

$$y_n = f(y_n, y_{n-1}, \ldots, x_n, x_{n-1}, \ldots) \qquad (4.1)$$

which cannot be readily solved for y_n. This problem was circumvented by predicting $y_n = p_n$ on the right side by some method not requiring y_n and then using Equation (4.1) to correct the estimate resulting in the predictor-corrector methods.

In Section 4.1 we discuss some techniques for solving nonlinear algebraic equations of this form.

211

In particular we find the roots of equations which can be written in the form f(x) = 0. Section 4.2 presents some schemes for the solution of systems of algebraic equations.

   Generally it is not necessary to have the solution to equations in a closed form formula, but only correct within a desired accuracy, say 0.001 inches or 20 grams. In fact, even with an exact expression, we know by now that we can enter 5.0000000 into the computer, do some manipulation on it which should result in the same number, and print out 4.9999999. Apparently we are satisfied with any method for approximately solving these equations if it results in solution to the desired accuracy. This chapter discusses a few of these methods in an attempt to develop an understanding of iterative techniques.

4.1   SOLUTION OF ALGEBRAIC AND TRANSCENDENTAL EQUATIONS

   To find the solutions for equations of the form f(x) = 0, consider first the problem of determining the square root of a number. Let's find $a = \sqrt{10}$ correct to three decimal places. $a = \sqrt{10}$ can be written in the form f(x) = 0 by taking $x - \sqrt{10} = 0$. Now as a first guess we can take the closest integer $a_1 = 3$. Define the error in this estimate

$$\alpha_1 = \sqrt{10} - a_1 = \sqrt{10} - 3$$

or

$$\sqrt{10} = 3 + \alpha_1$$

Squaring both sides of the equation gives

$$10 = 9 + 6\alpha_1 - \alpha_1^2$$

which can be solved for α_1 by the quadratic equation, giving the result $\alpha_1 = 3 - \sqrt{10}$. Now if we knew the square root of 10 we would have no need to look at this problem in the first place. Of course, we don't know it, so we come back to the equation

$$10 = 9 + 6\alpha_1 + \alpha_1^2$$

and observe that $\alpha_1 < 1$ so α_1^2 is even smaller. By neglecting the α_1^2 term we have

$$10 \approx 9 + 6\alpha_1$$

or

$$\alpha_1 \approx \frac{1}{6} \approx 0.165$$

So for our new estimate of a we can take

$$a_2 = a_1 + \alpha_1 = 3.165$$

Repeating the process above we obtain

$$\sqrt{10} = a_2 + \alpha_2$$

$$10 = a_2^2 + 2a_2\alpha_2 + \alpha_2^2$$

$$10 \approx a_2^2 + 2a_2\alpha_2$$

Solving for α_2 gives

$$\alpha_2 = \frac{10 - a_2^2}{2a_2}$$

$$= \frac{10 - (3.165)^2}{2(3.165)} \approx 0.00272$$

This result is correct to 3 decimal places. In general, given the estimate a_n we have

$$\sqrt{10} = a_n + \alpha_n$$

$$10 = a_n^2 + 2a_n\alpha_n + \alpha_n^2$$

By neglecting α_n^2 we have

$$10 = a_n^2 + 2a_n\alpha_n$$

or

$$\alpha_n = \frac{10 - a_n^2}{2a_n}$$

Thus

$$a_{n+1} = a_n + \alpha_n = a_n + \frac{10 - a_n^2}{2a_n}$$

or

$$a_{n+1} = \frac{2a_n^2 + 10 - a_n^2}{2a_n} = \frac{a_n^2 + 10}{2a_n}$$

To obtain the square root of a number B we take an initial guess a_1 and iterate using the recursion relation

$$a_{n+1} = \frac{a_n^2 + B}{2a_n} = \frac{1}{2}(a_n + \frac{B}{a_n}) \qquad (4.2)$$

EXAMPLE 4.1

Write a FORTRAN Program that will evaluate $\sqrt{500}$.

SOLUTION:

All that is necessary is an initial guess A and a DO loop around Equation (4.2)

```
B=500
A=20
DO 85 N=1,10
WRITE(6,28)N,A
85 A=0.5*(A+B/A)
STOP
28 FORMAT(I3,F10.5)
END
```

This program can be made more efficient by the inclusion of logic to stop the program when the value of A changes by less than the prescribed error. (See Problem 4.1.) ▼

Equation (4.2) for iteratively finding the square root of a number is equivalent to representing the geometric mean by the arithmetic mean.

$$a_{n+1} = \sqrt{a_n (\frac{B}{a_n})} \approx \frac{1}{2}(a_n + \frac{B}{a_n}) = \frac{a_n^2 + B}{a_n}$$

The error at the nth iteration is

$$\alpha_n = \sqrt{B} - a_n$$

$$= \sqrt{B} - \frac{a_{n-1}^2 + B}{2a_{n-1}}$$

$$= \frac{2a_{n-1}\sqrt{B} - a_{n-1}^2 + B}{2a_{n-1}}$$

$$= - \frac{(a_{n-1} - \sqrt{B})^2}{2a_{n-1}}$$

$$= - \frac{\alpha_n^2}{2a_{n-1}}$$

Thus if the error at any step of the iteration is known, the error at any other step can be calculated exactly. For $a_n > 0$, (this is the root we are looking for) we have all the errors negative except perhaps α_1 Thus the sequence $\{a_n\}$ converges to $\sqrt{B}$ from above. A pessimistic estimate of the error is that the errors are halved at each iteration. In actuality, as $\{a_n\} \to B$, the error decreases faster than α_1/n^2. (See Problem 4.2.) Thus after 10 steps the error is decreased at least by the factor $2^{10} = 1024 \approx 10^3$ and after 40 steps, $2^{40} \approx 10^{12}$.

Problem 4.3 demonstrates that the relative error sequence $\{\beta_n\}$ where

$$\beta_n = \frac{|\alpha_n|}{\sqrt{B}}$$

satisfies the inequality

$$\beta_{n+1} < \frac{\beta_n^2}{2}$$

demonstrating that as β_n gets smaller the successive relative errors decrease exponentially. If $\beta_n = 0.01$ then $\beta_{n+1} = 0.00005$ and $\beta_{n+2} = 0.00000000013 = 1.3 \times 10^{-10}$. As we get closer to the solution we essentially double the number of accurate places each iteration. So. if we have a good guess we need only take a few iterations to obtain an accurate result. Problem 4.4 discusses a recursion relation obtained in a similar manner for obtaining roots of the equation $x^k = a$. The evaluation of the roots of the equation $x^k = a$ is a special case of equations of the form $f(x) = 0$ [that is, $x^k - a = 0$]. Several schemes for finding the roots of general equations of the form $f(x) = 0$ are discussed below.

One possibility is to continue evaluating f while increasing the dependent variable x, and when f changes sign we know that either we have passed a root or $f(x)$ is discontinuous. This may require many steps. In fact, $f(x)$ may have no roots in the interval checked. Furthermore, if the steps are too large we might miss the roots of the equation, and if the steps are too small too many calculations are required. So we need some idea of how the function acts in order to pick intervals small enough so there are not too many calculations but not so large as to miss some roots. One possible search scheme is to keep incrementing x until f changes its sign, stop the procedure, and decrease the increment size and go back over the last interval. Then we do this with a smaller interval repeatedly until we obtain the desired accuracy. This technique has the danger of starting off with too large an interval and missing some roots.

It is desirable to have an idea about how to get a
good guess. If f(x) is very large, near x_0, then
there is no interest in the vicinity of x_0. The first
step in obtaining a good guess is to make a sketch of
the function.   From this sketch we obtain those x
which look like they might be in the vicinity of a
root of the equation f(x) = 0 - that is, those values
of x for which f is near the f = 0 axis.   With this
initial guess we obtain successively better estimates
of the root by the methods given below.

METHOD OF CHORDS
   If f(x) is continuous on the interval [a,b] and
f(a) and f(b) have opposite signs,   then   there is a
root of the equation f(x) = 0 in that interval.
Figure 4.1 shows a function having this property.

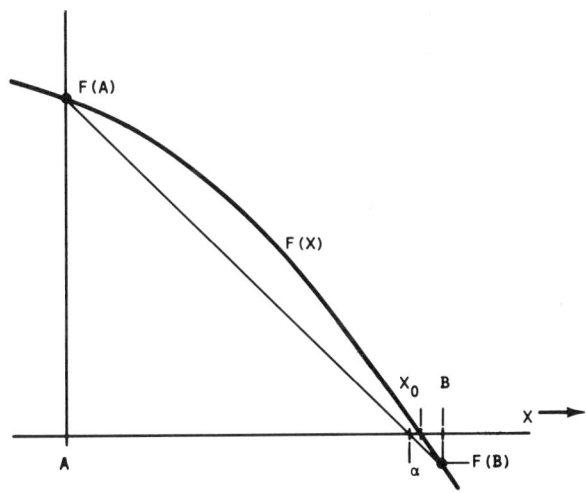

FIGURE 4.1 THE METHOD OF CHORDS FOR DETERMINING x_0 SUCH THAT F(x_0) = 0,

In Figure 4.1 we desire to find the value of $x = x_0$
for which $f(x_0)$ = 0. An estimate of x_0 is obtained by
taking the point that is the linear interpolation
between f(a) and f(b), that is, draw a   chord between

f(a) and f(b). Since the slope of the chord is

$$\frac{f(b) - f(a)}{b - a}$$

we have

$$\frac{0 - f(a)}{\alpha - a} = \frac{f(b) - f(a)}{b - a}$$

or

$$\alpha = a - f(a) \frac{b - a}{f(b) - f(a)} \qquad (4.3)$$

If $f(\alpha)$ has the same sign as $f(a)$ [that is, $f(a)f(\alpha) > 0$] we replace a by α, decreasing the size of the interval containing the root. Likewise if $f(\alpha)$ does not have the same sign as $f(a)$ [that is, it has the same sign as $f(b)$ or $f(a)f(\alpha) < 0$] we replace b by α. This procedure is repeated until b - a, the interval containing the root, is less than the prescribed error. If we choose $x_0 = (b+a)/2$ the error is certainly less than $(b-a)/2$. If in the vicinity of x_0, say in the interval [a,b] containing x_0, $\ddot{f}(x)$ and $\ddot{f}(x)$ have the same sign as shown in Figure 4.2, the

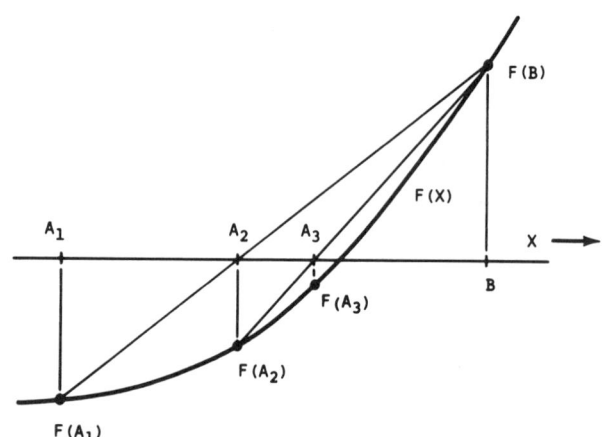

FIGURE 4.2   CONVERGENCE OF THE METHOD OF CHORDS IN THE VICINITY OF A SOLUTION TO THE EQUATION $F(X) = 0$.

chord will always intersect the f = 0 axis to the left
of x_0, thus we replace a by α each time and the
sequence a. converges to x_0 from the left.   Problem
4.17 shows two cases for $\dot{f}(x)$ and $\ddot{f}(x)$ of opposite
sign, for which the sequence $\{a_n\}$ converges to x_0 from
the right.   We see below that the Newton-Raphson
method has the opposite convergence properties, so we
can perform both methods and bracket the solution to
obtain an error estimate with which to terminate
calculations.

THE NEWTON-RAPHSON METHOD

If we do not have two values of x for which f(x)
has opposite signs we can still use the method of
chords to obtain an estimate of a solution as shown in
Figure 4.3.   The linear segment between f(a) and f(b)
is extended to the point where it intersects the f = 0
axis and is substituted for the closer of a or b.

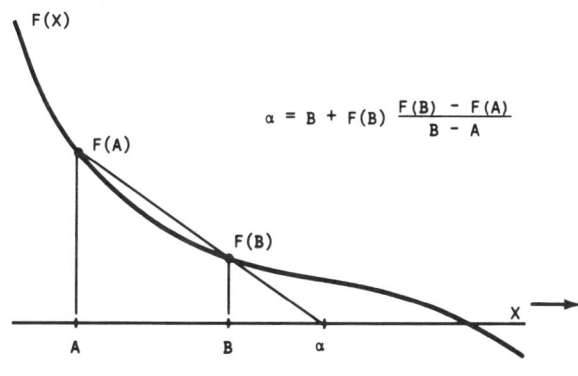

$$\alpha = B + F(B) \frac{F(B) - F(A)}{B - A}$$

FIGURE 4.3 EXTRAPOLATION BY THE METHOD OF CHORDS.

This addition allows convergence when it is difficult
to obtain two values of x for which f(x) has opposite
sign or when f(x) is either nonnegative or nonpositive
as shown in Figure 4.4.   This extrapolation is
particularly useful for finding the zeros of functions
that are represented by tabulated data.

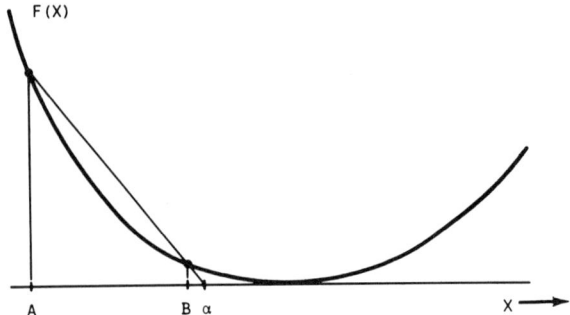

FIGURE 4.4 THE METHOD OF CHORDS FOR A NON-NEGATIVE FUNCTION.

If the derivative of f(x) can be readily obtained, the linear extrapolation in Figures 4.3 and 4.4 can be replaced by the extrapolation of f(a) along the derivative at point a giving the so-called *NEWTON-RAPHSON METHOD*, as shown in Figure 4.5.

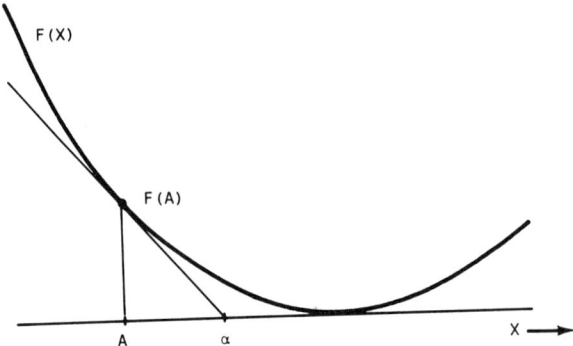

FIGURE 4.5 THE NEWTON-RAPHSON METHOD.

Since the slope of the linear extrapolation at f(a) is $\dot{f}(a)$ we can write

$$\dot{f}(a) = \frac{0 - f(a)}{\alpha - a}$$

or

$$\alpha = a - \frac{f(a)}{\dot{f}(a)}$$

Iteratively replacing a by α we have the recursion relation

$$a_{n+1} = a_n - \frac{f(a_n)}{\dot{f}(a_n)}$$

Figure 4.6 shows successive iterations of Equation (4.4) to obtain the solution to the equation $f(x) = 0$ by the Newton Raphson Method.

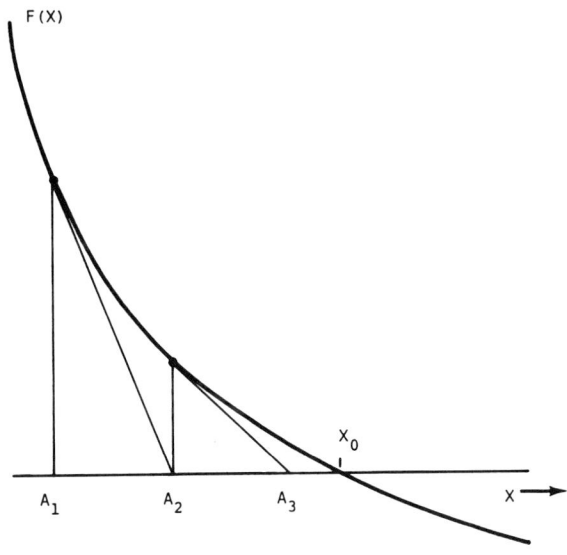

FIGURE 4.6   ITERATION BY THE NEWTON-RAPHSON METHOD
TO A SOLUTION OF THE EQUATION F(X) = 0.

EXAMPLE 4.2

Use the Newton Raphson Method to obtain a solution to the equation $f(x) = x^3 + x - 3 = 0$.

SOLUTION:

First we make a sketch of the function from a few tabulated values

x	0	1	2
f(x)	-2	-1	+6

Since we see that a solution exists between $x = 1$ and $x = 2$, we don't need a sketch; so let's start with $a_1 = 1$, then by Equation (4.4) and $f(x) = 3x^2 + 1$ we have. †

$$a_{n+1} = a_n - \frac{a_n^3 + a_n - 3}{3a_n^2 + 1}$$

$$a_1 = 1$$

$$a_2 = 1 - \frac{1 + 1 - 3}{3 + 1} = 1 - \frac{-1}{4} = 1.25$$

$$a_3 = 1.25 - \frac{(1.25)^3 + (1.25) - 3}{3(1.25)^2 + 1} = 1.220$$

$$a_4 = 1.220 - \frac{(1.220)^3 + (1.220) - 3}{3(1.220)^2 + 1} = 1.21344$$

Finally

$$a_5 = 1.21341166$$

is correct to eight significant digits. ▼

---

† Of course the FORTRAN statement
$$A = A - (A**3+A-3.)/(3.*A**2+1.)$$
accomplishes this operation.

There are convergence criteria for the method of chords and the Newton – Raphson Method but, unfortunately, they are generally of little utility. It is usually sufficient to keep in mind that these methods do not necessarily converge to the solution even if one exists. The methods are useful analysis tools which usually give favorable results. An aspect which makes them so useful is that if they do not work (that is, do not converge), this is known and erroneous results are not assumed to be correct. $f(a_n)$ can easily be compared with an error criterion at each iteration, and if $f(a_n)$ is less than the permissable error the calculation is terminated. Figure 4.7 shows some situations that can occasionally occur if the initial guess is not in the vicinity of a solution, vicinity meaning close enough for the method to converge.

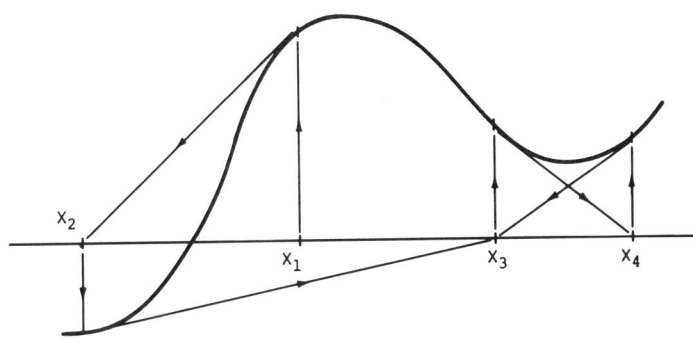

FIGURE 4.7   SOME CONVERGENCE PROBLEMS ENCOUNTERED WITH
THE NEWTON-RAPHSON METHOD.

HIGH ORDER METHODS

The Newton – Raphson formula for iteratively evaluating roots of equations can be obtained by truncating the Taylor series

$$f(x_{n+1}) = f(x_n) + \Delta x \dot{f}(x_n) + \frac{\Delta x^2}{2} \ddot{f}(x_n) + \ldots$$

to the first two terms on the right and setting $f(x_{n+1}) = 0$ (to solve the equation $f(x) = 0$)

$$0 = f(x_n) + x\dot{f}(x_n)$$

or

$$0 = f(x_n) + (x_{n+1} - x_n)\dot{f}(x_n)$$

which gives the previously obtained result

$$x_{n+1} = x_n - \frac{f(x_n)}{\dot{f}(x_n)} \qquad (4.4)$$

To obtain a second order formula, involving the second derivative of $f(x)$, we can take the first three terms of the Taylor series

$$0 = f(x_{n+1}) \approx f(x_n) + \Delta x\, \dot{f}(x_n) + \frac{\ddot{f}(x_n)\Delta x}{2}$$

$$(4.5)$$

To approximately solve (4.5) for x_{n+1}, use the estimate for Δx

$$\Delta x = \frac{f(x_n)}{\dot{f}(x_n)}$$

(Note: This is much like the predictor in a predictor corrector method.)

Then Equation (4.5) becomes

$$0 = f(x_{n+1}) \approx f(x_n) + (\dot{f}(x_n) - \frac{\ddot{f}(x_n)f(x_n)}{2\dot{f}(x_n)}\Delta x)$$

$$0 = f(x_n) + \dot{f}(x_n) - \frac{\ddot{f}(x_n)f(x_n)}{\dot{f}(x_n)}(x_{n+1}-x_n)$$

or

$$x_{n+1} = x_n - \cfrac{f(x_n)}{\dot{f}(x_n) - \cfrac{\ddot{f}(x_n)f(x_n)}{2\dot{f}(x_n)}} \qquad (4.6)$$

Equation (4.6) is called *NEWTON's SECOND ORDER FORMULA*. Note that if $f(x_n) = 0$ as assumed for the Newton - Raphson formula, Equation (4.6) reduces to Equation (4.4).

EXAMPLE 4.3

Give a recursion relation to iteratively obtain the square root of a number based on Newton's second order formula.

SOLUTION:

We desire to find a root of the equation

$$f(x) = x^2 + a = 0$$

Substitution of

$$f(x) = x^2 + a$$

$$\dot{f}(x) = 2x$$

$$\ddot{f}(x) = 2$$

Into Equation (4.6) gives

$$x_{n+1} = x_n - \cfrac{x_n^2 + a}{2x_n - \cfrac{2(x_n^2 + a)}{4x_n}}$$

$$x_{n+1} = x_n \left[1 - \frac{2(x_n^2 + a)}{(3x_n^2 - a)} \right] \blacktriangledown$$

GENERAL ITERATIVE TECHNIQUES

Iterative schemes for the solution of algebraic equations are recursion relations which state a rule by which past estimates of the solution are used to obtain a more accurate estimate. Any equation of the form $f(x) = 0$ can be written

$$x = \phi(x) \qquad\qquad (4.7)$$

in many ways. For example $f(x) = 0$ can be written

$$x = x + f(x)$$

or

$$x = x - \frac{f(x)}{\dot{f}(x)}$$

An iterative relation is obtained from Equation (4.7) by writing

$$x_{n+1} = \phi(x_n) \qquad\qquad (4.8)$$

Now if the sequence $\{x_n\}$ converges, it converges to a solution of the equation $f(x) = 0$. However, it may diverge, so the selection of the function $\phi(x)$ is critical for convergence. Equation (4.8) sequentially steps through values of x_n which hopefully converge to the solution $f(x_n) = 0$.

EXAMPLE 4.4

Find a solution to the equation $10x - 1 - \cos x = 0$ to slide rule accuracy.

SOLUTION:

One way to write the equation in the form $x_{n+1} = \phi(x_n)$ is

$$x_{n+1} = \frac{1 + \cos x_n}{10}$$

Try the initial guess $x_1 = 0$, then to slide rule accuracy

$$x_2 = \frac{1 + \cos 0}{10} = \frac{2}{10} = 0.2$$

$$x_3 = \frac{1 + \cos (0.2)}{10} = \frac{1.98}{10} = 0.198$$

$$x_4 = \frac{1 + \cos (0.198)}{10} = \frac{1.98}{10} = 0.198 \quad \blacktriangledown$$

To slide rule accuracy we get two successive identical answers so we can assume that the process converged and the solution is x = 0.198.

GEOMETRIC INTERPRETATION OF THE METHOD OF ITERATION

Figure 4.8 shows the two functions $\phi(x)$ and x plotted versus the dependent variable x. If the horizontal (independent variable) and vertical (dependent variable) axes have the same scale the plot of x versus x is a line with a slope of 45 degrees (+1). Now we choose an initial guess for x, say x_1, evaluate $\phi(x_1)$ and take $x_2 = \phi(x_1)$. We then evaluate $x_3 = \phi(x_2)$ and so on to obtain the solution. This sequence of operations representing one iteration is shown in Figure 4.8.

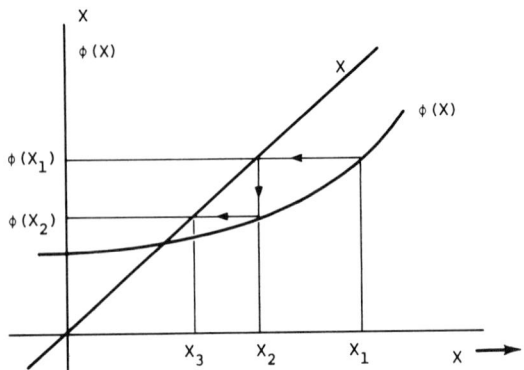

FIGURE 4.8 CONVERGENCE OF THE METHOD OF ITERATION FOR $|\phi'(x)| < 1$.

The function $\phi(x)$ shown in Figure 4.9 and the related iterations demonstrate that we cannot just blindly choose an iterative equation $x_{n+1} = \phi(x_n)$ and expect guaranteed convergence. (Show that the same convergence properties prevail if the initial guess is on the other side of the solution in both Figures 4.8 and 4.9.)

If the iteration $x_n = \phi(x_{n-1})$ diverges and the inverse function $\phi^{-1}(x_n) = x_{n-1}$ can be written, then the iteration follows through the same sequence $\{x_n\}$ but in the reverse direction. For example the iteration $x_n = 1/x_{n-1}^2$ diverges from the initial guess $x_0 = 2$ (check this, and make a sketch similar to Figures 4.8 and 4.9). However, interchanging the n and n-1 subscripts gives $x_{n-1} = 1/x_n^2$ or $x_n = 1/\sqrt{x_{n-1}}$ which converges. This concept is demonstrated in the example below.

EXAMPLE 4.5

Use the general method of iteration to find a solution to the equation

$$x + 2^x - 4 = 0.$$

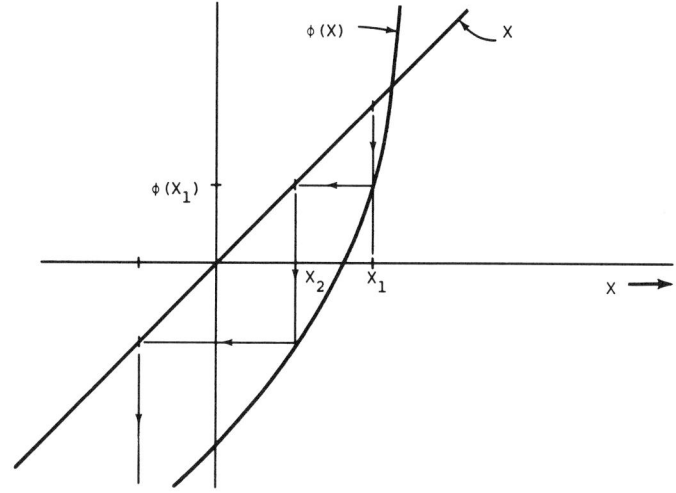

FIGURE 4.9 DIVERGENCE OF THE METHOD OF ITERATION FOR $|\phi'(x)| > 1$.

SOLUTION:

Taking $x_n = 4 - 2^{x_{n-1}}$ and $x_1 = 0$ yields

$$x_2 = 4 - 1 = 3$$

$$x_3 = 4 - 2^3 = 4 - 8 = -4$$

$$x_4 = 4 - 2^{-4} = 4 - 0.0625 = 3.9375$$

$$x_5 = 4 - 2^{3.9375} = 4 - 15.25 = -11.25$$

$$x_6 = 4 - 2^{-11.25} \approx 4$$

$$x_7 = 4 - 2^4 = 4 - 16 = -12$$

$$x_8 = 4 - 2^{-12} \approx 4$$

$$x_9 = -12$$

$$x_{10} = 4$$

and so on.

Clearly this process is divergent since the numbers do not become close to any value. In fact $\{x_n\}$ diverges in the sense that it oscillates between -12 and 4. Figure 4.10(a) shows this sequence of iterations.

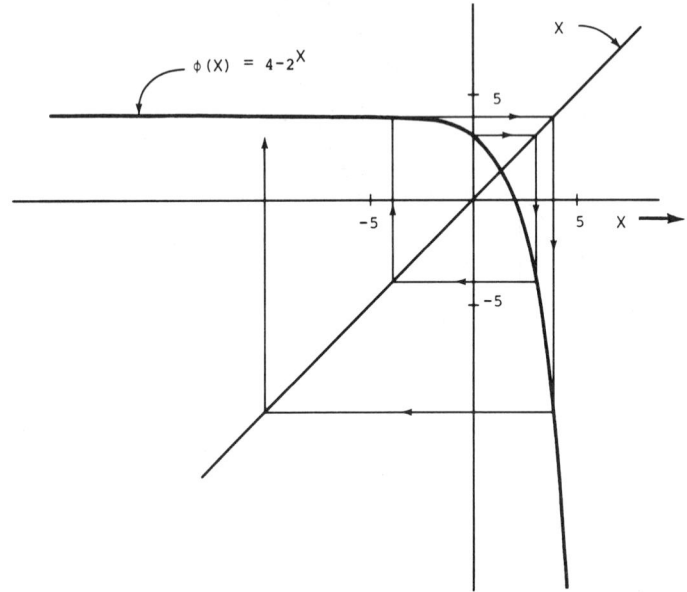

FIGURE 4.10(A)   EXAMPLE 4.5 $\phi(x) = 4 - 2^x$ (DIVERGENCE).

If we can make the iteration go in the opposite direction, that is take $x_{n-1} = 4 - 2^{x_n}$ or $x_n = \log_2(4-x_{n-1})$, we can obtain convergence to the desired solution. To show that it does indeed trace through the same values we start at $x_1 = -4$ and iterate

$$x_n = \log_2(4-x_{n-1})$$

$$x_1 = -4$$

$$x_2 = \log_2(4+4)\log_2(8) = 3$$

$$x_3 = \log_2(4-3) = \log_2(1) = 0$$

$$x_4 = \log_2(4-0) = 2$$

$$x_5 = \log_2(4-2) = 1$$

$$x_6 = \log_2(3) = 1.59$$

$$x_7 = \log_2(2.41) = 1.27$$

$$x_8 = \log_2(2.73) = 1.45$$

$$x_9 = \log_2(2.55) = 1.35$$

$$x_{10} = \log_2(2.65) = 1.41$$

$$x_{11} = \log_2(2.59) = 1.37$$

$$x_{12} = \log_2(2.63) = 1.39$$

$$x_{13} = \log_2(2.61) = 1.375$$

which is clearly converging to a number between 1.375 and 1.39. Figure 4.10(b) shows the functions x and $\phi(x_{n-1}) = \log_2(4-x_{n-1})$ (Trace out this convergent path on the figure.) ▼

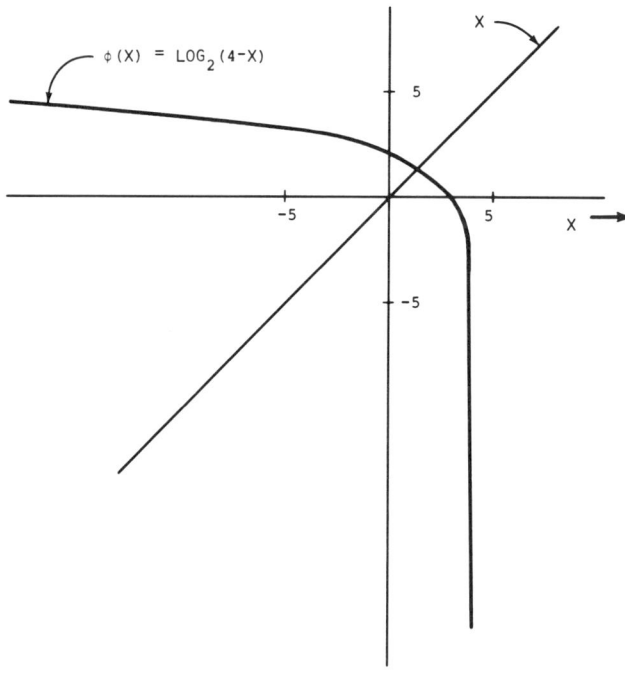

FIGURE 4.10(B)   EXAMPLE 4.5 $\phi(X) = \log_2(4-X)$ (CONVERGENCE).

A good initial guess speeds up convergence in most cases and sometimes gives convergence that cannot be obtained from a poor guess. However, if the iteration diverges when it is near the solution, the only hope is to change the iteration scheme as in the above example.

From Example 4.5 it appears that when the slope of $\phi(x)$ is small near the solution, the iteration converges and when the slope is large, the iteration diverges. We see below that this is indeed the case and that a sufficient condition for convergence is that the derivative $\dot\phi(x)$ be less than 1 in magnitude in the vicinity of the solution.

Let x be a solution to the equation

$$x = \phi(x) \tag{4.9}$$

and the initial guess is x_0. Then

$$x_1 = \phi(x_0) \tag{4.10}$$

subtracting Equation (4.10) from Equation (4.9) gives

$$x - x_1 = \phi(x) - \phi(x_0) \tag{4.11}$$

If $\phi(x)$ is differentiable in the interval $[x, x_0]$ we have, by the mean value theorem,

$$x - x_1 = (x - x_0)\dot\phi(\xi_0) \tag{4.12}$$

where ξ_0 is in the interval $[x, x_0]$. Similarly for the second iteration

$$x_2 = \phi(x_1) \tag{4.13}$$

Subtracting Equation (4.13) from Equation (4.9) gives

$$x - x_2 = \phi(x) - \phi(x_1) \tag{4.14}$$

and by the mean value theorem

$$x - x_2 = (x - x_1)\dot\phi(\xi_1) \tag{4.15}$$

For the nth iteration

$$(x-x_n) = (x-x_{n-1}) \dot\phi (\xi_{n-1}) \qquad (4.16)$$

By multiplying together the left sides of Equation (4.16) for all n, equating the result to the product of the right sides, and dividing out the common factors, we obtain

$$x - x_n = (x-x_0) \dot\phi (\xi_0) \dot\phi (\xi_1) \; \cdots \; \dot\phi (\xi_{n-1}) \qquad (4.17)$$

Equation (4.17) indicates that a sufficient condition for convergence of $\{x_n\} \to x$ is that all the derivatives $\dot\phi (\xi_0)$, $\dot\phi (\xi_1)$, ..., $\dot\phi (\xi_{n-1})$ are less than one in magnitude. In particular if $|\dot\phi (x)| < 1$ for all x in some interval containing the solution, the convergence is guaranteed by an initial guess in the interval. Also it is easy to show that if $|\dot\phi (x)| < 1$ for x in the interval [a,b] then there is at most one solution to the equation $x = \phi (x)$ in the interval. (See Problem 4.11.)

EXAMPLE 4.6

Discuss the convergence of the iterative equations given in Example 4.5 in terms of $\dot\phi (x)$.

SOLUTION:

(a) $x_n = \phi (x_{n-1}) = 4 - 2^{x_{n-1}}$

$\phi (x) = 4 - 2^x$

$\dot\phi (x) = -2^x \ln 2$

$\qquad\quad = -(1.693) 2^x$

$|\dot\phi (x)| = (0.693) 2^x < 1$ for $x < 0.53$

Thus we obtain convergence if there is a root for $x < 0.53$. There is none, so we obtain a sequence which moves out of this interval, either to $x = -\infty$ or $x > 0.53$, or becomes oscillatory as seen in Example 4.5.

(b) $x_n = \phi(x_{n-1}) = \log_2(4-x_{n-1})$

$$\phi(x) = \log_2(4-x)$$

$$\dot{\phi}(x) = \frac{-1}{(4-x)} \log_2 3$$

$$|\dot{\phi}(x)| = \left|\frac{\log_{\approx} e}{4-x}\right| = \left|\frac{1.443}{4-x}\right|$$

$$|\dot{\phi}(x)| < | \quad \text{for} \quad x < 2.557$$

Since this interval includes the interval found in part (a) the formulation in (b) is probably more likely to converge.▼

Frequently the functions are suffciently complicated that the intervals of convergence cannot be evaluated or the inverse function ϕ^{-1} cannot be determined to aid convergence. But these concepts used with a sketch of the functions usually give sufficient insight to obtain convergence after at most a few rearrangements of the equation. It is important to determine whether the printed answer from machine calculations is "the solution" of if the last iteration calculated before the program was terminated due to lack of convergence.

4.2 THE SOLUTION OF ALGEBRAIC SYSTEMS OF EQUATIONS

This section is devoted to the solution of systems of algebraic equations with particular emphasis on linear equations of the form

$$a_{11}x_1 + a_{12}x_2 + \cdots + a_{1n}x_n = b_1$$

$$a_{21}x_1 + a_{22}x_2 + \cdots + a_{2n}x_n = b_2$$

$$\cdot$$
$$\cdot$$
$$\cdot$$

$$a_{n1}x_1 + a_{n2}x_2 + \cdots + a_{nn}x_n = b_n \qquad (4.18)$$

where the coefficients a_{ij} are constants.

Writing Equations (4.18) in matrix notation we have

$$
\begin{bmatrix}
a_{11} & a_{12} & \cdots & a_{1n} \\
a_{21} & a_{22} & \cdots & a_{2n} \\
\cdot \\
\cdot \\
\cdot \\
a_{n1} & a_{n2} & \cdots & a_{nn}
\end{bmatrix}
\begin{bmatrix}
x_1 \\
\cdot \\
\cdot \\
\cdot \\
x_n
\end{bmatrix}
=
\begin{bmatrix}
b_1 \\
\cdot \\
\cdot \\
\cdot \\
b_n
\end{bmatrix}
$$

or

$$Ax = b \qquad (4.19)$$

where A is the $n \times n$ matrix of the coefficients a_{ij}, b is a known column of n elements, and x is the unknown vector of n elements.

Occasionally it is desirable to determine the solution x to Equation (4.19) for several different, say k, right hand side vectors b. Then the set of solutions $x_1, x_2, \ldots, x_k$ can be generated by the solution to the equation

$$AX = B \qquad (4.20)$$

where B is an $n \times 1$ matrix with columns $b_1, b_2, \ldots, b_k$ and X is the $n \times n$ solution matrix with columns $x_1, x_2, \ldots, x_k$.

For the special case $B = I$ (the identity matrix) we have the solution

$$AX = I \qquad\qquad (4.21)$$

or the solution matrix is the inverse of A (that is, $X = A^{-1}$).

Large systems of linear algebraic equations of the form (4.19) occur in many engineering problems, for example the solution of boundary value problems in ordinary and partial differential equations by finite difference methods as discussed in Chapter 6.

Two methods are generally studied in algebra courses for the solution of systems of linear algebraic equations - Cramer's rule and Gauss elimination. To evaluate x_j by *CRAMER's RULE* we replace the jth column of A with the vector b to obtain the matrix A_j Then the solution is obtained by the ratio of the determinants of A_j and A as given below.

$$x_j = \frac{|A_j|}{|A|} = \frac{\begin{vmatrix} a_{11} & a_{12} & \cdots & b_1 & \cdots & a_{1n} \\ a_{21} & a_{22} & \cdots & b_2 & \cdots & a_{2n} \\ \vdots & & & & & \\ a_{n1} & a_{n2} & \cdots & b_n & \cdots & a_{nn} \end{vmatrix}}{\begin{vmatrix} a_{11} & a_{12} & \cdots & a_{1j} & \cdots & a_{1n} \\ a_{21} & a_{22} & \cdots & a_{2j} & \cdots & a_{2n} \\ \vdots & & & & & \\ a_{n1} & a_{n2} & \cdots & a_{nj} & \cdots & a_{nn} \end{vmatrix}}$$

Although efficient algorithms exist for the evaluation of determinants, many multiplications are required $(1/3)(n^3 + 2n - 3)$ rendering Cramer's rule impractical for more than 10 or 20 equations.

Gauss elimination is performed by reducing the A matrix in Equation (4.19) to triangular form by adding together multiples of equations to eliminate selected variables from the equations. This procedure is demonstrated in the example below. Since the position of the elements of the A matrix specify for which x_i it is the coefficient, we can drop the x vector and work only with the augmented matrix

$$A_G = [A \quad b] = \begin{bmatrix} a_{11} & a_{12} & \cdot \cdot & \cdot & a_{1n} & b_1 \\ a_{21} & a_{22} & \cdot \cdot & \cdot & a_{2n} & b_2 \\ \cdot & & & & & \\ \cdot & & & & & \\ \cdot & & & & & \\ a_{n1} & a_{n2} & \cdot \cdot & \cdot & a_{nn} & b_n \end{bmatrix}$$

$$(4.22)$$

EXAMPLE 4.7

Solve the below equations by Gauss elimination.

$$2x_1 + 3x_2 - x_3 = 5$$
$$4x_1 + 4x_2 - 3x_3 = 3$$
$$2x_1 - 3x_2 + x_3 = -1$$

SOLUTION:

The augmented matrix is given by

$$A_a = [A \quad b] = \begin{bmatrix} 2 & 3 & -1 & 5 \\ 4 & 4 & -3 & +3 \\ 2 & -3 & 1 & -1 \end{bmatrix}$$

1) Multiplying row [1] by 1/2 (or $1/a_{11}$)

$$
\begin{array}{c}
[1] \\
[2] \\
[3]
\end{array}
\begin{bmatrix}
1 & 3/2 & -1/2 & 5/2 \\
4 & 4 & -3 & +3 \\
2 & -3 & 1 & -1
\end{bmatrix}
$$

2) Multiply row [1] by [4] (or a_{21}) and subtract it from row [2] to give the new row [2]' with $a_{12} = 0$. Also taking the new row [3]' = [3] - 2 [1]

$$
\begin{array}{c}
[1] \\
[2] \\
[3]
\end{array}
\begin{bmatrix}
1 & 3/2 & -1/2 & 5/2 \\
0 & -2 & -1 & -7 \\
0 & -6 & 2 & -6
\end{bmatrix}
$$

3) Dividing [2] by -2 (that is, [2]' = -1/2[2]) gives

$$
\begin{array}{c}
[1] \\
[2] \\
[3]
\end{array}
\begin{bmatrix}
1 & 3/2 & -1/2 & 5/2 \\
0 & 1 & 1/2 & 7/2 \\
0 & -6 & 2 & -6
\end{bmatrix}
$$

4) Now the new row [3]' = [3] + 6 [2] gives

$$
\begin{array}{c}
[1] \\
[2] \\
[3]
\end{array}
\begin{bmatrix}
1 & 3/2 & -1/2 & 5/2 \\
0 & 1 & 1/2 & 7/2 \\
0 & 0 & +5 & 15
\end{bmatrix}
$$

5) Finally taking [3]' = 1/5[3] we have

$$
\begin{array}{c}
[1] \\
[2] \\
[3]
\end{array}
\begin{bmatrix}
1 & 3/2 & -1/2 & 5/2 \\
0 & 1 & 1/2 & 7/2 \\
0 & 0 & 1 & 3
\end{bmatrix}
$$

Row   [3]   corresponds to the equation $x_3 = 3$ which can be substituted into the equation corresponding   to row [2] , $x_2 - (1/2)x_3 = -7/2$ to get $x_2 = 2$. Finally x_2 and x_3 are substituted into the equation corresponding to row   [1]   to obtain $x_1 = 1$. ▼

The matrix in the above example with elements $a_{ij} = 0$ for $i > j$ is said to be an *UPPER TRIANGULAR MATRIX*. Clearly a system of equations whose coefficient matrix is an upper triangular matrix can be easily solved by reverse substitution. Similarly if $a_{ij} = 0$ for $i < j$ we say the matrix is a *LOWER TRIANGULAR MATRIX*.

The manipulations to put the first n columns of the augmented matrix into upper triangular form can be continued to form the first n rows into a *DIAGONAL MATRIX* (that is, $a_{ij} = 0$ for $i \neq j$), and specifically into an *IDENTITY MATRIX* ($a_{ii} = 1$ and $a_{ij} = 0$ for $i \neq j$). In particular in Equation (4.23) we can form new rows [1] and [2] by taking $[2]' = [2] - (1/2)[3]$ and $[1]' = [1] + (1/2)[3]$ to obtain

$$
\begin{array}{c}
[1] \\
[2] \\
[3]
\end{array}
\begin{bmatrix}
1 & 3/2 & 0 & 4 \\
0 & 1 & 0 & 2 \\
0 & 0 & 1 & 3
\end{bmatrix}
$$

then to eliminate a_{12} we take $[1]' = [1] - (3/2)[2]$ to obtain

$$
\begin{array}{c}
[1] \\
[2] \\
[3]
\end{array}
\begin{bmatrix}
1 & 0 & 0 & 1 \\
0 & 1 & 0 & 2 \\
0 & 0 & 1 & 3
\end{bmatrix}
$$

which corresponds to the equations ($Ax = b$)

$$x_1 \qquad = 1$$

$$x_2 \quad = 2$$

$$x_3 = 3$$

If it had been desired to determine the solution matrix X consisting of the two solution vectors x_1 and x_2 to the equations

$$A x_1 = b_1 \qquad \text{and} \qquad A x_2 = b_2$$

we could have taken the $(n+2) \times n$ augmented matrix

$$[A \mid b_1 \mid b_2]$$

and performed the same row manipulation as in Example 4.7 to obtain

$$\begin{bmatrix} 1 & 0 & 0 & x_1^{(1)} & x_2^{(1)} \\ 0 & 1 & 0 & x_1^{(2)} & x_2^{(2)} \\ 0 & 0 & 1 & x_1^{(3)} & x_2^{(3)} \end{bmatrix}$$

where

$$X_1 = \begin{bmatrix} x_1^{(1)} \\ x_1^{(2)} \\ x_1^{(3)} \end{bmatrix} \qquad \text{and} \qquad X_2 = \begin{bmatrix} x_2^{(1)} \\ x_2^{(2)} \\ x_2^{(3)} \end{bmatrix}$$

Note that this requires no additional manipulation of the rows of the augmented matrix. It only requires that the row operations be performed on an additional column of the augmented matrix.

If we perform the same operation to transform the first n rows of the matrix

$$B = [A \mid b \mid I]$$

into the identity matrix we obtain the matrix

$$B' = [I \mid x \mid A^{-1}]$$

where x is the solution to the system $Ax = b$. (See Problem 4.20.)

For linear systems of equations of the form $Ax \doteq b$ with many nonzero elements in A there is no method known that is better than Gauss elimination for solution in terms of accuracy and computation time. There are many variants and special procedures for matrices with few nonzero elements and symmetry properties. There is insufficient space here to discuss them. However, the book by Forsythe and Moler.[9] discusses application of several algorithms to systems of equations primarily based on Gaussian elimination. Particular emphasis is placed on running time and storage requirements of the algorithms, and on an analysis of errors caused by various forms of limited-precision arithmetic in computers. Forsythe and Moler also provide programs in FORTRAN, ALGOL/60, and PL/1.

ANALOG COMPUTER SOLUTION OF LINEAR ALGEBRAIC EQUATIONS

With the development of digital computers there has been less interest in solving linear algebraic equations on analog computers than there was about 30 years ago. However, the analog computer solution of algebraic equations is still of some interest because it is similar to the methods used to solve simultaneous differential equations and gives some insight into the iterative solution of algebraic equations by digital methods.

Consider the system of equations

$$x_1 - x_2 = -1 \qquad (4.24a)$$

$$2x_1 + x_2 = 4 \qquad (4.24b)$$

Solving Equation (4.24a) for x_1 and Equation (4.24b) for x_2 we have

$$x_1 = x_2 - 1 \qquad (4.25a)$$

$$x_2 = -2x_1 + 4 \qquad (4.25b)$$

which corresponds to the block diagram shown in Figure 4.11.

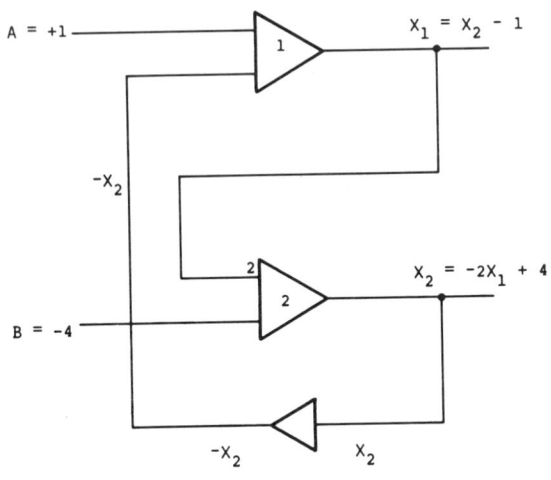

FIGURE 4.11 ANALOG COMPUTER SOLUTION OF LINEAR ALGEBRAIC EQUATIONS.

When the diagram shown in Figure 4.11 is patched up on a computer and the operate button depressed, the computer immediately overloads. If it is a 100 volt machine, and $A = 10$, $B = -40$, then x_1 and x_2 go to the limits of the machine, or approximately ± 110 to 150 volts.

After checking the equations and the patching we may be convinced that everything is fine but still decide to try solving Equation (4.24a) for x_2 and Equation (4.24b) for x_1. Doing this gives another equivalent diagram. However, patching up the new diagram results in outputs of $x_1 = 10$ and $x_2 = 20$ (assuming we take $A = 10$ and $B = -40$) and no overloads.

Although it happens almost instantly, this phenomenon is better understood by considering what happens in Figure 4.11 if x_1 and x_2 start off at some initial condition, say zero. Then the input to summer

1 is $1 - x_2 = 1$ and the input to summer 2 is $-4 + x_1 = -4$. Thus the outputs are $x_1 = -1$ and $x_2 = 4$, which gives new inputs to summers 1 and 2 of $1 - x_2 = -3$ and $-4 + 2x_1 = -6$ respectively and yield outputs $x_1 = 3$, $x_2 = 6$. Continuing the process we see that the summer outputs diverge and the computer overloads. At this point the reader should solve Equation (4.24a) for x_2 and Equation (4.24b) for x_1, and repeat the above iterative process starting from $x_1 = x_2 = 0$ and observe the convergence to the solution $x_1 = 1$ and $x_2 = 2$.

Even if the block diagram is stable, the high frequency dynamics of operational amplifiers can make up unstable feedback systems when connected in algebraic loops. This problem can be circumvented by solving a set of linear differential equations which have as a steady state solution the solution to the algebraic equations. For example, the solution to Equations (4.24) is also the steady state solution of the system

$$\dot{x}_1 = -x_1 + x_2 - 1 \qquad (4.26a)$$

$$\dot{x}_2 = -2x_1 - x_2 + 4 \qquad (4.26b)$$

However these equations represent an unstable system, so the solution diverges. To obtain a convergent solution it is best to have maximum negative feedback around each integrator. This is accomplished by constructing the differential equations so that in the jth equation

$$\dot{x}_j = \sum_{k=1}^{n} a_k x_k + b_j$$

a_{ij} is the largest coefficient among the a_{ij} values. Thus in Equation (4.24) we obtain a stable system which gives the desired steady state solution if we take

$$\dot{x}_1 = -2x_1 - x_2 + 4 \qquad (4.27a)$$

$$\dot{x}_2 = -x_1 + x_2 - 1 \qquad (4.27b)$$

Although algebraic equations are avoided in analog computation wherever possible, the above discussion and some experience with algebraic equations on the analog computer give insight into the stability of the numerical solution of algebraic equations on the digital computer.

THE METHOD OF SIMULTANEOUS DISPLACEMENTS (ITERATION)

Frequently, systems of linear equations in thousands of variables must be solved. If there are 2000 equations in 2000 unknowns the coefficient matrix contains 4,000,000 elements and there is little hope for solution unless most of the coefficients are zero. Fortunately these cases of many equations arise from physical situations where there are many variables but each is coupled to only a few other variables such as occurs in large electrical networks which can be considered to have each station directly connected only to its nearest neighbors. These equations also arise in the solution of boundary value problems or partial differential equations solutions as discussed in following chapters.

An iterative technique which is suitable for large sparse (few nonzero elements) systems of linear algebraic equations is the *METHOD OF SIMULTANEOUS DISPLACEMENTS*, sometimes referred to as the *JACOBI METHOD*. The equations are first arranged so that the diagonal elements of the coefficient matrix are as large as possible and none are zero. The nonzero constraint is necessary in order to use the method and, having the larger elements on the main diagonal, speeds convergence and improves stability as noted above in the analog solution. In the above discussion, when we write Equations (4.24a) and (4.24b) as $Ax = b$ in the form

$$\begin{bmatrix} 1 & -1 \\ 2 & 1 \end{bmatrix} \begin{bmatrix} x_1 \\ x_2 \end{bmatrix} = \begin{bmatrix} -1 \\ 4 \end{bmatrix}$$

the equations diverge by the iterative calculation, but when they are written

$$
\begin{bmatrix} 2 & 1 \\ 1 & -1 \end{bmatrix} \begin{bmatrix} x_2 \\ x_1 \end{bmatrix} = \begin{bmatrix} 4 \\ -1 \end{bmatrix}
$$

the iterative calculation converges.

The procedure of the method of simultaneous displacements after the above manipulation ($a_{ii} \neq 0$, and so on) is to solve the first equation for x, the second for x_2, and so on to give Equation (4.18) in the form

$$
x_1 = -\left(\frac{a_{12}}{a_{11}} x_2 + \frac{a_{13}}{a_{11}} x_3 + \ldots + \frac{a_{1n}}{a_{11}} x_n\right) + \frac{b_1}{a_{11}}
$$

$$
\vdots
$$

$$
x_n = -\left(\frac{a_{n1}}{a_{nn}} x_1 + \frac{a_{n2}}{a_{nn}} x_2 + \ldots + \frac{a_{nn-1}}{a_{nn}} x_{n-1}\right) + \frac{b_n}{a_{nn}}
$$

$$
(4.28)
$$

Clearly if $a_{kk} = 0$ the kth equation cannot be solved for x_k. Thus we require that $a_{kk} \neq 0$ for all k.

With appropriate definitions of A' and b' Equation (4.28) can be written as

$$
X = A'X + B' \tag{4.29}
$$

which is of the form $x = \phi(x)$ and can be iteratively evaluated as

$$
X^{(n+1)} = A'X^{(n)} + B' \tag{4.30}
$$

where the superscript on the vector $x^{(n)}$ is the iteration number. Starting with an initial estimate $x^{(0)}$ Equation (4.29) is iteratively solved by Equation (4.30), and if the sequence $\{x^{(n)}\}$ converges, it converges to a solution of Equation (4.29).

EXAMPLE 4.8

Solve the following equations by the simultaneous displacement algorithm.

$$2x_1 - x_2 = 0$$
$$x_1 + x_2 = 3$$

(4-36)

SOLUTION:

Solving the first equation for x_1 and the second equation for x_2 (Why this choice?) we have

$$x_1 = 1/2 \ x_2$$
$$x_2 = -x_1 + 3$$

writing the equations in the form of Equation (4.29)

$$\begin{bmatrix} x_1 \\ x_2 \end{bmatrix} = \begin{bmatrix} 0 & 1/2 \\ -1 & 0 \end{bmatrix} \begin{bmatrix} x_1 \\ x_2 \end{bmatrix} + \begin{bmatrix} 0 \\ 3 \end{bmatrix}$$

and in the iterative form of Equation (4.30)

$$\begin{bmatrix} x_1^{(n+1)} \\ x_2^{(n+1)} \end{bmatrix} = \begin{bmatrix} 0 & 1/2 \\ -1 & 0 \end{bmatrix} \begin{bmatrix} x_1^{(n)} \\ x_2^{(n)} \end{bmatrix} + \begin{bmatrix} 0 \\ 3 \end{bmatrix}$$

Taking the initial guess

$$\begin{bmatrix} x_1^{(1)} \\ x_2^{(1)} \end{bmatrix} = \begin{bmatrix} 0 \\ 0 \end{bmatrix}$$

we have

$$\begin{bmatrix} x_1^{(2)} \\ x_2^{(2)} \end{bmatrix} = \begin{bmatrix} 0 \\ 3 \end{bmatrix}$$

$$\begin{bmatrix} x_1^{(3)} \\ x_2^{(3)} \end{bmatrix} = \begin{bmatrix} 0 & 1/2 \\ -1 & 0 \end{bmatrix} \begin{bmatrix} 0 \\ 3 \end{bmatrix} + \begin{bmatrix} 0 \\ 3 \end{bmatrix} \doteq \begin{bmatrix} 1.5 \\ 3 \end{bmatrix}$$

$$\begin{bmatrix} x_1^{(4)} \\ x_2^{(4)} \end{bmatrix} = \begin{bmatrix} 0 & 1/2 \\ -1 & 0 \end{bmatrix} \begin{bmatrix} 1.5 \\ 3 \end{bmatrix} + \begin{bmatrix} 0 \\ 3 \end{bmatrix} = \begin{bmatrix} 1.5 \\ 1.5 \end{bmatrix}$$

$$\begin{bmatrix} x_1^{(5)} \\ x_2^{(5)} \end{bmatrix} = \begin{bmatrix} 0 & 1/2 \\ -1 & 0 \end{bmatrix} \begin{bmatrix} 1.5 \\ 1.5 \end{bmatrix} + \begin{bmatrix} 0 \\ 3 \end{bmatrix} = \begin{bmatrix} 0.75 \\ 1.5 \end{bmatrix}$$

Continuing the process we have

$$\begin{bmatrix} x_1^{(6)} \\ x_2^{(6)} \end{bmatrix} \begin{bmatrix} 0.75 \\ 2.25 \end{bmatrix} \begin{bmatrix} 1.125 \\ 2.25 \end{bmatrix} \quad \cdots \quad \begin{bmatrix} 1 \\ 2 \end{bmatrix}^{(\infty)}$$

which converges to the solution, but is not correct to 5 percent even after 10 iterations. [Rearrange the equations to obtain a divergent sequence.] ▼

Equation (4.30) can be programmed in FORTRAN using subroutines to perform the matrix operations with the following program.

```
CALL MULT(AP,X,APX,2,2,1)
CALL MADD(APX,BP,X,2,1)
```

```
SUBROUTINE MULT(A,B,C,N,M,K)
DIMENSION A(N,M),B(M,K),C(N,K)
DO 1 I=1,N
DO 1 J=1,K
C(I,J)=0.
DO 1 L=1,M
1 C(I,J)=C(I,J)+A(I,L)*B(L,K)
RETURN
END
```

```
SUBROUTINE MADD(A,B,C,N,M)
DIMENSION A(N,M),B(N,M),C(N,M)
DO 1 I=1,N
DO 1 J=1,M
1 C(I,J)=A(I,J)+B(I,J)
RETURN
END
```

A disadvantage of programming this method on the digital computer is the need to maintain the values of all the elements of $x^{(n)}$ while calculating the new estimates $x^{(n+1)}$. It is more convenient to label the kth element of x as $X(K)$ and program Equation (4.26) as

```
DO 50 K=1,M
X(K)=B(K)
DO 50 I=1,M
50 X(K)=A(K,I)*X(I)+X(K)
```

where $A(K,I) = a_{ki}$ is the element in the kth row and ith column of A' and $B(I) = b_1$ is the ith element of b' as given in Equation (4.29). ($a'_{ki} = a_{ki}/a_{kk}$, and $b'_i = b_i/a_{kk}$ and $a'_{kk} = 0$.)

But in the above program the most recently evaluated x_k's are used in the calculation of each new element of x. That is, $x_k^{(n+1)}$ is calculated using the most recent values of each element of x by the algorithm

$$x_i^{(k)} = \sum_{j=1}^{i-1} a_{ij}x_j^{(k)} + \sum_{j=i+1}^{n} a_{ij}x_j^{(k-1)} + b_i$$

(4.31)

(where $a_{ii} = 0$) instead of the simultaneous displacement algorithm given in Equation (4.32).

$$x_i^{(k)} = \sum_{j=1}^{n} a_{ij}x_j^{(k-1)} + b_i \qquad (4.32)$$

It turns out in this case that this procedure which is easier to program has more satisfactory convergence properties than the method of simultaneous displacements. The method of simultaneous displacements with the above modification is called the *GAUSS-SIDEL ALGORITHM* or the *METHOD OF SUCCESSIVE DISPLACEMENT*.

A sufficient condition for the convergence of the Gauss-Sidel method is that the equations are arranged so that the diagonal elements of A are the largest in the equation and

$$|a_{kk}| > |a_{k1}| + |a_{k2}| + \ldots + |a_{km}|$$

for k = 1, ..., m   (4.33)

Although this criterion is not necessary for convergence, and cannot usually be satisfied for all equations, it gives an indication of how to arrange the equations to improve the convergence properties. It is desirable to have the largest terms on the diagonal of A by arranging the equation to satisfy the inequality, Equation (4.33), as well as possible.

EXAMPLE 4.9

Repeat Example 4.8 using the Gauss-Sidel Algorithm.

SOLUTION:

Rewriting the equations in the appropriate form by solving each equation for the variable with largest coefficient as nearly as possible we have

$$x_1 = 1/2 \ x_2 \qquad\qquad (4.34a)$$

$$x_2 = -x_1 + 3 \qquad\qquad (4.34b)$$

Taking the initial guess $\begin{bmatrix} x_1^{(1)} \\ x_2^{(1)} \end{bmatrix} = \begin{bmatrix} 0 \\ 0 \end{bmatrix}$ gives

$$x_1^{(2)} = 0$$

$$x_2^{(2)} = 3$$

$$x_1^{(3)} = 1.5$$

$$x_2^{(3)} = 1.5$$

$$x_1^{(4)} = 0.75$$

$$x_1^{(5)} = 1.25$$

$$x_2^{(5)} = 1.875$$

Note that $x_2^{(3)}$ is calculated using $x_1^{(3)} = 4.316$ and that this method requires half as many iterations as required in Example 4.8. A geometrical interpretation of these two methods and a comparison of their convergence are investigated in Problems 4.14 and 4.15. ▼

An error criterion must be established to terminate calculations when convergence of the method has been

obtained. A commonly used criterion is obtained by
comparing the sum of the magnitudes of the elements
resulting from two successive iterations, and the
calculation terminated when B < ERROR where

$$B = |x_1^{(n)} - x_1^{(n-1)}| + |x_2^{(n)} - x_2^{(n-1)}| \cdots$$

$$+ |x_m^{(n)} - x_m^{(n-1)}| \qquad (4.35)$$

Also, to insure termination in cases of divergence,
the iterations should terminate after a sufficiently
large number of iterations or the value of the error
can be computed to ascertain that it is decreasing
with each iteration.

   The successive displacement algorithms are referred
to as relaxation methods in that the variables are
relaxed to their final value as a function of the
neighboring points in boundary value problems. A
technique that greatly speeds up convergence is
overrelaxation where an extrapolation is made based on
the past and new value of each of the elements of $\mathbf{x}$.
The assumption being that $x_i^{(k+1)} > x_i^{(k)}$ indicates
that x_i will increase with continued iterations and a
good estimate of x_i is obtained by taking

$$x_i^{(k)} = x_i^{(k-1)} + \omega[\alpha_i - x_i^{(k-1)}]$$

$$= (1-\omega)x_i^{(k-1)} + \omega\alpha_i \qquad (4.36)$$

where α_i is the value of $x_i^{(k)}$ obtained from Equation
(4.31). $\omega = 1$ corresponds to successive displacement
and $\omega = 2$ is usually unstable so a value of $\omega = 1.5$ to
1.6 generally speeds convergence nicely without
instability. Of course, the program can adaptively
adjust the weight ω to insure stability.

   If the equations are not linearly independent the
solution is not unique but if convergence is obtained
it is to a solution to the system of equations. If
the equations are linearly dependent some of the
equations are redundant and can be eliminated, leaving

fewer equations than unknowns. Also of interest are cases where the system of equations is said to be overspecified, that is there are more equations than unknowns, and the equations are inconsistent or have no solution. Other aspects of the iterative and "exact" methods for the solution of sets of linear algebraic equations plus other topics in linear algebra are discussed in detail in Faddeeva?

SYSTEMS OF NONLINEAR ALGEBRAIC EQUATIONS

The Newton–Raphson method can be extended to systems of nonlinear algebraic equations by writing the equations in vector form. Consider the set of nonlinear equations

$$f_1(x_1, \ldots, x_n) = 0$$

$$\vdots$$

$$f_n(x_1, \ldots, x_n) = 0 \qquad (4.37)$$

in vector form

$$F(X) = 0 \qquad (4.37)$$

The procedure is to proceed as in the Newton–Raphson iteration; however in place of the derivative of $F(x)$ with respect to x we must have the change in each component of $F(x)$ with respect to each component of x. This corresponds to the Jacobian matrix $J(x)$ of the equation $F(x)$ where

$$J(X) = \begin{bmatrix} \dfrac{\partial f_1}{\partial x_1} & \cdot & \vdots & \cdot & \dfrac{\partial f_1}{\partial x_n} \\ \dfrac{\partial f_n}{\partial x_1} & \cdot & \cdot & \cdot & \dfrac{\partial f_n}{\partial x_n} \end{bmatrix}$$

and the algorithm is

$$X^{(k+1)} = X^{(k)} - [J(X^{(k)})]^{-1} F(X^{(k)}) \quad (4.38)$$

which reduces to the scalar Newton-Raphson method Equation (4.4), for $n = 1$.

As in the one dimensional case we guess an initial value $x^{(0)}$ and iterate using Equation (4.35) until a convergence criterion is met or excessive iterations are performed without improvement of the error or residual $F(x^{(k)})$. A convenient measure for the error residual is

$$\text{ERROR} = \sum_{i=1}^{n} f_i(x^{(k)})$$

An alternative approach to the solution of nonlinear equations of the form of Equation (4.34) is to use optimization methods to obtain the minimum of the function

$$F(x_1, \ldots, x_n) = \sum_{i=1}^{n} |f_i(x, \ldots x_n)|^2 \quad (4.39)$$

a procedure which is explored in Chapter 7 along with other optimization methods.

PROBLEMS

4.1   (a) Add logic to the program in Example 4.1 that will automatically terminate calculation when the change in the result is less than 10^{-5}.

   (b) Use the program obtained in part (a) to evaluate $\sqrt{100}$ with a starting value of A = 5. Also determine $\sqrt{5}$ to r decimal places.

4.2   (a) Show that the error in the square root program satisfies the inequality

$$\alpha_{n+1} = (\frac{1}{2} - \frac{B}{2\alpha_n})\alpha_n < \frac{1}{2} \alpha_n$$

(b) Show that for $a_n \approx \sqrt{B}$, $\alpha/2$ is a pessimistic estimate and convergence is faster the closer a_n is to $\sqrt{B}$.

4.3 Define the relative error in the square root program to be

$$\beta_n = \frac{|\alpha_n|}{\sqrt{B}}$$

and show that

$$\beta_{n+1} < \frac{\beta_n^2}{2}$$

since after the first iteration $a_n > \sqrt{B}$.

4.4 Use the binomial expansion of $(a_{n+1} + \alpha_{n+1})^k = B$ and neglect the high order terms of α_{n+1} to obtain the recursion relation

$$a_{n+1} = \frac{B + (k-1) a_n^k}{k a_n^{k-1}}$$

to find a root of the equation in $a^k = B$.

4.5 Use a sketch to show that if f is constant the method of chords and the Newton-Raphson method converge to the solution to the equation $f(x) = 0$ from opposite sides.

4.6 Find a solution to the equation

$$x - \sin x - 0.5 = 0$$

(a) use the method of chords

(b) the Newton Raphson Method and

(c) a combined program which calculates (a) and (b) in parallel to bracket the result. Stop the iteration when you have the solution correct to 5 decimal places

4.7 Repeat Problem 4.6 for the function $x - e^{-x} = 0$.

4.8 Program the iterative solution of Example 4.5 and show that it does, indeed, cycle through the same sequence of values in opposite directions.

4.9 Write $f(x) = 0$ as $x_{n+1} = x_n + Kf(x_n)$ and choose K appropriately to allow convergence in the solution of the equation $x - e^{-x} = 0$.

4.10 Comment on the statement

"The Newton- Raphson method converges slowly for the function $f(x) = 0$, but near the solution $f(x)$ is essentially constant and equal to unity so an iterative equation of the form $x_{n+1} = x_n + Kf(x_n)$ will converge rapidly for $K = -1$ since

$$\frac{d}{dx}(x+Kf(x)) = \dot{\phi}(x) = 0$$

4.11 Show that if $\dot{\phi}(x) < 1$ for x in the interval [a,b] there is at most one α in [a,b] such that $\phi(\alpha) = \alpha$.

4.12 Set up an iterative scheme which will converge to a solution of the equation

$$\sin x + \cos x - 4x = 0$$

4.13 How many roots does the equation $xe^x = 2$ have? Find them.

4.14 Compare the iterative solution of the equations given in Examples 4.8 and 4.9 by constructing a sketch of the two equations (x_2 versus x_1) showing the iterative solutions obtained in the examples.

4.15 Interchange the labels on x_1 and x_2 in the equations given in Examples 4.8 and 4.9 and demonstrate that with this orientation the iterative algorithms diverge. Discuss this divergence in terms of the inequality demonstrated in Equation (4.24). Draw a sketch as outlined in Problem 4.14 comparing the two formulations (convergent and divergent).

4.16 Show that Equation (4.3) can be written

$$\alpha = b + f(b) \ \frac{b-a}{f(b) - f(a)}$$

4.17 Draw a sketch simialr to Figures 4.2 and 4.6 to show the convergence properties of the method of chords and the Newton-Raphson method for the solution of the equation $f(x) = 0$ if

(a) $\dot{f} < 0, \quad \ddot{f} < 0$ (b) $\dot{f} < 0, \quad \ddot{f} > 0$

(c) $\dot{f} > 0, \quad \ddot{f} < 0$ (d) $\dot{f} > 0, \quad \ddot{f} > 0$

4.18 Draw a sketch showing the convergence and divergence of the iterative solution of the equation $x^3 - 1 = 0$ by

(a) $x_n = [1/x_{n-1}]^2$, and

(b) the inverse equation obtained by interchanging x_n and x_{n-1} ($x_n = 1/\sqrt{x_{n-1}}$).

4.19 Write a program which will perform the interpolation indicated in Figure 4.12 for the iterative solution of $\phi(x) = x$.

4.20 Show that the Gauss elimination algorithm converts the matrix $[A \ b \ I]$ into $[I \ x \ A^{-1}]$ where $Ax = b$.

4.21 Write a program that will automatically interchange rows before the Gauss elimination algorithm so that the new a_{11} is the largest coefficient in the first row, and the 2nd through nth rows are rearranged making new diagonal term a_{22} the largest, and so on. It is not necessary to interchange the rows of $[A \ b \ I]$. An array of

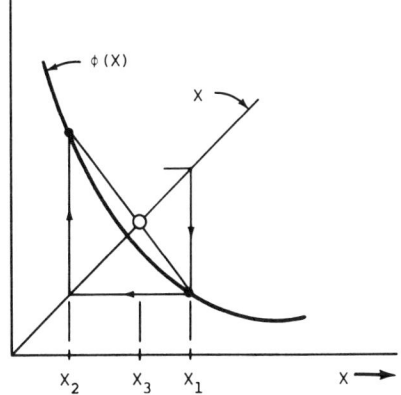

FIGURE 4.12   PROBLEM 4.19.

integers k_i $(i=1,...n)$ can be rearranged to replace the interchange by calling a_{k_ij} instead of a_{ij}.

4.22 Write a program that will automatically interchange rows and columns to make a_{11} the largest element of A, a_{22} the largest of the remaining $(n-1)^2$ elements, a_{33} the largest of the remaining $(n-2)^2$ elements, and so on. After a_{11} is chosen the first row and the first column cannot be moved.

4.23 Demonstrate the sensitivity of Gauss elimination to round off error when the diagonal elements are small by solving the following equations as shown and with the rows interchanged. Use 3 decimal floating arithmetic. Also give the true solution to 5 decimal places.

$$\begin{bmatrix} 0.0001 & 1.00 \\ 1.00 & 1.00 \end{bmatrix} \begin{bmatrix} x_1 \\ x_2 \end{bmatrix} = \begin{bmatrix} 1.0 \\ 2.0 \end{bmatrix}$$

4.24 Write a FORTRAN program that will solve a set of n simultaneous, linear algebraic equations by the Gauss-Sidel algorithm. Have the calculation terminate when the sum of the changes in the individual components is less than a number ERROR.

4.25 Use the program developed in Problem 4.24 to solve the following set of equations.

$$
\begin{bmatrix}
4 & -1 & 0 & 0 \\
-1 & 4 & -1 & 0 \\
0 & -1 & 4 & -1 \\
0 & 0 & -1 & 4
\end{bmatrix}
\begin{bmatrix}
x_1 \\
x_2 \\
x_3 \\
x_4
\end{bmatrix}
=
\begin{bmatrix}
1 \\
1 \\
1 \\
1
\end{bmatrix}
$$

4.26 Write a subprogram that will automatically arrange a set of linear algebraic equations so that the dominant terms (largest) are on the diagonal to precede your Gauss-Sidel algorithm program. (See Problems 4.21 and 4.22.)

4.27 Program an iterative program using a Newton-Raphson search for determining the solutions of the equation $f(x) = 0$ such that when a solution x_1 is determined the program repeats in an attempt to find a solution to the equation

$$
\frac{f(x)}{(x-x_1)} = 0
$$

The assumption is that the term $(x-x_1) = 0$ balances the zero of $f(x)$ at $x = x_1$ while preserving the other zeros of $f(x)$. Use your program to evaluate the roots of the equation

$$
x^3 - 2x^2 + x = 0
$$

4.28 Suppose an iterative algorithm results in the equation

$$
x_n = 0.9x_{n-1} + 10
$$

The steady state solution is obtained when $x_n = x_{n-1}$ or $(1-0.9)x_n = 10$. Thus the steady state solution is x = 100, which is obtained by iterating the given equation from any initial condition.

(a) Start from the initial value of $x_0 = 99$ and perform the iteration for two steps.

(b) Repeat part (a) with rounding such that each result is rounded to the nearest integer. What you experience is called the *DEADBAND EFFECT*.

(c) For this equation in part (b) evaluate the width of the deadband. That is, for what values of x does the calculation terminate?

(d) Give the width of the deadband as a function of the level of quantization in the result. In (b) the results are rounded to the nearest multiple of $E_0 = 1$. Evaluate the width in particular for $E_0 = .10^{-5}$.

5 hybrid computation

Hybrid computers are systems that utilize both analog and digital components and techniques. The main emphasis of the preceeding material in this text has been the utilization of analog and digital techniques for the solution of some problems, for which both computers might be applicable. Hopefully, sufficient experience has been gained from the text and the problems to appreciate the strong and weak points of the two systems. A hybrid system that connects together a digital computer and an analog computer has the good points of each. However, the weaknesses of each are also present. Not only does the system have the analog and digital errors but errors are also generated in the interface equipment that converts analog signals to digital signals (A/D converters) and digital signals to analog signals (D/A converters), making them compatable with the components they enter.

Time synchronization of the analog computer, digital computer, converters, and input-output devices can be controlled by clock pulses generated in either the digital logic section of the analog computer or the digital computer. In any event timing synchronization is extremely important and must be understood for successful programming. Along with block diagrams, flow charts and logic diagrams for mode control it is often necessary to maintain a timing chart for the overall program. Timing procedures are machine dependent and may differ for each installation, so a timing chart that is machine dependent is used here to demonstrate the timing relationships for the example programs.

This chapter begins with a brief discussion of analog, digital, and hybrid computation and equipment, Section 5.2 presents digital logic elements usually available on analog computers and demonstrates how these elements increase the flexibility of these machines. The increased capability of digital logic on an analog computer greatly expands the number of problems that can be solved on the analog computer. Section 5.3 presents the interface elements for hybrid computation and some programming techniques for hybrid computers.

5.1 ANALOG, DIGITAL AND HYBRID COMPUTATION

A computer is a system designed to carry out a logical procedure or represent a mathematical model. The digital computer performs operations on discrete quantized variables in a sequential (serial) manner. Some special purpose digital computers, such as digital differential analyzers, include both sequential operation and parallel operation within different portions of the computer. However, most general purpose digital computers used for scientific computation are serial in operation and step sequentially through the program. On the other hand, an analog computer is designed to represent a mathematical model relating continuous variables. General purpose analog computers are parallel machines which are primarily applied to the rapid solution or simulation of differential equations. The basic units

of general purpose digital and analog computers are
the *ARITHMETIC UNIT* and the *INTEGRATORS* respectively.
Note that the digital computer usually has a single
arithmetic unit while the analog computer has many
integrators. This difference is fundamental to their
radically different characteristics.

The parallel operation of the analog computer
allows the high speed solution of differential
equations, perhaps at thousands of solutions per
second, independent of the complexity of the program.
A higher order system merely requires a more complex
block diagram and more integrators. The fast solution
and the opportunity for "hands-on" operation allow the
designer to develop a feel for the system during the
design process. Digital time shared installations
provide this feel to some extent but lack the intimacy
of a cathode ray tube (CRT) display of an analog
computer in REPOP performing the high speed simulation
of the system under study. All digital simulations
that combine increased computing speed and improved
display input and output devices will allow a "hands-
on" operation like the analog computer. However,
systems of which a significant portion is described by
ordinary or partial differential equations will
continue to have them simulated on the analog portion
of hybrid computers since the speed restrictions on
the digital computer can only be overcome by the
addition of multiprocessors.

The serial operation on discrete variables in the
digital computer is particularly suitable for event
oriented problems in discrete variables. Such
problems include queueing problems, inventory control,
and general nondifferential systems involving record
keeping of alpha-numeric data and the storage of
information. Generally most problems can be solved on
digital computers, including those for which the
analog computer is better suited. So there is a
tendency to perform an analog type of computation on
the digital computer since this doesn't require the
interface of the programmer with any different
equipment, even though it may be expensive and time
consuming. This effect, coupled with the
unsuitability of the analog computer for some

problems, has led operators who are trained
exclusively in digital techniques to say that analog
computation is a dying art. Of course, this is not
so.

Hybrid installations combine the speed of the
analog computer with the accuracy of the digital
computer. They have the convenience of efficient
input and output of data in both discrete and
continuous form. The digital portion can perform
operations for which it is best suited and that are
inconvenient or impossible in the analog portion, like
mode control logic based on complex arithmetic
results, the realization of variable time delays by
function storage and playback, and the generation of
nonlinear functions of several variables from complex
expressions or table look-up.

If a hybrid system is operated with a 1:1 time
scale the simulation can contain actual hardware from
the system being simulated. A dynamic simulation of
space flight might contain the space vehicle with the
dynamics of flight being generated in reaction to
adjustments within the vehicle. The analog signals in
the simulation are both real and simulated variables.
An element of the space flight vehicle system is a
digital computer which can best be simulated by a
digital computer. So the overall simulation is some
analog-digital combination that interfaces with itself
and the actual hardware.

The timing circuitry of the analog computer can be
utilized to perform modifications and control of the
digital program. The extreme case being an analog
controlled digital simulation. Also the electronic
switches and relays of the analog computer can be controlled by the digital program and arithmetic results
to control and alter the analog computer and its mode
control.

A major area requiring trade off in a hybrid system
involves the accuracy of the analog computer, which is
generally between 0.01 percent and 2 percent of full
scale independent of computing speed but not of equipment cost. On the other hand, the accuracy of the
digital computer can be set at any practical value,
but at the expense of increased computation time. If

there are portions of the simulation that are particularly sensitive to errors and must be calculated accurately these portions can be integrated on the digital portion of the system. Thus the high frequency portions can still be integrated on the analog portion and only the sensitive areas integrated on the digital portion of the hybrid computer.

The following section presents what might be called a first generation hybrid computer. The addition of digital logic elements to an analog computer does indeed give it some hybrid characteristics.

5.2 LOGIC AND LOGIC CONTROL OF ANALOG COMPUTERS

The addition of just a few digital logic elements to an analog computer greatly increases the flexibility of the computer. Gates, flip-flops, counters, and timers allow the computing of digital signals to control the logical operations and mode control of an analog simulation.

The signals used in logic operations are of two levels which may differ with each machine and which are represented by the states '0' and '1'. Generally '0' corresponds to zero volts and '1' corresponds to a positive voltage – the voltage level you have been using for patching mode control on your computer. The signals are used to perform control operations and mode control switching. The operation of the logic circuits is in terms of the '0' and '1' states of the input and output signals. Some commonly used terms for '1' and '0' are "up and down", "yes and no", and "true and false."

LOGIC COMPONENETS

The basic elements for digital logic on analog computers are the AND and OR gates, and flip-flops. Also included in the digital portion of analog computers are a clock to generate timing pulses, counters, and interval timers.

The *AND GATE* has two or more inputs. The output of the AND gate is '1' if and only if all of the inputs

are '1'. If any of the inputs is '0' then the output
is '0'.

The *OR GATE* also has two or more inputs. The
output of the OR gate is '1' if any of the inputs is
'1'. If all of the inputs are '0' then the output is
'0'.

Clearly the names of the gates correspond to the
operations they perform. With inputs a and b the AND
gate has output '1' if and only if a and b are '1'.
The OR gate has output '1' if either a or b is '1'.
If both inputs to an OR gate are '1' the output is
also '1'. The AND operation is written (a·b) while
the OR operation is written (a+b). Figure 5.1 shows
the programming symbols for the AND and OR elements.

On most computers the *COMPLEMENT* (opposite logic
signal) of the signal a denoted $\bar{a}$ is also available as
an output. $\bar{a}$ is called 'a bar' or 'not a'. A *LOGIC
INVERTER* changes a signal '1' to '0', changes '0' to
'1', and has the small triangle shown in Figure 5.1
for its programming symbol. The inverter operation is
usually obtained as the inverted output of a single
input AND gate.

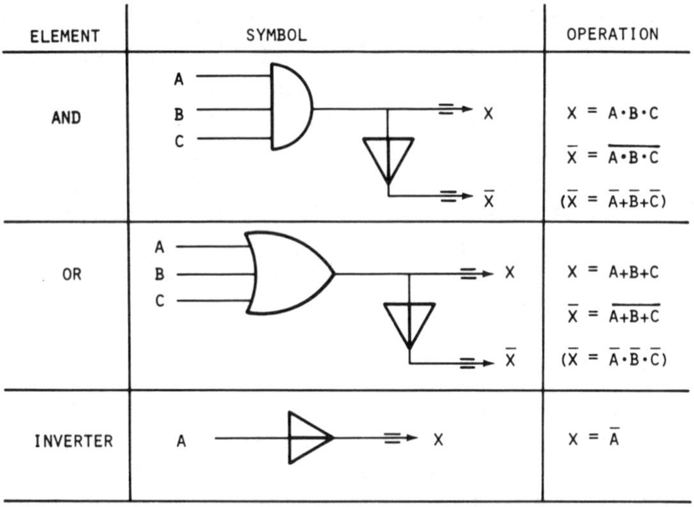

ELEMENT	SYMBOL	OPERATION
AND	A B ⟹ X C ⟹ $\bar{X}$	$X = A \cdot B \cdot C$ $\bar{X} = \overline{A \cdot B \cdot C}$ $(\bar{X} = \bar{A} + \bar{B} + \bar{C})$
OR	A B ⟹ X C ⟹ $\bar{X}$	$X = A + B + C$ $\bar{X} = \overline{A + B + C}$ $(\bar{X} = \bar{A} \cdot \bar{B} \cdot \bar{C})$
INVERTER	A ⟹ X	$X = \bar{A}$

FIGURE 5.1 SOME DIGITAL LOGIC ELEMENTS FOR ANALOG COMPUTERS.

Most analog computers have only one type of gate – usually AND gates with built-in inverters. The other type of gate can then be constructed by the following theorem.

DEMORGAN'S THEOREM

An OR gate with inputs a and b has output '1' if either a or b is '1', and output '0' if both a and b are '0'. Thus the complement of the output is '1' if both a and b are '1', which is a statement of DeMorgan's theorem

$$a + b = \overline{\overline{a} \cdot \overline{b}} \qquad (5.1)$$

since $\overline{a + b} = \overline{a} \cdot \overline{b}$.

Thus all logic operations can be obtained with only AND gates and inverters, a combination sometimes called NAND gates or NAND logic.

If some of the inputs of an OR gate are unpatched the output is '1' if any of the patched inputs is '1'. The AND gate operates electronically as an OR gate by DeMorgan's theorem; that is, if any input is '0' the output is '0'. Thus the AND gate also ignores unpatched inputs.

EXAMPLE 5.1

Give a logic diagram using only AND gates and inverters which has output '1' if (a is '0' or b is '0') or (c is '1' and d is '0').

SOLUTION:

The statement of the desired function is more easily written in the notation given in Figure 5.1. We desire the function

$$(\overline{a} + \overline{b}) + (c \cdot \overline{d}) = \overline{(a \cdot b)} + (c \cdot \overline{d}) \qquad (5.2)$$

(Check this with the statement above and with DeMorgan's theorem.) ▼

The function in Equation (5.2) is accomplished with the diagram shown in Figure 5.2(a). Figure 5.2(b) shows the OR gate replaced by an AND gate preceeded and followed by inverters to obtain the OR operation. In Figure 5.2(c) the inverters are combined to obtain the final diagram. Note that the inverter for signal d can be obtained as a single input AND gate, or if d is the output of another gate we just take the inverted output d from that gate.

If the output signal is used to RESET integrator 5 in a simulation it can be labeled as shown in Figure 5.2(c).

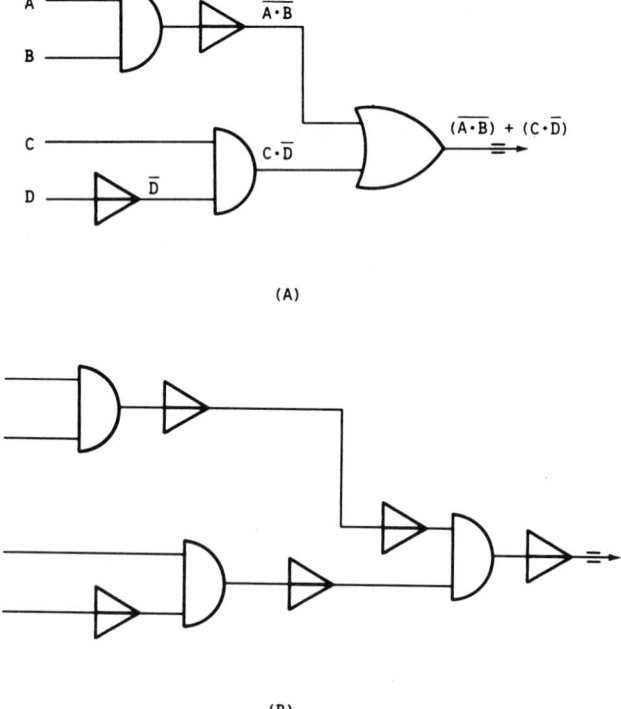

(A)

(B)

FIGURE 5.2   EXAMPLE 5.1 AND AND OR GATES TO OBTAIN
THE LOGIC FUNCTION $(\overline{A \cdot B}) + (C \cdot \overline{D})$.

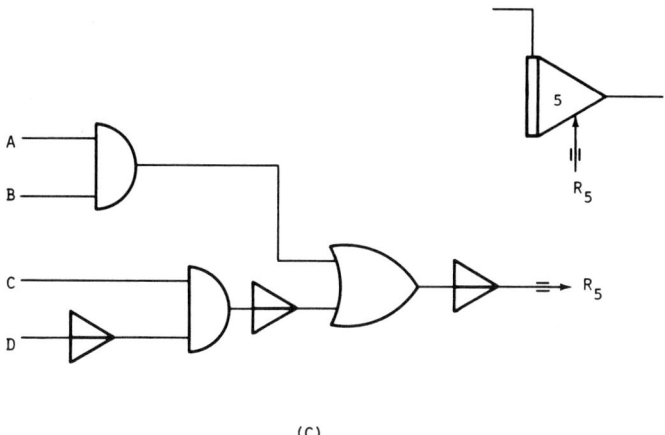

(C)

FIGURE 5.2 EXAMPLE 5.1

THE FLIP-FLOP

Most analog computers with digital logic have *FLIP-FLOPS* which are bistable elements having output '0' or '1'.   There are many types of flip-flops, but perhaps the most common is one with three inputs called SET, RESET and TRIGGER as shown in Figure 5.3.   The output

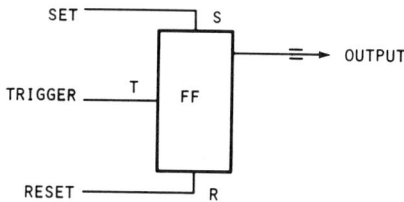

FIGURE 5.3 FLIP-FLOP.

of the flip-flop remains in a constant state of '0' or '1' until an input occurs.   An input '1' into SET

switches the output to '1' while an input '1' into
RESET switches the output to '0'. If the output is
already '0' when the RESET occurs the output just
remains '0'. Simultaneous inputs into both S and R
are not permitted. The input TRIGGER always changes
the state of the flip-flop. If TRIGGER goes to '1'
the flip-flop switches state (that is, goes from '0'
to '1' or '1' to '0' depending on the state before the
trigger occurred.)

The output of the flip-flop is '0' or '1' depending
on the state of the flip-flop, thus its output is a
level type of signal. The inputs SET, RESET and
TRIGGER can be either levels or pulses, which are
levels of short duration.

CLOCKS, COUNTERS, AND TIMERS
In the discussion of mode control we discussed
REPOP control in which the mode of operation of the
computer alternately switches between two (or more)
seates. Frequently it is desirable to have different
timing sequences in different parts of the analog
program in addition to different integrators in
different modes at the same time, as in REPOP.
Essentially independent control of each integrator can
be obtained using the logic elements discussed above.
The timing of the operation is then obtained using
counters and timers that are synchronized by a
sequence of *CLOCK PULSES*.

The *CLOCK* of the computer is a sequence of pulses
generated by an internal oscillator, say at a frequency of 10 kHz or some other rate. Also available
may be pulses that are counted down from the basic
clock. So a typical computer might have pulses that
occur every 100 μ sec (10 kHz), 1 ms, 10 ms, 100 ms,
and 1 second all in synchronization. Each pulse of
the 1 ms interval coincides with every tenth pulse of
the 100 μ sec clock, and so on.

There are also patchable counters that count input
pulses. Only one of the outputs of the counter is
'1'. Each time a pulse appears on the RUN or COUNT
input the '1' is shifted to the next output of the
counter as shown in the timing diagram in Figure 5.4.

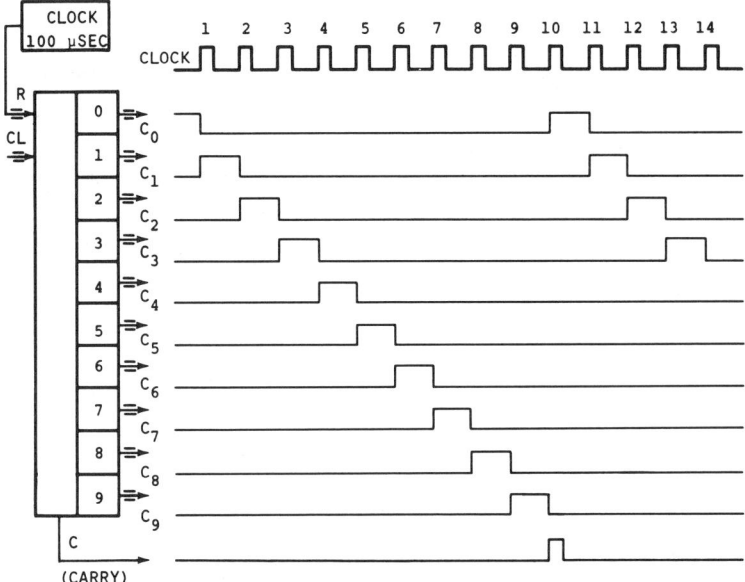

FIGURE 5.4 COUNTER OPERATION.

Also available are the input CLEAR, which switches the counter back to its initial state (C_0 = '1', all other C_k = '0'), and the output CARRY which occurs when the counter goes from C_9 = '1' to C_0 = '1' and can be used to further "count down" the pulses.

Counting to other than base 10 can be accomplished by clearing the counter before it completes the total count as shown ik the following example.

EXAMPLE 5.2

Use a counter to count down a sequence of pulses to obtain an output on every fourth clock pulse.

SOLUTION:

The count can be obtained by using a counter that is cleared when the output C_4 becomes '1'. However, usually a counter cannot be cleared by one of its inputs. (Why not?) So we can set a flip-flop with C_4 and use the output of the flip-flop to clear the counter. Once the counter is cleared ($C_0 = '1'$) we can reset the flip-flop from C_0. The resultant circuit is shown in Figure 5.5.▼

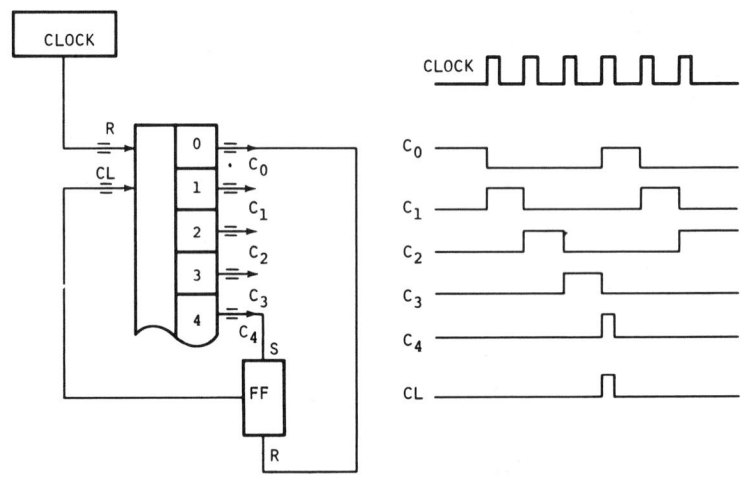

FIGURE 5.5   EXAMPLE 5.2 COUNT DOWN BY FOUR.

The *TIMER* is a logic element having a single input line and two output lines as shown in Figure 5.6. In the absence of an input pulse both outputs remain in the '0' state. When the input goes to state '1' for a short (pulse) duration, output T_0 becomes '1' and remains in the '1' state for a preset and adjustable interval T_0 After the T second interval line T. returns to the '0' state and the completion line C has a short pulse output. Figure 5.6 shows a pulse on the

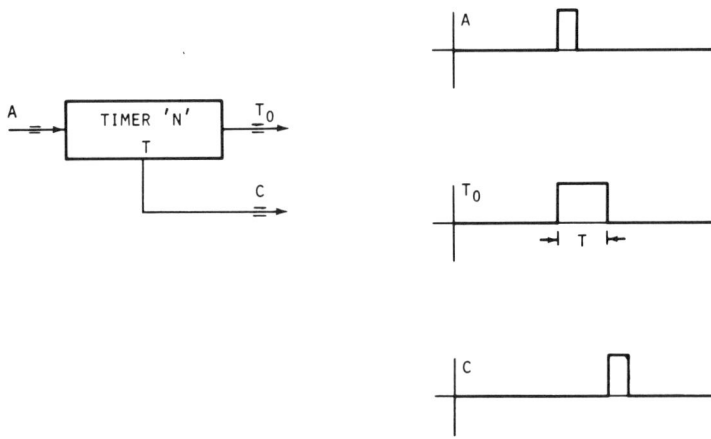

FIGURE 5.6 TIMER OPERATION.

line of a timer and the corresponding output
waveforms. "N" is the number of the timer and T is
the interval of the timer output.

Several timers can be connected in cascade to
generate timing intervals for mode control, delay, or
sequential operations. Figure 5.7 shows three cascade
timers forming a sequential ring. The completion
pulse of each timer is used to trigger the next time.
Timer 1 triggers timer 2, timer 2 triggers timer 3,
and timer 3 triggers timer 1. Thus the ring of timers
runs in sequence with each timing interval being
independently adjustable.

An analog computer memory can be constructed by
using integrators in the so-called *TRACK-STORE* mode.
The function to be stored is connected into the RESET
jack of the integrator so the output is equal to
(minus) the input when the integrator is in RESET.
(When the integrator goes to HOLD it keeps the value
of the output at that instant.) The following example
demonstrates the idea.

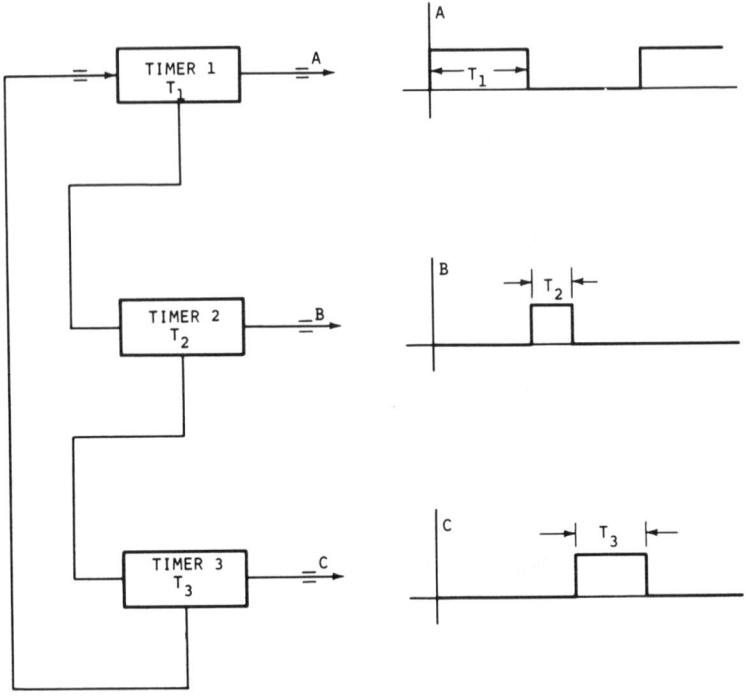

FIGURE 5.7 A RING OF TIMERS.

EXAMPLE 5.3

Give an analog computer memory which will store four values of the function x(t).

SOLUTION:

Figure 5.8(a) shows the block diagram and Figure 5.8(b) gives a waveform x(t) and corresponding outputs of the memory integrators. The timing pulses to reset the integrators are generated by counting clock pulses. ▼

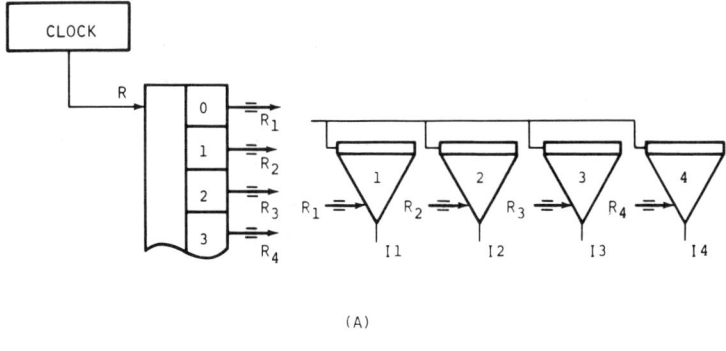

(A)

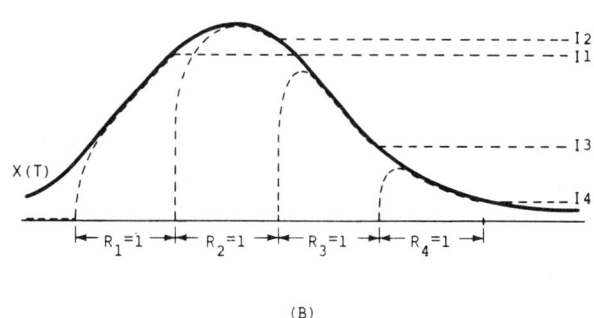

(B)

FIGURE 5.8 EXAMPLE 5.3 ANALOG COMPUTER MEMORY.

We conclude this section with two elements which are fundamental to analog controlled digital operations - the comparitor, and digitally controlled analog operations - the D/A switch.

The *COMPARITOR* is an element with a digital output of '0' or '1' depending on whether the sum of the analog inputs is negative or positive. The comparitor symbol and operation are shown in Figure 5.9.

The *D/A SWITCH* is an open circuit for analog signals when the digital input is '0' and a short circuit for a digital input of '1'. The D/A switch symbol and operation are shown in Figure 5.9.

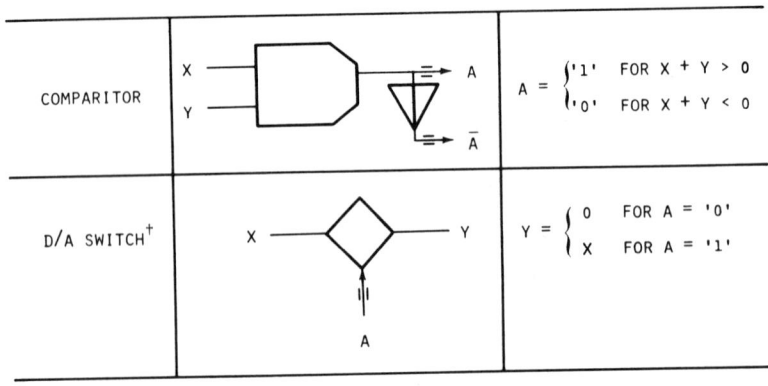

FIGURE 5.9 COMPARITOR AND D/A SWITCH. †

The D/A switch can be used to switch the outputs of the memory integrators in Example 5.3 to obtain a delayed reconstruction of the signal x(t) in Figure 5.8.

EXAMPLE 5.4

Use D/A switches to construct a delay of the function x(t) in Example 5.3.

SOLUTION:

Consider the block diagram shown in Figure 5.10(a). If S_1 = '1' when R_2 = '1', S_2 = R_3, and so on, then the output y(t) will simply be constant and equal to the immediate past value, giving the output labeled as w(t) in the figure. If S_2 = R_1, S_3 = R_2, and so on as shown, then when integrator 2 is in RESET switch S_3 will be closed, giving constant output equal to the value sampled three periods ago as indicated by y(t) in the figure.▼

---

† When both analog and digital signals are represented on the same diagram the digital line can be identified by the convention given below.

ANALOG

DIGITAL

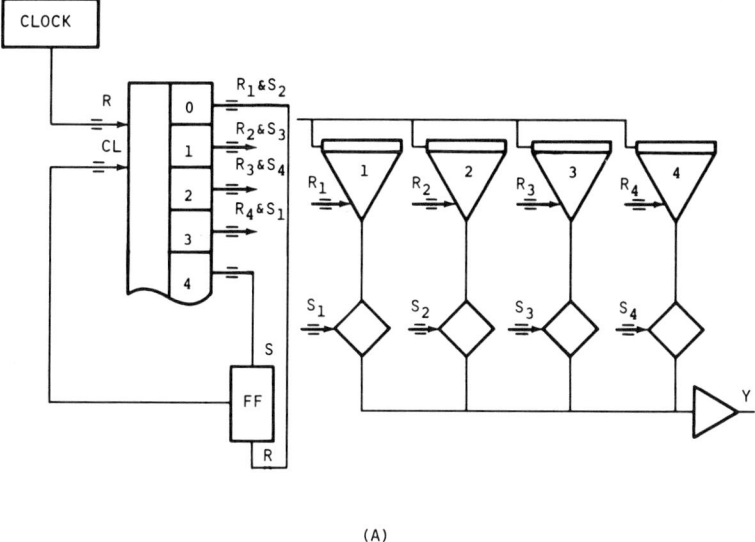

(A)

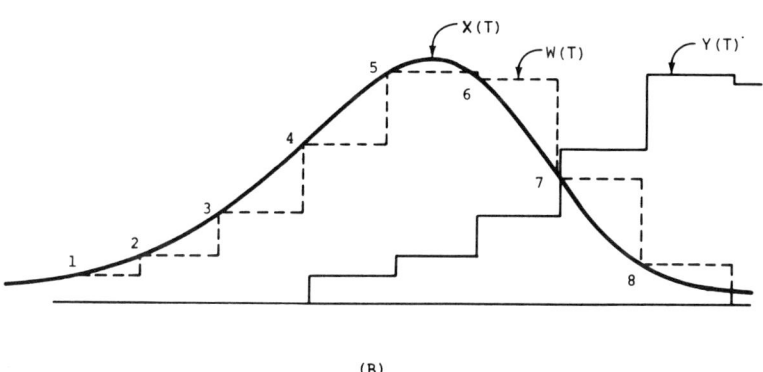

(B)

FIGURE 5.10 EXAMPLE 5.4 ANALOG MEMORY AND RECONSTRUCTION USED FOR DELAY.

EXAMPLE 5.5

In the analog computer solution of boundary value problems it is frequently desirable to stop the simulation when certain criteria are met. For example, if a projectile satisfying the equation

$$M\ddot{y} + B|\dot{y}|\dot{y} = -Mg \qquad (5.3)$$

is given initial vertical velocity $\dot{y}(0) = V_0$ it may be desired to calculate the time before it hits the ground, perhaps to determine the horizontal position, correct the initial velocity to decrease the error, and rerun the simulation.

Give a block diagram that will perform the simulation and put the simulation in HOLD when y = -0.5.

SOLUTION:

The system block diagram for the equation

$$\ddot{y} = \frac{B}{M}|\dot{y}|\dot{y} - g$$

appears in Figure 5.11. The comparitor is "biased" such that when (y + 0.5) > 0 or y > -.5 the comparitor output '1' goes to C_1 and C_2 keeping integrators 1 and 2 in COMPUTE. When (y + 0.5) becomes just less than zero the comparitor switches to '0', dropping the integrators into HOLD. If HOLD must be patched or a different operation is desired at this transition the inverted output from the comparitor can be used. Note in the figure that for a larger value of B the limiting velocity is smaller and a longer time elapses until y = -0.5. However, the comparitor still will not switch the simulation to HOLD until y < -0.5 regardless of how long it takes. This example is considered in greater detail in the next chapter.▼

5.3 HYBRID COMPUTERS

The speed of the digital computer will probably not increase to the point of being competitive with the analog computer for the simulation of dynamic systems without the introduction of multiprocessor systems containing many parallel arithmetic units working in parallel.

The modern analog computers (parallel hybrid computer), as discussed in the last section, have good

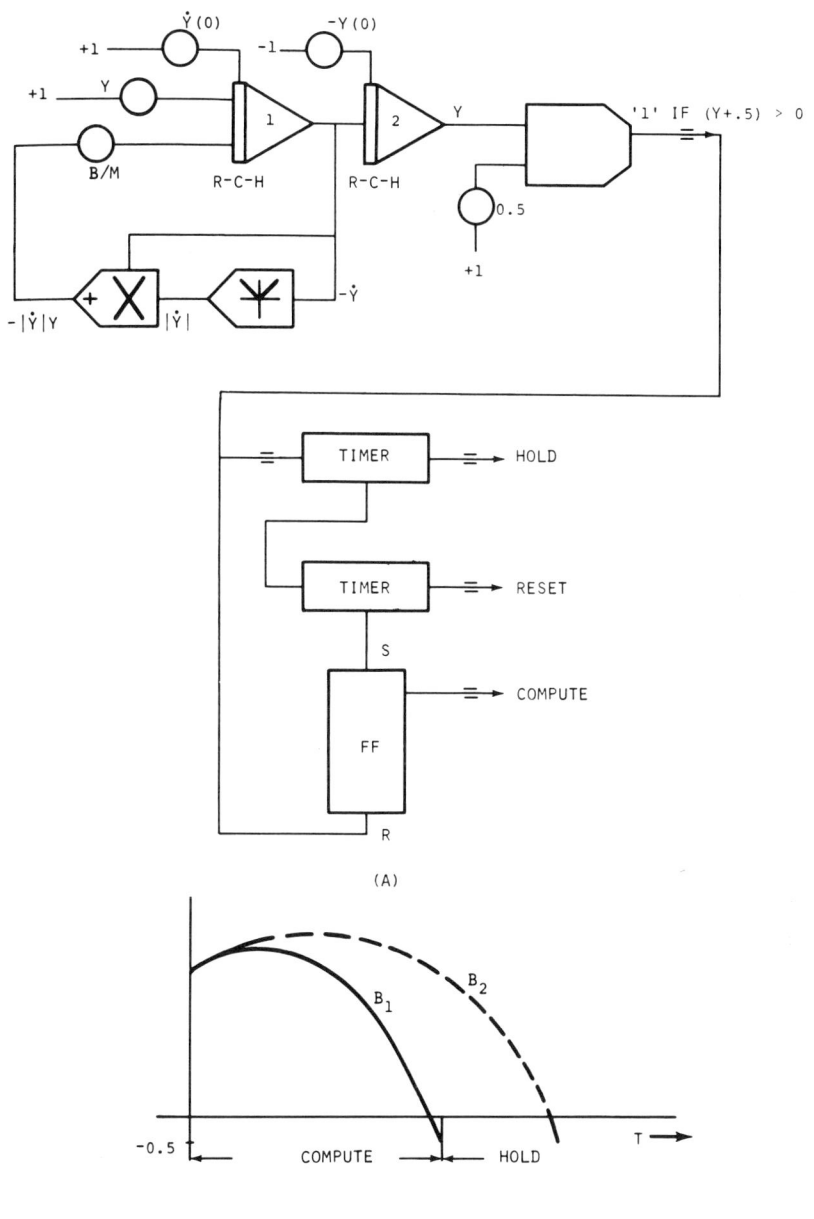

FIGURE 5.11 EXAMPLE 5.5 USE OF COMPARITOR FOR MODE CONTROL ($B_2 > B_1$).

dynamic   performance,   low   power   requirements,
reasonable cost, reasonable accuracy, good logic, and
fast   electronic   switches.   Frequently,   when
comparisons are made between analog and digital
simulation, the modern digital computer is compared
with the analog computer of the 1950's which of course
cannot compete.   However, the modern analog computer
is a powerful computational tool in its own right.

The second generation hybrid computer resulted from
the connection of a small general purpose digital computer to an analog computer to perform such functions
as analog computer setup with digitally set POTs,
automatic checkout of analog programs, and function
storage and playback using the digital memory.

A full hybrid system uses the complete capabilities
of both the analog and digital equipment to obtain
efficient problem solution and simulation with high
speed and good accuracy.   The availability of the
digital computer for multichannel function storage and
playback, the generation of nonlinear functions of
several variables, and other uses to be discussed
below requires multichannel high speed analog to
digital and digital to analog converters. Figure 5.12
is a block diagram for a general hybrid computer
containing an analog computer, a digital computer,
interface equipment and input-output devices.   The
organization of the machine is such that the program
is controlled through the software of the digital computer and the patching on the patchboard. Logical
operations can be performed either in the digital program or on the parallel logic of the analog computer.
The *SENSE LINES* monitor the state of the entire system
to insure execution of the program in proper sequence.
The *CONTROL LINES* and *INTERRUPT LINES* provide control
of the operation based on the program and the status
of the sense lines.   Of major importance in
programming is the assurance that digital calculations
are completed before the results are converted or
transferred to analog operations.   When a digital
calculation is completed a sense line can indicate
that the result is available.   Also, the digital
computer can be stopped temporarily, by means of an
interrupt, to allow the operator to change parameter

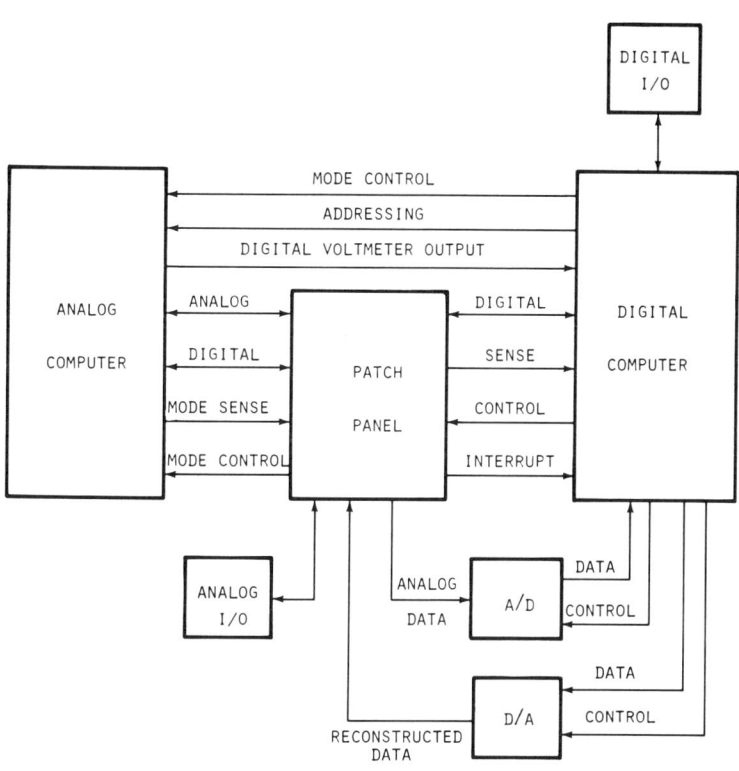

FIGURE 5.12 HYBRID COMPUTER BLOCK DIAGRAM.

values or perform logical operations. These control, sense, and interrupt signals are usually "one bit" signals like the logic levels of the analog computer.

Along with the monitoring and control functions, the interface must also control the high speed interchange of data between the two computers through analog to digital converters and digital to analog converters. This data conversion is discussed in some

detail because it is fundamental to hybrid
computation, and facilitates an understanding of the
relation between the time or parallel oriented analog
signals and the discrete or sequentially oriented
digital signals.

DIGITAL TO ANALOG CONVERTERS (DAC)
   The D/A switch is a one bit digital to analog converter. If the analog input is the machine unit
reference REF = 1, then the output is an analog signal
a = 1.0 when the digital signal n = 1, and a = 0.0
when n = 0. Thus the one bit digital word is
converted with a D/A switch. Higher resolution
digital numbers are maintained in registers in the
digital computer in binary form. Assuming the numbers
are machine unit scaled so that the number n is in the
range $-1 < n < +1$, or between 0 and 1 with sign.
Any number n between 0 and 1 can be represented in the
BINARY form by

$$n = \frac{1}{2} u_1 + \frac{1}{4} u_2 + \frac{1}{8} u_3 + \ldots$$

$$= \sum_{k=1}^{\infty} 2^{-k} u_k \qquad (5.4)$$

where u_k = 0 or 1 depending on the value of that bit.
In the digital computer only a finite number of bits
can be used for the representation of the number, so
we have the digital numbers n_D where

$$n_D = \sum_{k=1}^{N} 2^{-k} u_k \qquad (5.5)$$

n_D can be stored in a digital register by maintaining
the u_k as either 0 or 1. If all the u_k are 1 then
$n_D = (1 - 2^{-N})$ or one minus the LEAST SIGNIFICANT BIT
(LSB).

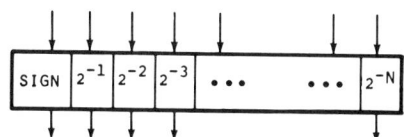

FIGURE 5.13 REGISTER CONTAINING DIGITAL NUMBER N_D PLUS SIGN.

A digital register containing n_D plus a sign bit can be represented by the block shown in Figure 5.13. The inputs are digital lines which set numbers in the register, and the output lines are digital lines with values either 0 or 1.

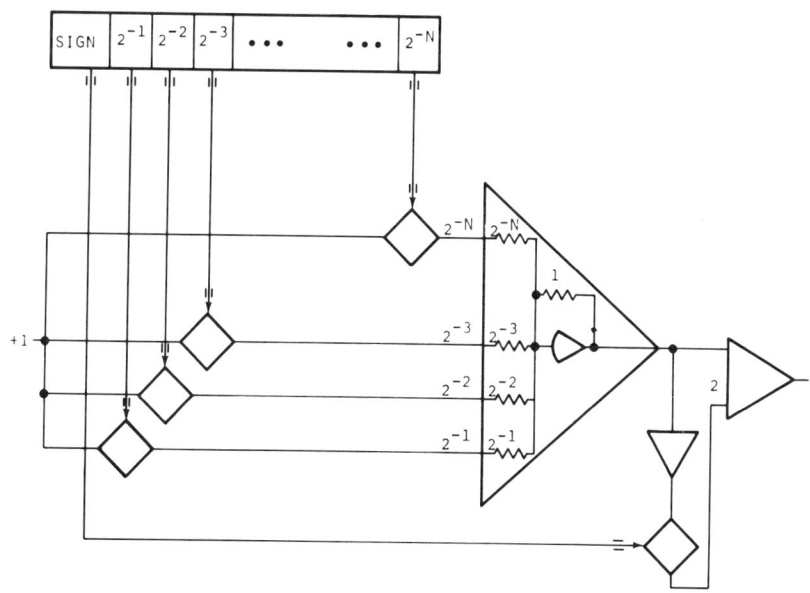

FIGURE 5.14 A DIGITAL TO ANALOG CONVERTER.

The outputs of the register can be used in conjunction with an amplifier with appropriate gain to obtain fast digital to analog conversion as shown in Figure 5.14. To operate the DAC, the register weights the reference into the amplifier by means of the gains of the significance of each bit. This type of DAC is commonly called a *WEIGHTED RESISTOR DAC* since the gains are obtained by different input resistors for the operational amplifier. A disadvantage of the weighted resistor DAC arises because of the wide range of resistance values needed to obtain the 12–14 bit resolution that is usually required. A more easily realized DAC can be made with the addition of double throw D/A switches. These switches have two analog inputs, with the one switched to the output depending on the state of the digital input. A symbol for the double throw D/A switch is shown in Figure 5.15.

Since the double throw D/A switch allows the line to be grounded with a 0 input, an R - 2R resistor

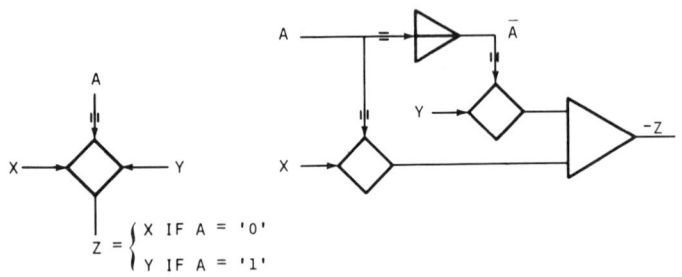

(A) DOUBLE THROW D/A SWITCH   (B) EQUIVALENT DIAGRAM

FIGURE 5.15 THE DOUBLE THROW D/A SWITCH.

ladder can be utilized, as shown in Figure 5.16. To simplify the figure, a four bit DAC is shown, but the principle is, of course, applicable for any wordlength. It can be shown that if all of the inputs are 0 except that from the kth (corresponding to 2^{-k}), then the output of the DAC is in fact 2^{-k}.† Thus by superposition the output for a general digital number n_D is as specified by Equation (5.5).

If an analog signal v_a is applied to the reference input of the DAC the output is v_a weighted by the digital word. Thus the DAC multiplies the analog signal by the digital number and is in effect a digitally controlled potentiometer. In this configuration the converter is called a *MULTIPLYING*

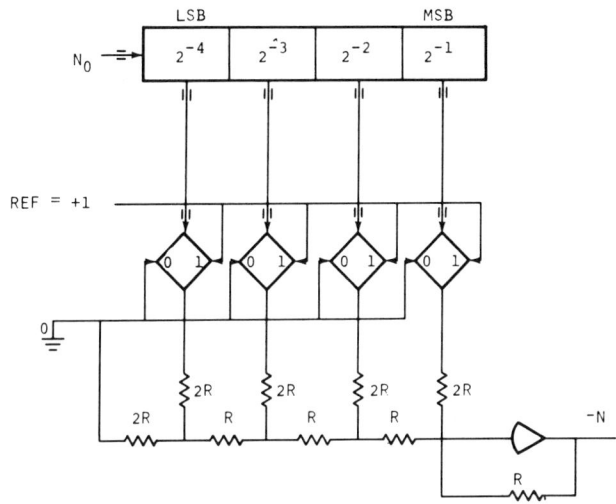

FIGURE 5.16 LADDER NETWORK DAC.

DIGITAL TO ANALOG CONVERTER (MDAC). Typical modern DACs have output conversions correct to the least significant bit in less than 5 μsec.

---

† Assume an input of one bit at the kth place. Then the effective resistance of the less significant bits to ground is 2R. The equivalent of this 2R and the 1 bit and its 2R is a voltage 1/2 and series resistor R. Continuing, we get a total equivalent voltage 2^{-k} and series resistor R. Thus the output is -2^{-k}.

ANALOG TO DIGITAL CONVERTERS

An *ANALOG TO DIGITAL CONVERTER (ADC)* has a digital output that is the binary representation of the input analog signal at the time the input is sampled. The input signal is maintained constant during the conversion time with a track store element or an integrator in HOLD.

The comparitor is the simplest analog to digital converter. It can be biased such that the output is 1 if the analog input $v_a > 0.5$, and the output 0 if $v_a < 0.5$. The addition of more comparitors and some logic gives the very fast *PARALLEL ADC*. To obtain a two bit converter we need the binary output given in Table 5.1, where b_1 is the most significant bit and b_2 is the least significant bit. We require that a

TABLE 5.1 A TWO BIT ADC TABLE.

ANALOG INPUT	BINARY OUTPUT	
V_A	B_1	B_2
$0.75 < V_A$	1	1
$0.5 < V_A < 0.75$	1	0
$0.25 < V_A < 0.5$	0	1
$V_A < 0.25$	0	0

circuit give $b_1 = 1$ when $v_a > 0.5$, and $b_1 = 1$ for $v_a < 0.75$ or $0.25 < v < 0.5$. The three comparitors shown in Figure 5.17 provide the signals C_1, C_2 and C_3 consistent with these values of v_a, and the logic elements perform the desired operations $b_1 = C_1$ and $b_2 = (C_2 \cdot C_3) + C_1$. Thus this system performs the two bit conversion. The parallel ADC is very fast but also very expensive because of the requirement for many comparitors.

An ADC that uses fewer components is the *RAMP ADC* which operates by measuring the time for a ramp func-

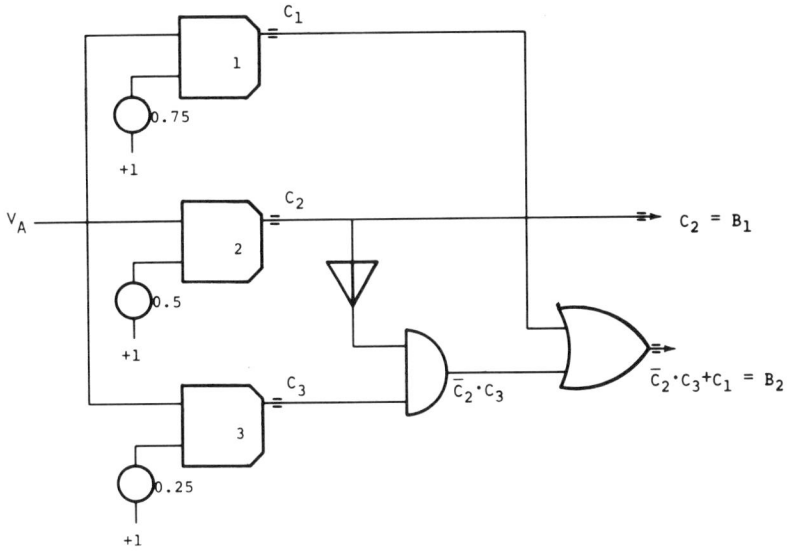

FIGURE 5.17   A PARALLEL ANALOG-TO-DIGITAL CONVERTER.

tion v_r to become equal to the analog signal. The elapsed time is measured with a counter and a standard clock. The output of the counter is then the digital number. Figure 5.18(a) shows a ramp ADC and associated logic. Conversion is initiated by putting the ramp generator into COMPUTE and resetting the counter. Figure 5.18(b) shows an analog input signal v_a and the ramp v_r, (c) shows $v_a + v_r$ which crosses zero, thereby changing the comparitor output a to '0' and stopping the clock input to the counter as shown in (d). The number in the counter is then proportional to v_a at the time when $v_a + v_r = 0$. The accuracy of the ramp ADC can be increased by increasing the complexity of the design giving the so-called dual ramp ADC and triple ramp ADC.

Probably the most commonly used ADC is based on successive approximation where the digital output is fed into a DAC whose output is compared to the analog input v_a, and the digital number is successively adjusted to minimize the difference between the two

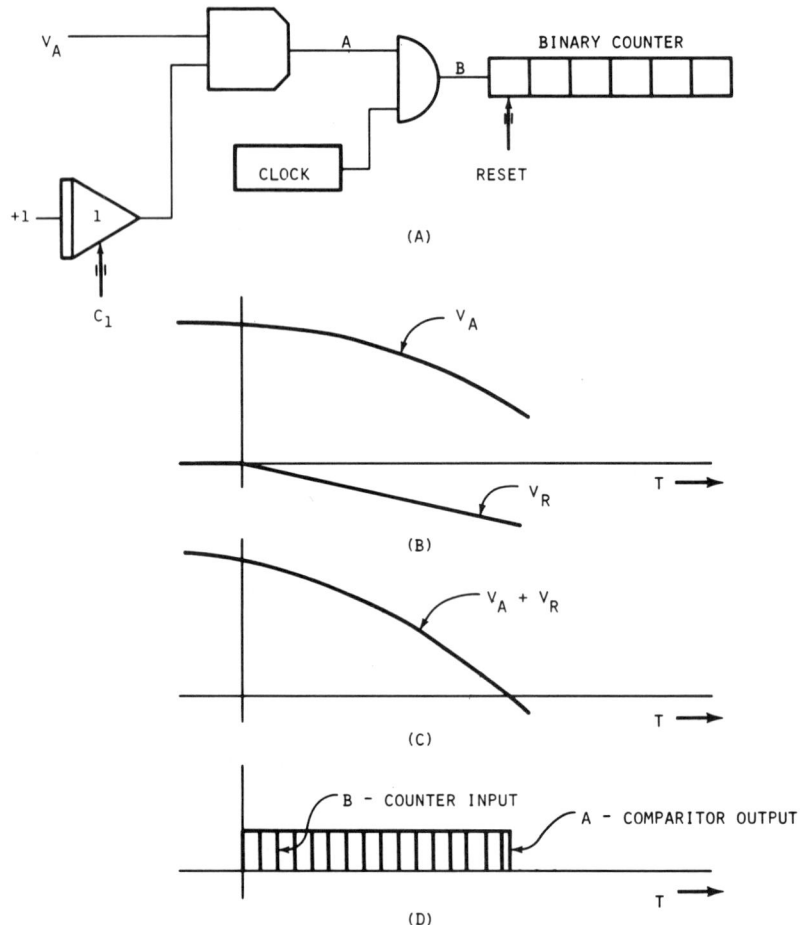

FIGURE 5.18 RAMP ADC.

values. Figure 5.19 shows a block diagram for a *SUCCESSIVE APPROXIMATION ADC*. The logic and control is consistent with the desired method of successive approximation.

There are two basic strategies for adjusting the bits of the ADC. The method of successive approximation adjusts the most significant bit first

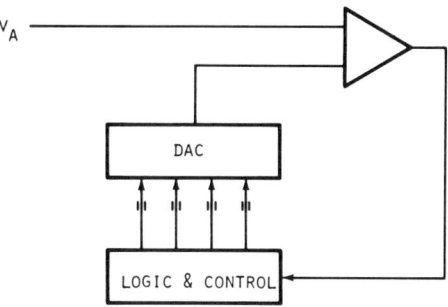

FIGURE 5.19 SUCCESSIVE APPROXIMATION ADC.

at the start of conversion and proceeds to the least
significant. The incremental method works at the
other end and adjusts the least significant bit only,
with carry, to change the output one quantization step
at a time. The result is that the successive
approximation method follows large jumps well but must
readjust all bits at each new conversion, while the
incremental method can follow smooth changes in v_a
with relatively little error. Figure 5.20 shows a

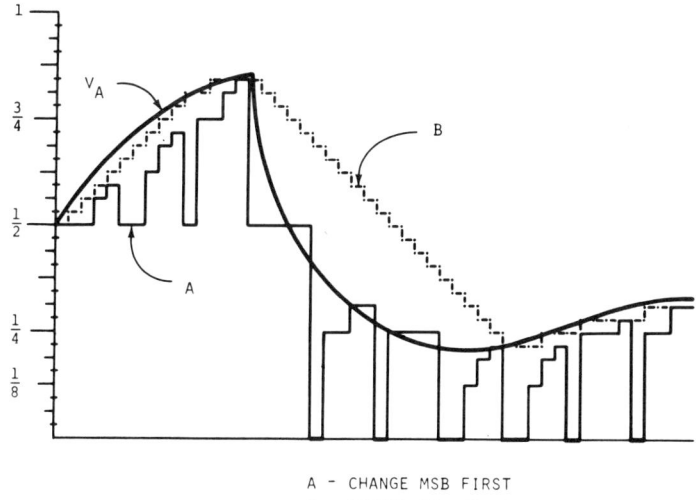

A - CHANGE MSB FIRST
B - CHANGE BY LSB AT EACH STEP

FIGURE 5.20 TWO STRATEGIES FOR A SUCCESSIVE APPROXIMATION ADC.

signal v_a and the resultant output of successive approximation ADCs using these two strategies.

DATA RECONSTRUCTION

Since the output of the DAC must take on some value between sample values, an interpolation scheme must be employed. Perhaps the most common is simply to maintain the output constant with a track store integrator, which results in piecewise constant outputs as obtained in Figure 5.10 for the output of the delay reconstructor. Alternatively the output points can be fit by a piecewise linear interpolation as shown in the following example.

EXAMPLE 5.6

Give an analog computer block diagram that gives a linear interpolation (first order HOLD) between the sample values of a piecewise constant signal.

SOLUTION:

Figure 5.21 shows the block diagram and the reconstruction. Integrator 1 resets to the sampled value $y_n - x_n$ which is held constant and integrated for T seconds with a gain of $1/T$ changing $x(t)$ linearly by $y_n - x_n$ to the value y_n.▼

Note that the output of the above system is the linear interpolation of the data points with a delay of one sample interval T. The constant interpolator (zero order HOLD) has an effective delay of $T/2$ as discussed in Problem 5.12. Since $y(t)$ in Figure 5.21 is sampled only when integrator 1 is in RESET it need not be a piecewise constant function, so the system gives a linear interpolation between the samples regardless of the character of $y(t)$.

A typical hybrid computer might have 24 channels each of A-to-D and D-to-A conversion, the digital number being represented by 13 or 14 bits plus a sign to just exceed the best analog accuracy normally obtained. 13 bits corresponds to 0.01 percent full scale accuracy. The ADC channels may be switched sequentially to a single ADC through a coding network

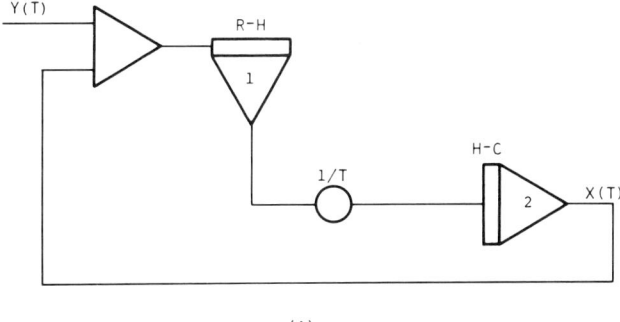

(A)

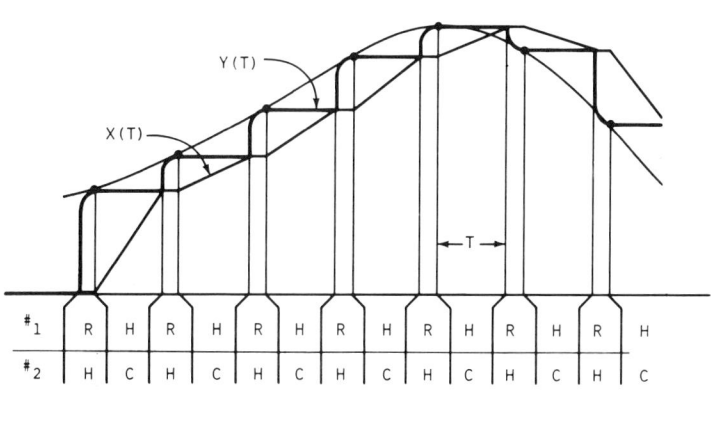

(B)

FIGURE 5.21 EXAMPLE 5.6 LINEAR DATA RECONSTRUCTION.

called a *MULTIPLEXER*. The distribution of the digital numbers to the correct channel is performed by a demultiplexer. Similarly the DAC channels may also be multiplexed through a single DAC. Of course, several channels sharing an ADC or DAC reduces the rate at which conversions can be performed on each channel. ADCs have conversion times in the range of 1 – 50 μsec with about a 10:1 ratio of price for the faster speed. DACs are somewhat faster with conversion time in the range 0.1 - 10 μs depending on cost. Thus the converters can perform conversions upward to a million per second depending on cost. Multiplexers can switch from channel to channel effectively in less than a μsec.

PROGRAMMING AND SOFTWARE

Hybrid computing centers generally have a scientific compiler like FORTRAN IV that contains subroutines for logical operations, mode control, and analog subroutines. The compiler also includes subroutines for setup, checkout and debugging of the program. Some typical subroutines are the following

ADSEL(I,J)   operates the analog computer address selector connecting POTs, AMPs, or TRUNKs to the selector output. The arguments are codes for the element type and number. CALL ADSEL(2,5) might select amplifier No. 5.

POTSET(I,A)   sets POT No. 1 to value A where $0 \leq A \leq 1.0$.

RESET   puts the analog computer into RESET.

HOLD   puts the analog computer into HOLD.

COMPUT   puts the analog computer into COMPUTE.

MDAC(I,J)   multiplies the analog input to MDAC No. 1 by the fixed point number J. CALL MDAC(6,K) reads the digital number K into MDAC No. 6 (or channel 6 of MDACs).

ADC(I,J)   reads the input of ADC No. 1 into the variable J. CALL ADC (12,I5) converts the input to ADC 12 and puts the result in I5.

LOGAD(I,J)   reads logic signals from analog to digital machine.

LOGDA(I,J)   reads logic signals from digital to analog machine.

This brief list gives the general idea of the role of hybrid software. It is important to maintain documentation of the software package for each installation. Of course the programmer uses additional subprograms unique to his problem in programming the machine while the library subprograms are of general applicability to hybrid computation.

ERRORS

The errors of analog and digital calculation discussed in Chapter 1 are present in hybrid computation along with additional errors in the interface. The analog accuracy depends on the accuracy of the computing elements and the larger the simulation, the greater the chance of an error buildup. Furthermore, dynamic errors exist due to frequency limitations of the amplifiers and other components, so the error increases with the frequency of the simulation variables. Since the major application of hybrid computers is the fast repetitive simulation of dynamic systems these, dynamic errors are of utmost concern, particularly since we want to time scale solutions to run as fast as possible while maintaining accuracy.

Dynamic errors are characterized as phase shift (or delay), changes in gain with frequency and velocity limiting. The fastest rate of change the output of an element can follow is called the slewing rate of the element.

Table 5.2 lists some comparisons of analog and digital computers. A hybrid computer simulation makes use of the good points of each machine but has the disadvantage of being somewhat more difficult to program than either the analog or digital machines because of the complexity of the system and the lack of standardization of notation and programming. It is of course necessary that the programmer understand both analog and digital computer programming and simulation.

With the increased versatility and decreased cost of large scale integrated (LSI) circuits, analog computers may soon have accurate digital integrators that can be conveniently patched on an analog patch

TABLE 5.2 SOME COMPARISONS OF ANALOG AND DIGITAL COMPUTERS.

	ANALOG	DIGITAL
SPEED	FAST SOLUTION OF DIFFERENTIAL EQUATIONS THROUGH PARALLEL OPERATION	SLOWER THAN ANALOG FOR DIFFERENTIAL EQUATIONS (10-100 TIMES) DUE TO SEQUENTIAL OPERATION
ACCURACY	COMPONENTS ARE 0.01 PERCENT FOR FAST SOLUTIONS, ABOUT 1 PERCENT DYNAMIC ERROR	ANY DESIRABLE ACCURACY AT COST OF COMPUTER TIME
FLOATING POINT ARITHMETIC	NO	YES
MEMORY	WEAK	GOOD
SOFTWARE	LITTLE	GOOD
SETUP & DEBUGGING	INDIVIDUAL, EXPENSIVE AND TAKES EXPERIENCE	SOMEWHAT STANDARDIZED BUT ALSO TAKES EXPERIENCE
COST/RUN	LOW	HIGH
INTER-ACTION	EXCELLENT	POOR BUT IMPROVING WITH TIME SHARED GRAPHIC DISPLAY TERMINALS. A SMALL OPEN SHOP DIGITAL COMPUTER IS BETTER FOR HANDS-ON OPERATION
COST	$10,000 - $30,000 SMALL MACHINE $200,000 LARGE INSTALATION	$10,000 - $30,000 MINI-COMPUTER $2,000,000 - LARGE MACHINE
DISPLAYS	GOOD	IMPROVING (ANALOG TYPE OUTPUTS) CRT XY PLOTTER

board. The resultant (digital) simulation can have parallel processing units, a total digital facility operating with a hands on patching convenience. In this case the analog type of integrator blocks would be ordered to any desired accuracy, and each computer can have an abundance of reasonably accurate, and few very accurate, integrators together with logic and nonlinear functions that can be preprogrammed and patched into the analog equipment. In the near future, digital computers will no doubt be on the market which are made up of parallel processing units that can be interconnected (patched) through programming instructions and special software. The block manipulation programming required for such machines will be closer to hybrid computation than either analog or digital techniques alone. A knowledge of hybrid computation is an excellent background for understanding and developing programming techniques for these parallel digital computers. In particular, one is not constrained to think exclusively in terms of an analog oriented or digitally oriented problem approach.

Since the signals in the real world are continuous and analog in character there are many areas of application for hybrid computation, such as the solution of boundary value problems in ordinary and partial differential equations, optimization and stochastic processes. The use of random number generators facillitate the application of hybrid computers for statistical analyses by Monte Carlo methods. Hybrid computers are also used for data reduction, and signal analysis and processing. The remaining chapters of the book are devoted to some of these applications. The following example demonstrates the applicability of using the digital portion of a hybrid computer for function storage and playback.

EXAMPLE 5.7

L. Stark † has shown that a reasonable model for the pupil response mechanism is

$$H(s) = \frac{0.16e^{-0.18s}}{(1+0.1s)^3}$$

---

† L. Stark, "Stability Oscillations, and Noise in the Human Pupil Servo-mechanism," *Proc. IRE*, Vol. 47, Nov. 1959, pp. 1925-1939.

This response mechanism is measured by shining a narrow beam of light into the center of the pupil so that the light entering the pupil is unaffected by the iris area, and results in an open loop operation. A closed loop (feedback) system is obtained if light is spread over the total area of the eye so that dilation and constriction of the pupil effect the amount of light entering the pupil. A high gain feedback system is obtained when a narrow beam of light is focused on the border of the iris so that small movements of the iris result in large changes in light intensity at the retina.

A feedback model for the pupil servomechanism with variable gain K is given by the equation

$$(0.1D+1)^3 y(t) = K[x(t-0.18)-y(t-0.18)] \quad (5.6)$$

Give a computer simulation to investigate the stability of the system as a function of the gain K.

SOLUTION:

The hybrid computer is particularly well suited for the simulation of delay differential equations like Equation (5.6) because the differential equation can be solved rapidly on the analog computer while the delay is generated with an ADC, the digital memory, and a DAC. Figure 5.22 is a block diagram representing the system for constant excitation. (See Problem 5.18.)

The block diagram was programmed on a hybrid computer system that had a 100V reference analog machine, with full scale MDACs of 16,383. So the machine unit scale on the analog has REF = 1 patched to +100, while on the digital computer it corresponds to 16,383, $(2^{15} - 1)$. However, the ADCs have a full scale of 2047, $(2^{12} - 1)$; thus full scale on the analog computer is converted to 2047 in the digital computer. These scale changes are compensated for by a single FORTRAN statement multiplying incoming signals by 16383/2047, thus insuring that the value returned to the DAC is within the range of the device and that the output is scaled properly for the analog

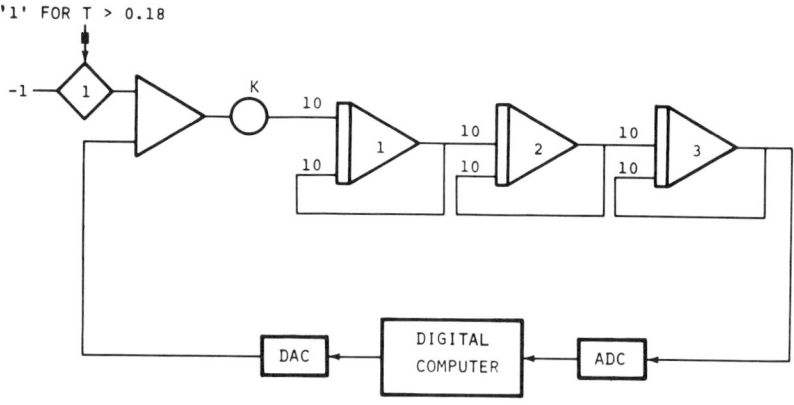

'1' FOR T > 0.18

FIGURE 5.22 MODEL OF THE HUMAN PUPIL SERVOMECHANISM.

computer.

For this example, the COMPUTE, HOLD, and RESET subroutines in the digital computer were used to control analog computation. The hybrid program does the following

1. Samples the output of integrator 3, which is the input to ADC 1.

    ```
    CALL ADC(1,IY)
    ```

2. Scales the sample and stores it in array JY(N).

    ```
    JY(N)=(16383/2047)*IY
    ```

3. Reads out delayed sample through DAC 1, which is patched to an input of summer 1.

    ```
    KY=JY(N-KT)
    CALL DAC(1,KY)
    ```

The simulation is initially run for 10 seconds with 10^3 sample increments, or 100 samples per second. Thus the delay of 0.18 seconds corresponds to 18 samples or KT = 18. Of course the first 18 values of JY(N) must be set equal to zero.

Following is an abbreviated program incorporating these statements and the necessary mode control for the program. Several machine dependent details have been omitted in order to place the emphasis on the computational procedures. A delay is inserted after

mode control statements to allow time for the change
in mode. If each execution of a DO loop uses 10
microseconds then going through a DO loop I times
gives 10I microseconds delay using the subroutine

```
          SUBROUTINE DELAY(I)
          DO 10 J=1,I
       10 CONTINUE
          RETURN
          END


          DIMENSION JY(1018)
C
C   RESET ANALOG COMPUTER.
C
          CALL RESET
C
C   ALLOW 100 MILLISECONDS FOR RESET.
C
          CALL DELAY(10000)
C
C   SET INITIAL VALUES IN ARRAY JY.
C
          DO 11 K=1,18
       11 JY(K)=0
C
C   BEGIN BY PUTTING ANALOG IN COMPUTE.
C
          CALL COMPUT
C
C   DO LOOP TO TAKE AND RETURN SAMPLES.
C
          DO 12 N=19,1018
C
C   DELAY 10 MILLISECONDS FOR NEXT SAMPLE.
C
          CALL DELAY(1000)
          CALL ADC(1,IY)
          JY(N)=(16383/2047)*IY
          KY=JY(N-18)
          CALL DAC(1,KY)
       12 CONTINUE
          CALL HOLD
```

Accurate timing of other operations can be obtained
by an evaluation of the time needed to perform each
operation in the digital program or by putting the
analog portion of the computer in HOLD during the
digital calculations for which the times are not
accurately calculated.

Figure 5.23 gives plots of the response y(t) from
integrator 3 for various values of gain K. (For what
K is the system stable?) Also presented is the
response y(t) for K = 2 with and without delay (that
is, KT = 18 and KT = 0). The delay is apparently a
significant factor in causing the instability

connected with the gain K. The simulation model demonstrates the instability at high gain which is called "induced pupillary hippus." This program has been run in real time and can of course be time scaled to run faster. Note that the limiting factor is not the speed of the analog computer, since it can complete a total solution in less than 100 ms. ▼

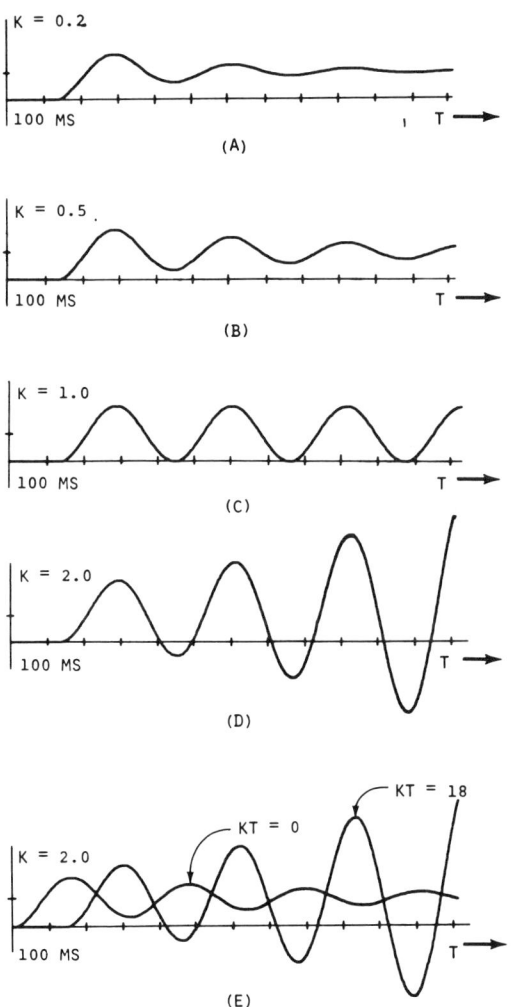

FIGURE 5.23 INDUCED PUPILLARY HIPPUS SIMULATION.

The example above uses the timing of the digital computer for control of the simulation. The following example shows the use of *INTERRUPT CALLED SUBROUTINES* in a program controlled by the analog (logic) portion of the computer.

An interrupt called subroutine is initiated when a logic signal on an interrupt line signals an interrupt. The digital computer completes the operation in progress, and then calls the appropriate subroutine associated with that interrupt line. Upon completion of the called subroutine the digital computer resets the interrupt line and returns to the instruction it would have performed next if the interrupt had not occured.

EXAMPLE 5.8

Analog integrators can be used to obtain an XY plot from an XY array in the digital computer using linear interpolation. Assume the arrays X(N), Y(N) and IMD(N) corresponding to the XY values for the plot and a move draw bit (logic signal) to raise and lower the pen of the X-Y plotter.

Figure 5.24 shows a block diagram for the required patching.

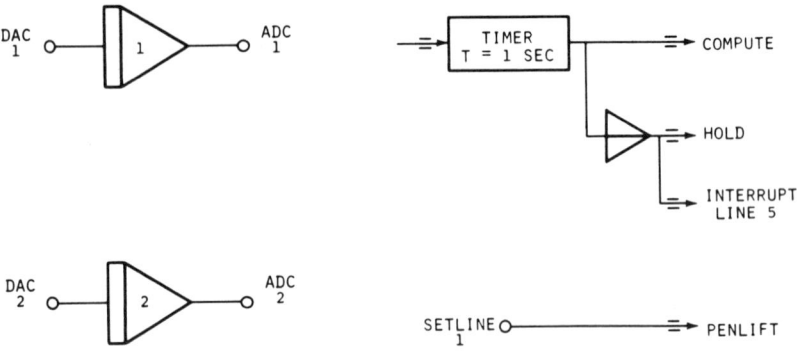

FIGURE 5.24 LINEAR INTERPOLATOR.

CALL COMPUT puts the analog portion in COMPUTE and triggers a timer which is set for 1 second. The timer output is patched so it puts the computer in HOLD at the end of the 1 second interval. The HOLD signal is also used to set the interrupt to call the subroutine PLOT XY. If the arrays X(N), Y(N), and IMD(N) are assumed for NPTS points, the main program loops with the statement

```
20 IF((INDEX-NPTS).NE.0) GO TO 20
```

When the interrupt occurs the subroutine is called. Upon completion of the subroutine, the program returns to statement 20 and loops until the next interrupt. The plotting continues until all points are plotted.

```
SUBROUTINE PLOTXY(X,Y,INDEX)
DIMENSION X(300),Y(300)
CALL ADC(1,PASTX)
CALL ADC(2,PASTY)
DX=PASTX-X(INDEX)
DY=PASTY-Y(INDEX)
CALL DAC(1,DX)
CALL DAC(2,DY)
CALL SETLIN(5,IMD(INDEX))
CALL COMPUT
INDEX=INDEX+1
RETURN
END
```

The program block diagram shown in Figure 5.25 represents the operation of the system. The incremental input DX is the difference between the new value and the past value with a sign reversal to account for the inversion in the integrator. Setline 5 controls the lifting of the pen. (Check the statements in the program to verify proper operation.)▼

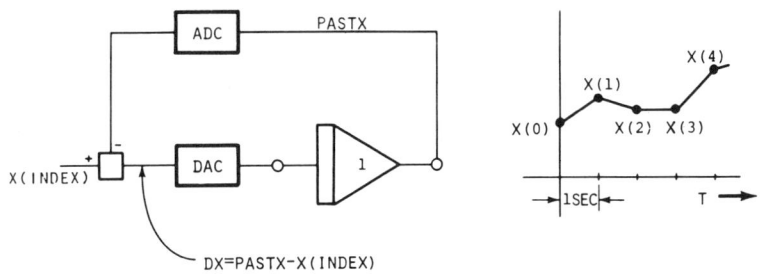

FIGURE 5.25   EXAMPLE 5.8 LINEAR INTERPOLATOR USING INTERRUPT CALLED SUBROUTINE.

PROBLEMS

5.1 Give a logic Diagram using AND gates, OR gates, and inverters to realize the following logic functions.

(a) $a + b + (c \cdot d)$

(b) $(\bar{a}+b) \cdot c \cdot \bar{d}$

(c) $(a \cdot \bar{b}) + (\bar{a} \cdot c)$

(d) $(a \cdot b) + (c \cdot d)$

(e) $(a \cdot \bar{b}) + (\bar{a} \cdot b)$

[(e) is called an *EXCLUSIVE OR* normally denoted $a \oplus b$.]

5.2 Give a block diagram using only AND gates and inverters to realize the logic functions given in Problem 5.1.

5.3 Give a logic diagram using only AND gates and inverters to realize a decoder which has logic inputs a and b, and the four outputs $w = a \cdot b$, $x = a \cdot \bar{b}$, $y = \bar{a} \cdot b$, and $z = \bar{a} \cdot \bar{b}$.

5.4 (a) Give a block diagram using a 10 KHz (10^4 pulses per second) clock, a counter, AND gates, inverters, and flip-flops to control two integrators with the following strategy during each ms (10^{-3} seconds): integrator 1 is in RESET and 2 is in COMPUTE for the first 0.3 ms, 1 is in COMPUTE and 2 is in HOLD for the next 0.4 ms, both 1 and 2 are in RESET for the last 0.3 ms, and then the cycle repeats.

(b) How can you make the above sequence run one tenth as fast? That is, how do you time scale the logic and mode control when you

time scale the integrator? [If you have a
1 KHz clock or another counter, it is easy.]

5.5 Patch up the block diagram obtained in Problem
5.4 and verify its operation. Time scale it sufficiently that you can plot the mode control signals.

5.6 Give an analog block diagram including one or
more synchronized analog tape recorders to
average the signals

$$r_i(t) = y(t) + n_i(t)$$

Include mode control to start, stop, and reset
the tape recorders. Comment on what type of
errors and inaccuracies might occur in this
system.

5.7 Give a block diagram, including comparitors and
appropriate logic, to perform 3 bit parallel
analog-to-digital conversion where the input
signal is in the range $-1 < v_a < +1$ and the
quantization steps are equally spaced. Make a
sketch of the output v_o as a function of v_a (the
static characteristic of the ADC).

5.8 (a) Show that the parallel conversion into an n
bit binary number requires 2^{n-1} comparitors.

(b) If the wordlength is chosen at the accuracy
of normal analog tolerance, 12 - 14 bits are
required. How many comparitors would be required for this resolution in a parallel ADC?

5.9 A ramp ADC gives output that is the digital equivalent of the unknown signal v_a at the completion
of the conversion. Since the conversion time depends on v_a the exact time of conversion is asyn-
chronous even if the convert commands are syn-
chronous. Include a tracking integrator and
appropriate mode control to hold v_a constant at
the value it has when the convert command (start
pulse) occurs.

5.10 Make a sketch of the response of a 4 bit ADC for successive approximation and incremental approximation to the inputs (a) $v_a = \sin t$ and (b) $v_a = t - (t-1) \, u(t-1)$. Assume that the 4 bit converter can make a comparison and change every 100 ms.

5.11 Use the circuit shown in Figure 5.21 to obtain the error in trapezoidal integration. Plot the maximum error as a function of the frequency of excitation. (See Problem 2.24.) (Hint: Don't forget to account for the delay when comparing it with the exact integral.)

5.12 An ADC has an effective time delay of the con-version time since the completion of the conversion occurs at some time τ_1 later than v_s is sampled. Also a DAC has an effective delay of one half the D/A conversion interval τ_2 as a result of the zero order data reconstruction. Figure 5.26(a) shows the signals v_s, the digital signal v_s^* and the output of a DAC v_s' for the system shown in Figure 5.26(b). During data reconstruction v_s' contains a delay from v_s^* of between zero and τ_2, so an "average" value is $\tau_2/2$.

Use a digital z-transform simulation to simulate the equation $\ddot{y} + y = 0$, assuming y is piecewise constant, and generate a sine wave. Then simulate the effective delay by introducing delays in the right hand side of the recursion relation. Make a plot of some measure of response error versus delay from the ADC-DAC combination. (Don't forget that the piecewise constant assumption introduces an effective delay of $\tau/2$.)

In real time systems the introduction of a digital computer in the loop introduces an additional time delay since the continuous portion of the program cannot be put into HOLD during the digital calculation.

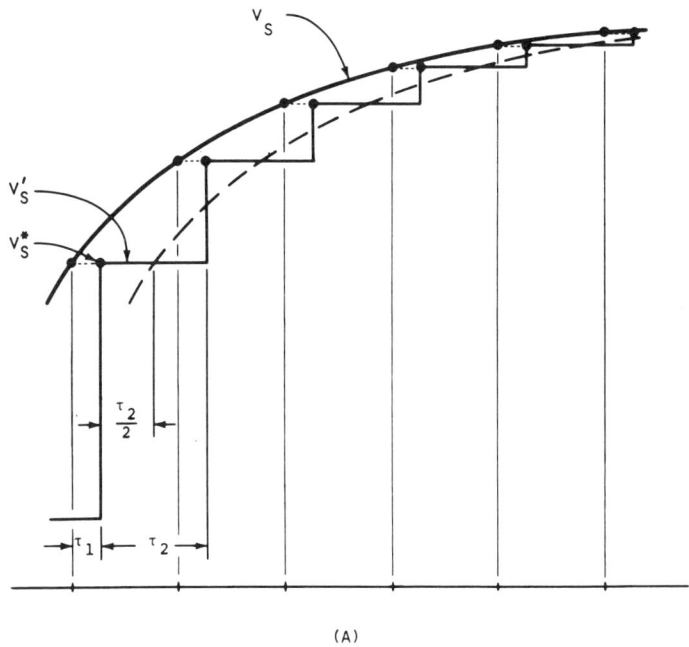

(A)

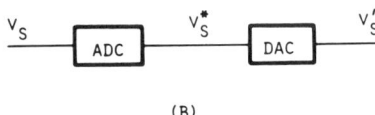

(B)

FIGURE 5.26   PROBLEM 5.12.

5.13 Time delay errors such as those discussed in Problem 5.12 are generally compensated for by the introduction of some prediction scheme which hopefully just balances the delay. Using a z-transform, suppose the delay in the signal y(t) is τ and we take z-transforms with increment size τ. Then the ADC-DAC combination corresponds to a pulse transfer function $H(z) = z^{-1}$ which can be completely compensated for by an ideal prediction $H(z) = z$. Since we have no ideal predictors we can approximate one by extrapolating the derivative of the input signal $\dot{y}$ to obtain

$$y_{n+1} \approx y_n + \tau \dot{y}_n$$

So in place of y(t), for a first order approximation we can use $y + \tau \dot{y}$ which can be obtained by a feed forward path from the last two integrators of the analog portion of the program.

(a) Simulate an analog sine wave generator as described in Problem 5.12 and compensate for the time delay by taking the output of the analog as $y + \tau \dot{y}$ rather than y and evaluate the effect of this compensator.

(b) The above result can be obtained by a Taylor series in y about the point y(nτ). Give a block diagram for the sine wave simulation including a second derivative term in the compensation.

5.14 (a) With reference to the sine wave generator of Problem 5.13, draw a block diagram to implement the following procedure and comment on its effectiveness as time delay compensation.

(b) Test the procedure by a simulation similar to that described in Problem 5.13 for a sine wave generator.

   Since an input to the digital computer of $y + \tau \dot{y}$ compensates for a time delay of τ, then an alternative compensation is obtained by assuming that the ADC-DAC combination has output $y - \tau y$ rather than y and this input $y - \tau \dot{y}$ can be patched into the analog computer appropriately to eliminate the delay.

5.15 Describe how the time delay discussed in Problems 5.12 - 5.14 might be compensated for in the digital computer portion of the program. Show that the block diagram in Figure 5.2] satisfies Equation (5.6).

5.16 Write a hybrid program (using the digital computer) for function storage and playback for the solution of the delay differential equation

$$\dot{y}(t) + y(t) - 0.2y(t-0.1) = \cos 10t$$

Assume it takes 30 μs for one step of a DO loop and 10 ms for the analog computer to RESET. The initial condition function is

$$y(t) = 10t + 1 \quad \text{for } -0.1 < t < 0$$

5.17 Show that the block diagram for the model of the human pupil servomechanism in Figure 5.22 does indeed satisfy Equation 5.6.

5.18 B. A. Chubb † has presented three case studies indicating the costs of performing design simulations on the analog computer, on the digital computer using FORTRAN and the Continuous System Modeling Program CSMP, and on a hybrid computer. Assume a cost of $200/hour for the digital computer, $80/hour for the analog computer, $80/hour for the hybrid computer and $18/hour for labor including overhead. Assume the FORTRAN program compiles (translates) in negligible time and the CSMP program compiles in 0.5 minutes; evaluate the following for each case by each method of simulation used.

(a) The cost to obtain the first run

(b) The time to obtain the first run

(c) Cost per run by each method after the first run

(d) The total cost to obtain 10, 50, 100, 200 runs

---

† B. H. Chubb, "Economic Evaluation of the CSMP digital Simulation Language," *Simulation*, March 1970, pp 101-103.

CASE 1   CSMP FORTRAN

Number of coding statements   102   340

Programming time in hours   6   28

One run execution time in min.   1   1

CASE 2   CSMP FORTRAN ANALOG

Number of coding statements   41   167   —

Programming time in hours   2   24   12

One run execution time in min.   5.4   5.4   0.3

CASE 3   CSMP HYBRID

Number of coding statements   35   55+patching

Programming time in hours   1   3

One run execution time in min.   13   1

6 computer solution of boundary value problems

Thus far in the solution of differential equations by analog or digital computation we have assumed that initial conditions are available on all of the variables. In many problems of ordinary or partial differential equations, boundary conditions are given at two or more values of the independent variable, some perhaps at the initial time and the remainder at the final time. Those problems in which boundary conditions are specified at two points are called *TWO POINT BOUNDARY VALUE PROBLEMS* (TPBVP) and are somewhat more complex than initial value problems. Those problems for which the boundary conditions are given at more than two values of the independent variable are called multipoint boundary value problems and, because they are significantly more difficult, they are not discussed in this book.

An nth order TPBVP is generally solved by one of the following methods

1. If the equations are linear they are solved by the solution of n initial value problems

309

2. Shooting or trial and error procedures

3. Successive linearization of nonlinear problems

4. The differential equation is replaced by a set of finite difference equations which is solved algebraically by the procedures of Chapter 4.

Boundary value problems in partial differential equations are usually solved by the approximation of some or all of the partial derivatives by finite differences. Section 6.2 is devoted to the solution of TPBVP and boundary value problems in partial differential equations.

6.1 BOUNDARY VALUE PROBLEMS IN ORDINARY DIFFERENTIAL EQUATIONS

A linear TPBVP can be solved analytically or computationally with little difficulty. This discussion is presented as an introduction to techniques for the solution TPBVP. Nonlinear TPBVPs are solved by successive linearization of the nonlinear equations.

Consider a linear system described by the state equation

$$\dot{X} = AX \tag{6.1}$$

with known solution

$$X(t) = \Phi(t)X(0) \tag{6.2}$$

where $\Phi(t) = e^{At}$ and $X(t)$ is an n dimensional state vector.

If there are n independent initial conditions specified, Equation (6.1) has a unique solution. However, some of the conditions may be given at the initial time $t = 0$ and some at the final time $t = t_f$. Suppose we have k initial conditions and n-k final conditions. Then the remainder of the elements of $X(0)$ and $X(t_f)$ are related by Equation (6.2). $\Phi(t)$ is

known since it depends only on A and not the boundary
conditions. So we have

$$X(t_f) = \Phi(t_f) X(0) \qquad\qquad (6.3)$$

which is n equations in the 2n initial and final
values, n of which are given.

If more than n conditions are given for an nth
order system, the conditions clearly are not
independent and there may or may not be a solution.
If less than n boundary conditions are specified, the
solution may not be unique and the general solution
contains arbitrary constants consistent with the
degrees of freedom allowed by the unspecified boundary
values. Even in the case of n conditions there may
not be a solution to the TPBVP, in which case the
boundary conditions are said to be inconsistent.

EXAMPLE 6.1

Obtain the solution to the equation $\ddot{y} + y = 0$ that
satisfies the boundary conditions $Y(0) = 0$ and
$y(\pi/4) = 1$

SOLUTION:

The general solution to the differential equation
is

$$y(t) = A \cos t + B \sin t$$

Substitution of $t = 0$ and $t = \pi/4$ allows the
evaluation of the constants A and B. However, we
complete the solution in state notation to demonstrate
the procedure used for higher order systems.

Writing the equation in state notation we have
($x_1 = y$ and $x_2 = \dot{x}_1$)

$$\begin{bmatrix} \dot{x}_1 \\ \dot{x}_2 \end{bmatrix} = \begin{bmatrix} 0 & 1 \\ -1 & 0 \end{bmatrix} \begin{bmatrix} x_1 \\ x_2 \end{bmatrix}$$

The fundamental matrix $\Phi(t)$ can be obtained from an analog computation, digital computation, or the known solution as outlined in Section 3.4. The first column of $\Phi(t)$ can be obtained as the response to the initial condition (first initial value problem)

$$\begin{bmatrix} x_1(0) \\ x_2(0) \end{bmatrix} = \begin{bmatrix} 1 \\ 0 \end{bmatrix}$$

which gives the response

$$\begin{bmatrix} x_1 \\ x_2 \end{bmatrix} = \begin{bmatrix} \cos t \\ -\sin t \end{bmatrix}$$

in the block diagram shown in Figure 6.1. (See Example 3.11.)

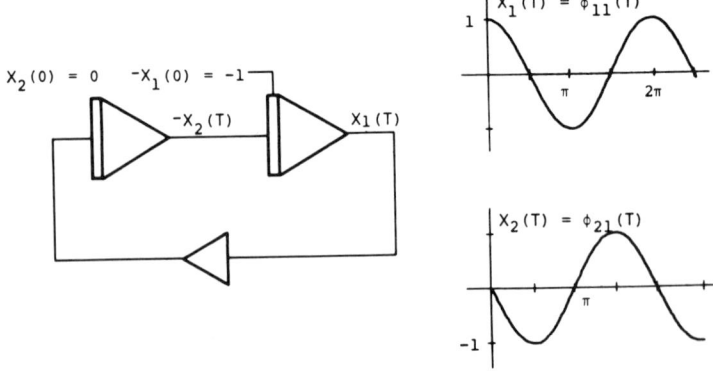

FIGURE 6.1   EXAMPLE 6.1 GENERATION OF $\phi(T)$ FOR THE SOLUTION OF A LINEAR TWO POINT BOUNDARY VALUE PROBLEM (TPBVP).

Obtaining the second column of $\Phi(t)$ in a similar manner (the second initial value problem) we have the state equations

$$\begin{bmatrix} x_1(t) \\ x_2(t) \end{bmatrix} = \begin{bmatrix} \cos t & \sin t \\ -\sin t & \cos t \end{bmatrix} \begin{bmatrix} x_1(0) \\ x_2(0) \end{bmatrix}$$

at $t = t_f = \pi/4$ ($\sin \pi/4 = \cos \pi/4 = 1/\sqrt{2}$)

$$x(\pi/4) = \Phi(\pi/4) x(0)$$

or

$$\begin{bmatrix} x_1(\pi/4) \\ x_2(\pi/4) \end{bmatrix} = \begin{bmatrix} 1/\sqrt{2} & 1/\sqrt{2} \\ -1/\sqrt{2} & 1/\sqrt{2} \end{bmatrix} \begin{bmatrix} x_1(0) \\ x_2(0) \end{bmatrix}$$

$$\begin{bmatrix} 1 \\ x_2(0) \end{bmatrix} = \begin{bmatrix} 1/\sqrt{2} & 1/\sqrt{2} \\ -1/\sqrt{2} & 1/\sqrt{2} \end{bmatrix} \begin{bmatrix} 0 \\ x_2(\pi/4) \end{bmatrix}$$

Substituting the given boundary conditions $x_1(0) = 0$ and $x_1(\pi/4) = 1$ we have two equations in the two unknown boundary values.

Thus

$$x_2(\pi/4) = \sqrt{2}$$

and

$$x_2(0) = 1$$

The desired initial conditions to match up the solution to the given boundary conditions are

$$\begin{bmatrix} x_1(0) \\ x_2(0) \end{bmatrix} = \begin{bmatrix} 1 \\ 1 \end{bmatrix}$$

and the solution is

$$\begin{bmatrix} x_1(t) \\ x_2(t) \end{bmatrix} = \begin{bmatrix} \cos t & \sin t \\ -\sin t & \cos t \end{bmatrix} \begin{bmatrix} 1 \\ 1 \end{bmatrix}$$

$$= \begin{bmatrix} \cos t + \sin t \\ -\sin t + \cos t \end{bmatrix} \blacktriangledown$$

Generally an nth order differential equation (or n first order equations) are solved n times on either an analog or digital computer with appropriate initial conditions to obtain the fundamental matrix which is evaluated at the final time. Then the solution at the final time t_f

$$X(t_f) = \phi(t_f - t_0) X(t_0)$$

and the unknown elements of $x(t_f)$ are obtained. This procedure is equivalent to choosing a computation increment of $(t_f - t_0)$ and evaluating the discrete state transition matrix which takes $X(t_0)$ into $X(t_f)$ in one step.

As mentioned above, the boundary values may be inconsistent and a solution may not exist regardless of the number of boundary values specified (greater than one, of course). For the system in Example 6.1 the boundary conditions $y(0) = 0$ and $y(\pi) = 1$ are inconsistent since we have

$$\begin{bmatrix} 1 \\ x_2(\pi) \end{bmatrix} = \begin{bmatrix} -1 & 0 \\ 0 & -1 \end{bmatrix} \begin{bmatrix} 0 \\ x_2(0) \end{bmatrix}$$

which is clearly inconsistent since the first row corresponds to the equation $1 = 0$. The boundary conditions $y(0) = 0$ and $y(\pi) = 1$ require a sinusoid of period 2π to have values at $t = 0$ and $t = \pi$ which are clearly impossible. The solution to the differential equation $\ddot{y} + y = 0$ with the initial condition $y(0) = 0$ is

$$y(t) = A \sin t$$

which requires that $y(\pi) = 0$. Thus the given final value cannot be achieved.

SHOOTING METHODS

The trial and error method is perhaps the first and most obvious technique for the solution of TPBVP. The procedure is to guess the unknown initial conditions, obtain a solution, and check how well the final values compare with the desired final values. Based on this comparison the initial condition guesses are adjusted and another solution obtained. Hopefully the process is convergent and the final values get closer to the desired values. However, in nonlinear problems, convergence is not guaranteed and in high order equations it is difficult to determine if each of the guessed initial conditions should be increased or decreased. A good first guess greatly improves the possibility of obtaining a solution and if only one of the initial conditions is adjusted for each solution an indication of its effect on the final values becomes more evident. However, this procedure required many time consuming solutions.

The procedures for adjusting the initial conditions as a function of the result must be tailored for each problem, a technique that is particularly suitable for analog computers, small "hands-on" digital computers, or hybrid computers. The program can then be adjusted during the problem solution to improve convergence. The essence of the method is demonstrated in the following example.

EXAMPLE 6.2

Using a shooting method to solve the boundary value problem

$$\ddot{y} + \dot{y} + y = 0 \quad \left\{ \begin{array}{l} y(0) = 1 \\ y(1) = 0 \end{array} \right\} \qquad (6.4)$$

SOLUTION:

To solve the equation on either the analog or digital computer we need initial conditions for $y(0)$ and $\dot{y}(0)$. Since $y(0)$ is specified we can guess $\dot{y}(0) = 0$ and plot a solution as obtained in the previous chapters. Either analog or digital simulations give the solution which has final value $y_1(1) = 6.6$, as shown in Figure 6.2. It appears from that solution that we want $\dot{y}(0)$ to be negative to decrease $y(1)$. Trying $\dot{y}(0) = -1$ we have a new solution with final value $y_2(1) = -4.0$. These results can be used to obtain a new estimate for $\dot{y}(0)$ by linearly interpolating the results of the two previous solutions to that value of $\dot{y}(0)$ which gives $y(1) = 0$, or

$$\dot{y}_3(0) = \dot{y}_2(0) - y_2(1) \frac{\dot{y}_2(0) - \dot{y}_1(0)}{y_2(1) - y_1(1)}$$

(Verify that this is the correct expression for linear interpolation.)

To obtain an iterative scheme with which the digital computer can automatically iterate to obtain the required final value $y(1) = 0$ we take

$$\dot{y}_{n+1}(0) = \dot{y}_n(0) - y_n(1) \frac{\dot{y}_n(0) - \dot{y}_{n-1}(0)}{y_n(1) - y_{n-1}(1)} \qquad (6.5)$$

Figure 6.2 shows the solution for two initial conditions $\dot{y}(0) = 0$ and $\dot{y}(0) = -1$, and a third solution based on the initial condition iteration, Equation (6.5), that converges exactly to $y(1) = 0$, using the z-transform substitution, assuming that $y(t)$

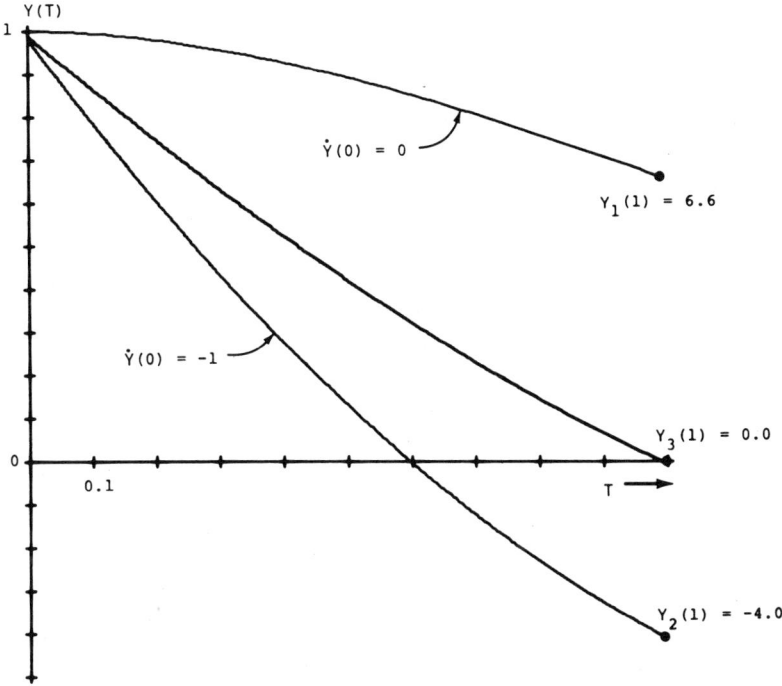

FIGURE 6.2   TRIAL AND ERROR SOLUTION OF LINEAR TPBVP BY DIGITAL SIMULATION.

is piecewise linear, and having an increment size $\tau = 0\ 10.$ ▼

It is not surprising that the linear interpolation of Equation (6.5) obtains the correct initial condition on the first evaluation since the system Equation (6.4) is linear and can be written

$$X(t) = \begin{bmatrix} y(t) \\ \dot{y}(t) \end{bmatrix} = e^{At} \begin{bmatrix} y(0) \\ \dot{y}(0) \end{bmatrix}$$

$$X(1) = \begin{bmatrix} y(1) \\ \dot{y}(1) \end{bmatrix} = e^{A} \begin{bmatrix} y(0) \\ \dot{y}(0) \end{bmatrix}$$

Thus $y(1)$ depends linearly upon $\dot{y}(0)$ and $y(0)$, so a linear interpolation is exact.

For nonlinear equations, the final value is a non-linear function of the initial conditions and the solution must iterate to the desired final value. Furthermore, for nonlinear equations it is not clear if a solution exists or if convergence is assured if one does exist. For some so-called *ILL-CONDITIONED SYSTEMS* small changes in initial conditions give large changes in response and final value, which render the simulation very sensitive to roundoff errors and essentially eliminating the possibility of hitting the desired final value. These equations can be approached by shooting solutions in reverse time (that is, by solving the equations backwards; see Problem 2.38) or by the finite difference methods of the next section.

Some of the difficulties encountered in solving nonlinear TPBVPs are demonstrated by the following example where a parameter of the equation is to be adjusted to make the equation fit the boundary conditions. †

EXAMPLE 6.3

Solve the characteristic value problem by adjusting ε in the van der Pol equation

_____

† Example 6.3 is not strictly a TPBVP but a characteristic value (eigenvalue) problem in which the characteristic value ε must if possible be evaluated to satisfy the boundary conditions.

$$\ddot{y} + \varepsilon(y^2-1)\dot{y} + y = 0$$

to meet the boundary conditions

$$y(0) = 0.0, \quad \dot{y}(0) = 0.5, \quad y(2) = 2.0$$

SOLUTION:

Using a straightforward strategy of incrementing the characteristic value ε proportional to the error in the final value, we obtain successive trajectories $y(t)$ as shown in Figure 6.3(a). The corresponding phase trajectories are shown in Figure 6.3(b). Note that although ε continues to increase the change in the final value, $y(2)$ changes less with each solution. It can be shown that the condition $y(2) = 2$ is satisfied as $\varepsilon \rightarrow \infty$. Thus the problem cannot be solved computationally. Clearly the final condition $y(2) = 1$ can be obtained, although $y(2) > 2$ cannot be obtained by adjustments in ε.▼

A similar procedure to that utilized in the example above can be employed for the iterative solution of boundary-value problems on the analog computer by using mode control switching and setting the unknown initial conditions as a function of the obtained final values.

Figure 6.4 shows a block diagram for the analog solution of Example 6.2 which turns out to be the same as that obtained by digital simulation, as shown in Figure 6.2.

Since it appears that when we decrease the initial value $\dot{y}(0)$ it decreases the final value $y(1)$ (in the range we have seen, anyway), we can feed back $-y(1)$ to the initial condition $\dot{y}(0)$ to further decrease it to the desired value. However, when we RESET integrator 1, the new value for $\dot{y}(0)$ must be maintained. This can be accomplished by accumulating $-y(1)$ in integrator 3 as shown in Figure 6.4.

While the differential equation solver (integrators 1 and 2) is in COMPUTE, the accumulator (integrator 3) is in HOLD, maintaining the initial value of zero.

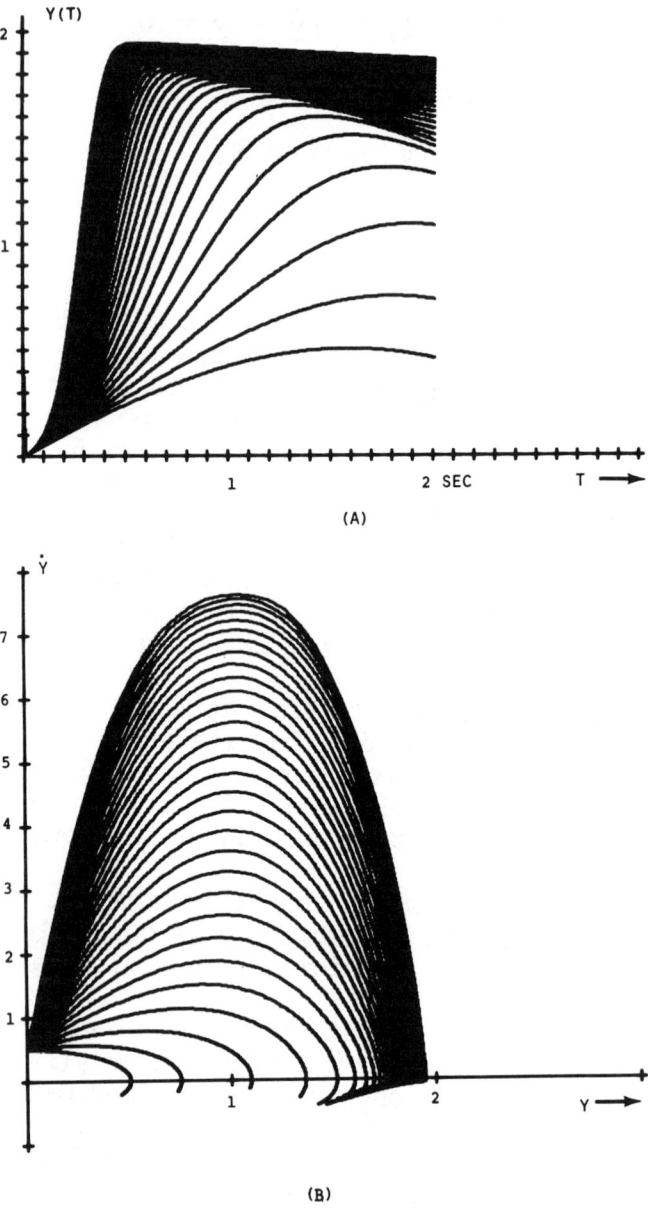

(A)

(B)

FIGURE 6.3   SOLUTION OF A NONLINEAR CHARACTERISTIC VALUE PROBLEM.

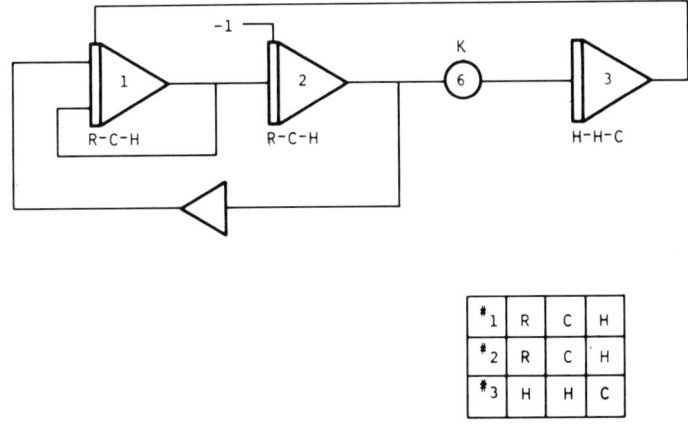

FIGURE 6.4   ADJUSTING $\dot{Y}_{(0)}$ AS A FUNCTION OF $Y_{(1)}$ FROM THE PREVIOUS RUN
TO SOLVE THE TPBVP OF EXAMPLE 6.2.

After one solution integrators 1 and 2 go to HOLD, maintaining the error, and integrator 3 to COMPUTE, integrating the error, to change the value to that at which integrator 1 resets. The amount that the initial condition is changed as a function of the final value depends on the gain k of POT 6 and the time that integrator 3 is in COMPUTE. Generally if k is small the solution converges slowly and k can be increased to speed up convergence. However, if the gain is too great the solutions diverge and become successively worse on opposite sides of the desired final value. If the solution diverges for even small k then the initial condition is perhaps being adjusted incorrectly as a function of the output. If this happens, convergence may be obtained by patching an inverter after integrator 3 or taking the input for integrator 3 as -y rather than y. Generally, if the solution does not converge, try something else.

Figure 6.5 shows the solutions obtained from the syseem in Figure 6.4 for the mode control shown. Note

that the error is integrated to drive the unknown
initial condition ẏ(0) to that value which drives the
error to zero. The plot of y(t) in Figure 6.5 shows
the response during all modes. If the plotter were
reset with the pen up when integrators 1 and 2 are in
RESET, then the plots would be in the format of
Figures 6.2 and 6.3.

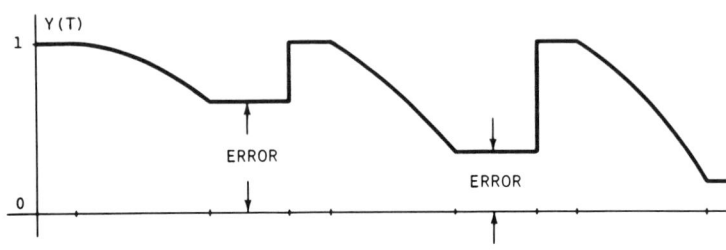

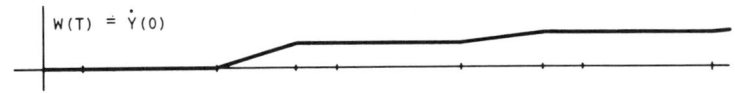

$^\#1_\&\,^\#2$	R	C		H	R	C		H	R	C
$^\#3$	H	H		C	H	H		C	H	H

FIGURE 6.5 ANALOG COMPUTER SOLUTION OF THE TPBVP IN EXAMPLE 6.2.

If three mode operation is not available in REPOP
from digital logic using counters and timers, the

accumulator integrator can be preceded by a track-store element as presented in Figure 6.6(a). While the system (integrators 1 and 2) is in COMPUTE, the tracking integrator (No. 3) is in RESET tracking on the solution $-y(t)$. When the final time t. of the TPBVP is reached and the output of the tracking integrator is the error, in the most recent solution. At this time integrator 3 goes to HOLD maintaining the error while the accumulator integrator (No. 4)

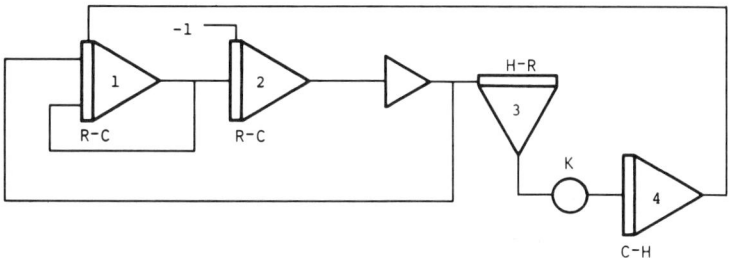

(A)

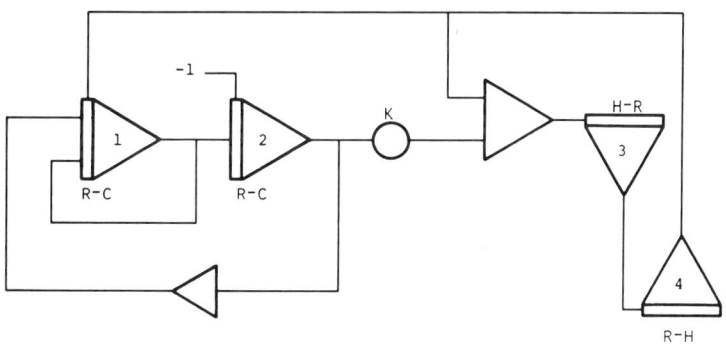

(B)

FIGURE 6.6 ALTERNATIVE TWO MODE SOLUTION OF A TPBVP.

integrates the error to change the initial condition
$\dot{y}(0)$. During this time the system is in RESET so
integrator 1 tracks on this corrected value of $\dot{y}(0)$.

The interval of time that integrator 4 is in
compute effects the rate of correction. A faster
operation is obtained with the track-store pair wired
up as shown in Figure 6.6(b). It is left as an
exercise to describe the operation of this circuit,
including sketches of the initial condition $\dot{y}(0)$ and
the response $y(t)$ similar to Figure 6.5. (See Problem
6.7.) On some analog computers the initial condition
jacks can be used for summing initial conditions, in
which case the summer in Figure 6.6(b) can be omitted.

MULTIPARAMETER PROBLEMS

In the above discussion we have considered
adjusting a single parameter or initial value to
achieve a desired final result. TPBVPs that require
the determination of several parameters or initial
conditions are more difficult. If each of the
parameters effects only one of the end conditions,
then the problem reduces to several single parameter
problems which can be solved successively. However,
usually all the parameters effect all the errors.
Actually, it is unknown how each effects each error,
so rather sophisticated search schemes must be
utilized to insure that the errors go to zero as the
solution proceeds. One procedure consists of
continuously determining the effect of each parameter
on the total error (some function of the final values
and the trajectory of the solution) in terms of the
maximum rate of minimizing the error. Then this
gradient vector is used to correct the parameters for
another run. These procedures are discussed in some
detail in the next chapter. The following example
gives the solution to a two parameter problem with a
reasonable well defined strategy for adjusting the
parameters. If the parameter adjustment involves more
than the most elementary algebraic calculations the
problem must be solved on a hybrid computer if it is
to be done efficiently. The analog computer simulates
the system while the digital computer performs the
optimization and parameter adjustments.

EXAMPLE 6.4

Find the periodic solutions with period $T = 2\pi$ of the equation

$$\ddot{y} + \dot{y} + y^3 = \cos t \qquad (6.6)$$

SOLUTION:

It is desired to find initial conditions $y(0) = a$ and $\dot{y}(0) = b$ such that

$$y(0) = y(T) = a$$

and

$$\dot{y}(0) = \dot{y}(T) = b$$

A reasonable strategy for adjusting $y(0)$ and $\dot{y}(0)$ is to set the new values of the initial condition as the average of the initial and final values for the previous run. That is

$$y(0)_{new} = \frac{1}{2}[y(0) + y(T)] \qquad \text{last run}$$

and

$$\dot{y}(0)_{new} = \frac{1}{2}[\dot{y}(0) + \dot{y}(T)] \qquad \text{last run}$$

Figure 6.7(a) shows plots in parameter space of the values of $y(0)$ and $\dot{y}(0)$ for successive runs using the above strategy for adjustment after each run, starting from various initial estimates a and b. The parameter space plots are constructed from the linear interpolation between successive values of the iterative solution of the boundary vale problem. Figures 6.7(b) and (c) show a particular iterative sequence of the approximate solutions to Equation (6.6) in the phase plane and time domain respectively.

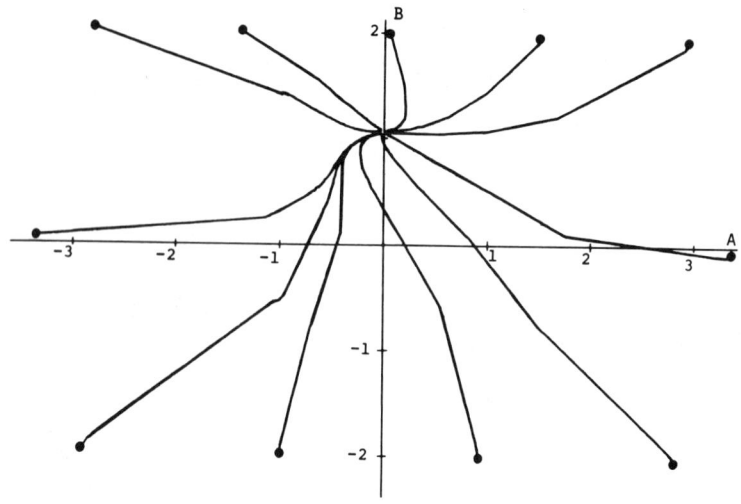

(A) TRAJECTORIES OF INITIAL CONDITIONS

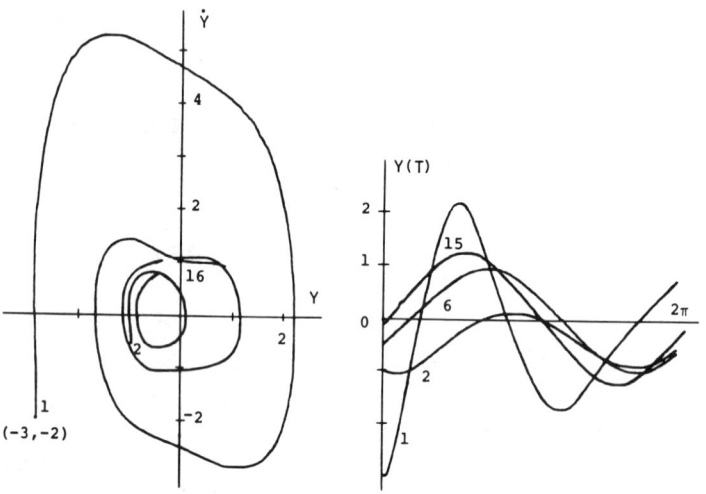

(B) PHASE TRAJECTORIES FOR
    ITERATIONS 1, 2, AND 16
    STARTING FROM
    $(Y(0), \dot{Y}(0)) = (-3,-2)$.

(C) TIME RESPONSE FOR ITERATIONS
    1, 2, 6, AND 15 STARTING FROM
    $(Y(0), \dot{Y}(0)) = (-3,-2)$.

FIGURE 6.7   EXAMPLE 6.4 SEARCH FOR PERIODIC SOLUTIONS OF THE EQUATION
$\ddot{Y} + \dot{Y} + Y^3 = \cos \tau$ FROM VARIOUS STARTING VALUES.

The mode control shown in Figure 6.8 accomplishes the desired initial condition adjustment on $y(0)$.

$$y(0)_{new} = (y(0)_{old} - y(T))/2$$

A similar circuit is used for adjustment of $\dot{y}(0)$. ▼

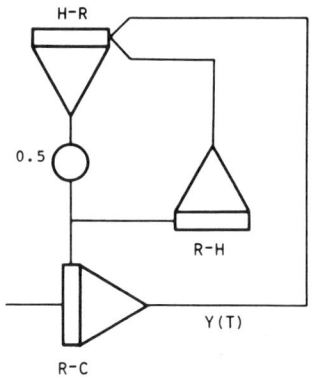

FIGURE 6.8   ITERATIVE SEARCH FOR THE PERIODIC SOLUTIONS IN EXAMPLE 6.4.

FINITE-DIFFERENCE METHOD

Boundary value problems can be solved by finite-difference methods to obtain a set of algebraic equations at discrete values of the independent variable. The resultant algebraic equations are solved by the methods discussed in the previous chapter to obtain the values of the dependent variable (the solution). This method usually works best for linear differential equations because they yield linear algebraic equations which are particularly suitable for iterative solution. The following example demonstrates the method.

EXAMPLE 6.5

Solve the boundary value problem

$$\ddot{y} + \dot{y} + y = 0 \qquad \left\{ \begin{array}{l} y(0) = 1 \\ y(1) = 0 \end{array} \right\} \qquad (6.7)$$

by a finite difference method.

SOLUTION:

First divide the interval of solution [0,1] into equal increments, and write an approximating difference equation at the discrete points by taking

$$y(t_n) = y_n$$

$$\dot{y}(t_n) \approx \frac{y_{n+1} - y_{n-1}}{2\Delta t}$$

and

$$\ddot{y}(t_n) \approx \frac{\dot{y}(t_n + \frac{\Delta t}{2}) - \dot{y}(t_n - \frac{\Delta t}{2})}{\Delta t}$$

$$\approx \frac{(y_{n+1} - y_n)/\Delta t - (y_n - y_{n-1})/\Delta t}{\Delta t}$$

$$\approx \frac{y_{n+1} - 2y_n + y_{n-1}}{\Delta t^2}$$

The differential equation $\ddot{y} + \dot{y} + y = 0$ can be approximated at each point by the difference equation

$$\frac{y_{n+1} - 2y_n + y_{n-1}}{\Delta t^2} + \frac{y_{n+1} - y_{n-1}}{2\Delta t} + y_n = 0$$

or

$$(2-\Delta t)y_{n-1} - (4-2\Delta t^2)y_n + (2+\Delta t)y_{n+1} = 0$$

$$(6.8)$$

If the independent variable is divided into 10 equal increments $\Delta t = 0.1$ we have the 9 equations in 9 unknown

$$1.9y_0 - 3.98y_1 + 2.1y_2 = 0$$

$$1.9y_1 - 3.98y_2 + 2.1y_3 = 0$$

$$\cdot$$
$$\cdot$$
$$\cdot$$

$$1.9y_8 - 3.98y_9 + 2.1y_{10} = 0$$

where $y_0 = 1$ and $y_{10} = 0$ by the boundary conditions. In matrix notation we have

$$\begin{bmatrix} -3.98 & 2.1 & 0 & 0 & \cdots \\ 1.9 & -3.98 & 2.1 & 0 & \cdots \\ 0 & 1.9 & -3.98 & 2.1 & \cdots \\ & & & & \end{bmatrix} \begin{bmatrix} y_1 \\ y_2 \\ y_3 \\ \cdot \\ \cdot \\ \cdot \\ y_9 \end{bmatrix} = \begin{bmatrix} -1.9 \\ 0 \\ 0 \\ \cdot \\ \cdot \\ \cdot \\ 0 \end{bmatrix}$$

$$(6.9)$$

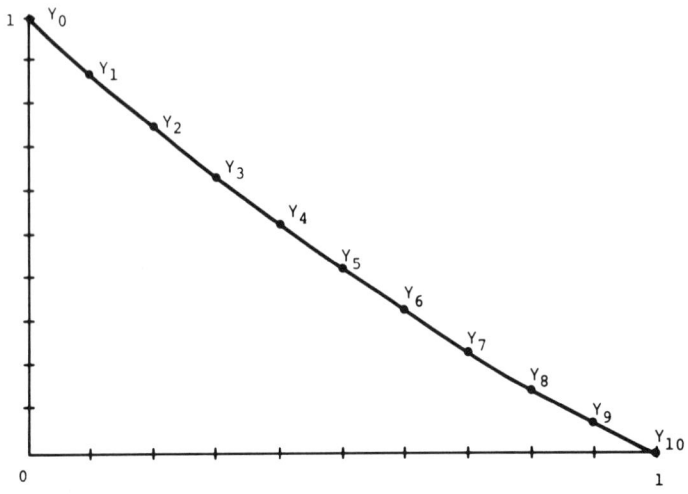

FIGURE 6.9   EXAMPLE 6.5 FINITE DIFFERENCE SOLUTION OF TPBVP.

which is a system of linear algebraic equations in a form convenient for solution by successive displacement. Solution of these equations gives the approximate solution to the differential equation shown in Figure 6.9. ▼

The system matrix in Equation (5.6) contains many zero elements, and is said to be a *SPARCE MATRIX* in contrast to a *DENSE MATRIX* which has mostly non-zero elements. Sparce matrices generally result when differential equations are approximated by a finite difference methods since each of the unknowns is coupled to only a few of the other unknowns. In particular as in the above example the relationship between adjacent variables is the same for all unknowns and the matrix itself need not be stored. It is only necessary to store the recursion relation in general form. (For instance, as in Equation (6.8)). A matrix of this form is called a *BAND MATRIX*), because the only non-zero elements of the matrix are in a band about the main diagonal. The finite difference equations are not only band matrices but the elements along each diagonal are equal. Thus, their values need only be stored once, and correspond to the coefficients of the recursion relation.

The procedure for programming the finite difference equations is to set the boundary conditions and then range through the variables, adjusting them so as to be consistent with Equation (6.8) and repeating the process until convergence is obtained. This procedure is equivalent to solving Equation (6.9) by successive displacement. Special procedures for Gaussian elimination are particularly efficient for the solution of systems characterized by band matrices (see, for example, Forsythe and Moler[4]).

6.2 BOUNDARY VALUE PROBLEMS IN PARTIAL DIFFERENTIAL EQUATIONS

The finite-difference method discussed in the last section can be extended to the solution of partial differential equations, such as *LAPLACE'S EQUATION*,

$$\nabla^2 v = \frac{\partial^2 v}{\partial x^2} + \frac{\partial^2 v}{\partial y^2} = 0 \qquad (6.10)$$

POISSON'S EQUATION,

$$\nabla^2 v = \frac{\partial^2 v}{\partial x^2} + \frac{\partial^2 v}{\partial y^2} = f(x,y) \qquad (6.11)$$

the *HEAT FLOW EQUATION*,

$$\nabla^2 v = K \frac{\partial v}{\partial t} \qquad (6.12)$$

and the *WAVE EQUATION*

$$\nabla^2 v = \alpha^2 \frac{\partial^2 v}{\partial t^2} \qquad (6.13)$$

To solve partial differential equations by finite-differences, the derivatives are replaced by finite difference approximations as in the last section. The region of interest for the solution is divided into a mesh or grid over which the approximate solution is obtained, as shown in Figure 6.10.

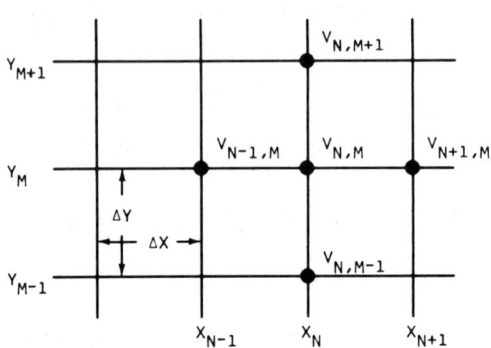

FIGURE 6.10   GRID FOR THE FINITE DIFFERENCE SOLUTION OF PARTIAL DIFFERENTIAL EQUATIONS BY FINITE DIFFERENCES.

Over the grid we can make the central difference approximation

$$\left.\frac{\partial v}{\partial x}\right|_{x_n, Y_m} \approx \frac{v_{n+1,m} - v_{n-1,m}}{2\Delta x}$$

In a similar fashion

$$\left.\frac{\partial v}{\partial x}\right|_{x_n, Y_m} \approx \frac{v_{n,m+1} - v_{n,m-1}}{2\Delta y}$$

The second derivatives can be approximated by the second difference as in Example 6.5.

$$\left.\frac{\partial^2 v}{\partial x^2}\right|_{x_n, y_m} \approx \frac{v_{n+1,m} - 2v_{n,m} + v_{n-1,m}}{\Delta x^2}$$

$$\left.\frac{\partial^2 v}{\partial y^2}\right|_{x_n, y_m} \approx \frac{v_{n,m+1} - 2v_{n,m} + v_{n,m-1}}{\Delta y^2}$$

To solve Laplace's equation we merely take the above approximations and substitute them into Equation (6.10).

EXAMPLE 6.6

Give a finite difference approximation for Laplace's equation in two dimensions assuming a grid which is equally spaced in both dimensions (that is, $\Delta x = \Delta y$).

SOLUTION:

Substitution of the finite difference approxmations into Equation (6.10) gives, at the grid point (x_n, y_m)

$$\frac{\partial^2 v}{\partial x^2} + \frac{\partial^2 v}{\partial y^2} = 0$$

$$\frac{v_{n+1,m} - 2v_{n,m} + v_{n-1,m}}{\Delta x^2} + \frac{v_{n,m+1} - 2v_{n,m} + v_{n,m-1}}{\Delta y^2} = 0$$

If $\Delta x = \Delta y$ we have

$$v_{n+1,m} + v_{n-1,m} + v_{n,m+1} + v_{n,m-1} = 4v_{n,m}$$

which is equivalent to stating that the solution (potential) at each point is equal to the average value of the potential at the four adjacent points. Thus we can set up the iterative procedure

$$v_{n,m} = \frac{1}{4}(v_{n+1,m} + v_{n-1,m} + v_{n,m+1} + v_{n,m-1})$$

$$(6.14)$$

and range over the grid until no point changes by more than the prescribed error. Figure 6.11 shows a region over which we might desire the solution and the corresponding selection of grid points. The boundary points are set at the specified boundary values and the solution is obtained by successive application of Equation (6.14) at each of the interior points. This procedure corresponds to the solution of the simultaneous algebraic finite difference equations.▼

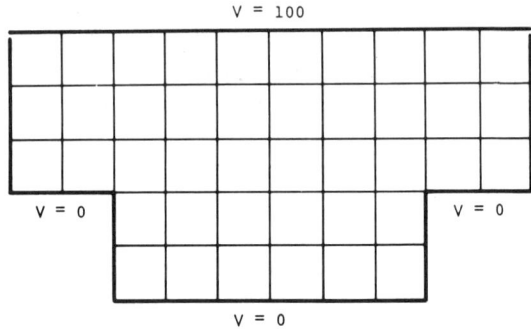

V = 100

V = 0   V = 0

V = 0

FIGURE 6.11   GRID FOR THE FINITE DIFFERENCE SOLUTION OF
LAPLACE'S EQUATION WITH A PRESCRIBED BOUNDARY.

A difficulty occurs in the solution of partial differential equations over unbounded regions. For example, the wave and heat-flow equations have time for one of the independent variables, so we must obtain a sequence of solutions corresponding to each time increment. But since time continues independently we lack boundary conditions on one side of the region, as shown in Figure 6.12.

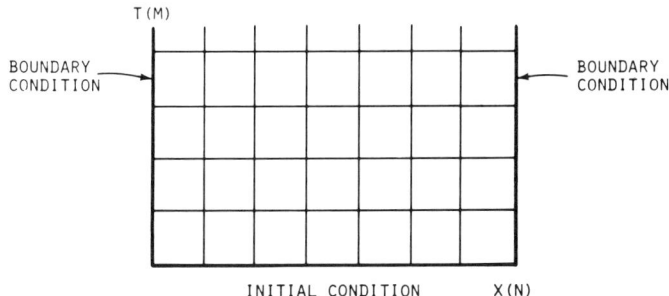

FIGURE 6.12 PARTIAL DIFFERENTIAL EQUATIONS IN AN UNBOUNDED REGION.

Consider the heat flow equation in one dimension

$$\frac{\partial^2 v}{\partial x^2} = K \frac{\partial v}{\partial t} \qquad (6.12)$$

The space derivative can be approximated by the central difference

$$\frac{\partial^2 v}{\partial x^2} \approx \frac{v_{n-1,m} - 2v_{n,m} + v_{n+1,m}}{\Delta x^2}$$

and the time derivative can be approximated by either the forward difference

$$\frac{\partial v}{\partial t} \approx \frac{v_{n,m+1} - v_{n,m}}{\Delta t}$$

or the back difference

$$\frac{\partial v}{\partial t} \approx \frac{v_{n,m} - v_{n,m-1}}{\Delta t}$$

which can be substituted into Equation (6.15) to obeain the finite difference equations

$$v_{n,m+1} = v_{n,m} + \frac{\Delta t}{K \Delta x^2} v_{n-1,m} - 2v_{n,m} + v_{n+1,m}$$

(6.15)

and

$$v_{n,m} = \frac{1}{2 + \frac{\Delta x^2}{\Delta t} K} v_{n-1,m} + v_{n+1,m} + \frac{\Delta x^2}{\Delta t} K v_{n,m-1}$$

(6.16)

respectively. The n and m subscripts correspond to space and time increments respectively.

If we approximate Equation (6.12), by using the forward difference, we obtain an explicit formula for $v_{n,m+1}$ based on the solution at grid points corresponding to the past time increment. Thus there is no iterative solution involved. Unless the time increment is very small (less than $\Delta x^2/2K$, in fact) Equation (6.16) is unstable and particularly sensitive to round-off errors.

Equation (6.17) is an implicit formula involving all the grid points in the time increment under consideration and has more favorable convergence properties. Usually, the larger the coefficient of the unknown element in the solution of equations by finite difference methods, is the better the stability properties. This is consistent with the requirement to maximize the magnitude of the diagonal elements since this is merely the Seidel algorithm of successive displacements.

Finite difference methods applied to all space and time variables are called method of *DISCRETE-SPACE-DISCRETE-TIME* or merely DSDT. For an exhaustive treatment of DSDT methods for the solution of boundary value problems in partial differential equations, see Forsythe and Wassow[5]).

EXAMPLE 6.7

Give a flow chart for a DSDT solution to the one dimensional heat flow Equation (6.15) for $0 \le x \le 1$ with $K = 1$, initial condition $v(x,0) = 100°$ and boundary conditions $v(0,t) = v(1,t) = 100°$.

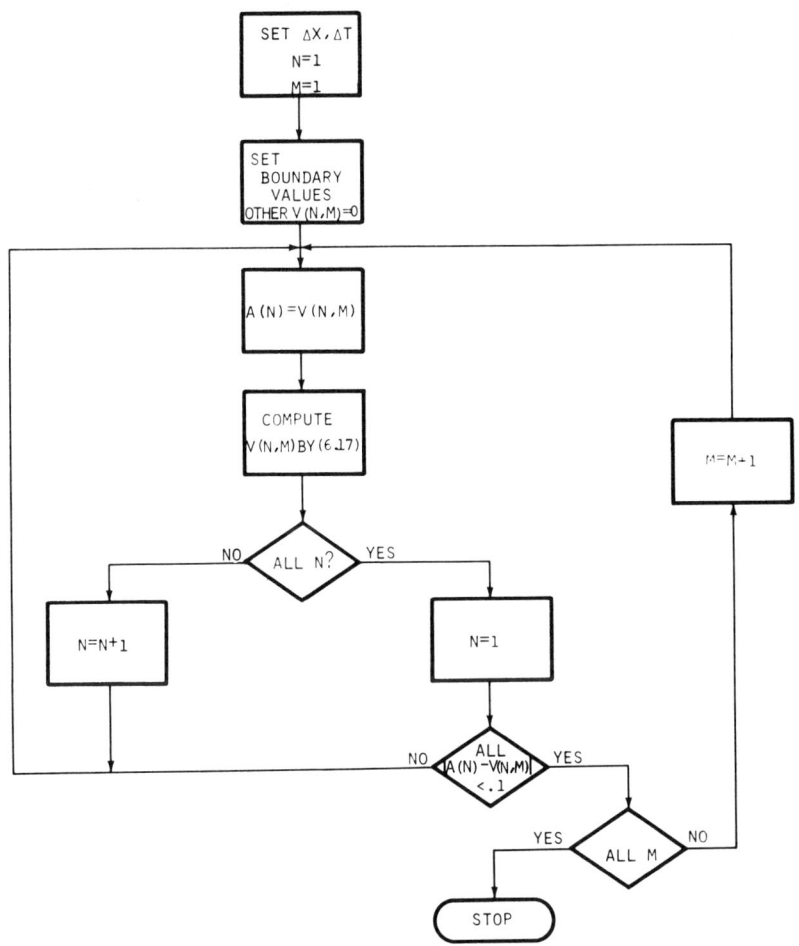

FIGURE 6.13 EXAMPLE 6.7 FLOW CHART FOR THE DSDT SOLUTION
OF THE HEAT FLOW EQUATION.

SOLUTION:

We choose $\Delta x = -0.05$ for 20 increments in x and $\Delta t = 10^{-4}$ (to insure $\Delta x^2 > 2K\Delta t$) and iterate each row of constant time t by Equation (6.17) until no value changes by more than $0.1°$. A flow chart for the program is given in Figure 6.13, on the previous page. Plots of the solution to this example are given in the following examples. The implementation of this program is left as an exercise. (See Problem 6.23.)▼

DISCRETE-SPACE-CONTINUOUS-TIME (DSCT)

An analog computer solution to the boundary value problems in partial differential equations can be obtained by replacing the derivatives in the space variables by finite differences and integrating the time derivatives with analog computer integrators. For example, the heat flow equation in one dimension

$$\frac{\partial^2 v}{\partial x^2} = K \frac{\partial v}{\partial t} \qquad (6.12)$$

is approximated by taking an increment size Δx in the space variable to obtain

$$\frac{v_{x-\Delta x} - 2v_x + v_{x+\Delta x}}{\Delta x^2} \approx K \frac{dv_x}{dt}$$

thus

$$\frac{dv_x}{dt} \approx \left(\frac{\Delta x^2}{K}\right) v_{x-\Delta x} - 2v_x + v_{x+\Delta x} \qquad (6.17)$$

Equation (6.17) can be implemented on the analog computer by assuming the variable $v_{n\Delta x}$ is known for each n as the output of an integrator. Then the inputs to the integrators are given by Equation (6.17). Figure 6.14 is a block diagram for the discrete space points $x - \Delta x$, x, and $x + \Delta x$ and the corresponding values of $v_{x-\Delta x}$, v_x, and $v_{x+\Delta x}$. The block diagram can be timed scaled by $K/\Delta x^2$ to remove the POTs. If the solution is acceptable with alternating signs on successive

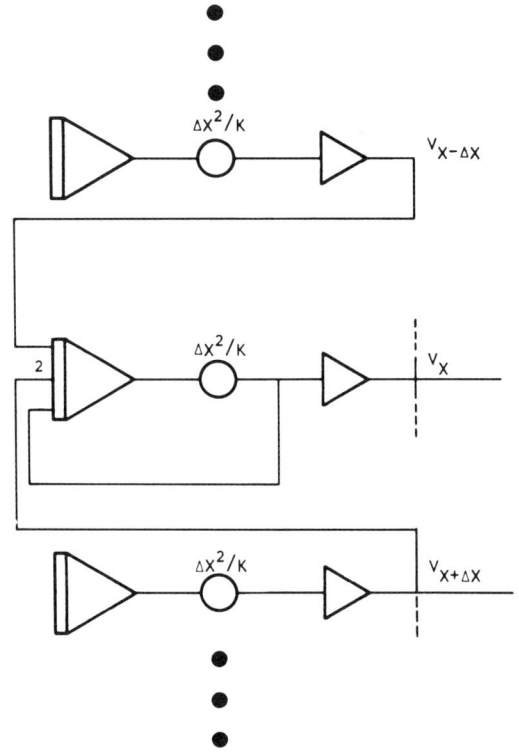

FIGURE 6.14 DSCT SOLUTION OF THE HEAT FLOW EQUATION
ON AN ANALOG COMPUTER.

variables, the inverters can be eliminated to obtain
the block diagram in Figure 6.15. The initial
condition placed on each integrator is the
corresponding value of the initial value of $v(x,0)$.
The remaining boundary conditions are patched in as
$v_{x+\Delta x} = v_0$ on the first integrator and $v_{x+\Delta x} = v_N$ on
the last integrator with appropriate sign if there are
an odd number of integrators. The DSCT solution is
obtained from a single run of the simulation for as
long as a solution is desired. Increased accuracy is
obtained by a finer division of the space variable
(that is, decrease Δx) resulting in additional
discrete values and consequently additional analog

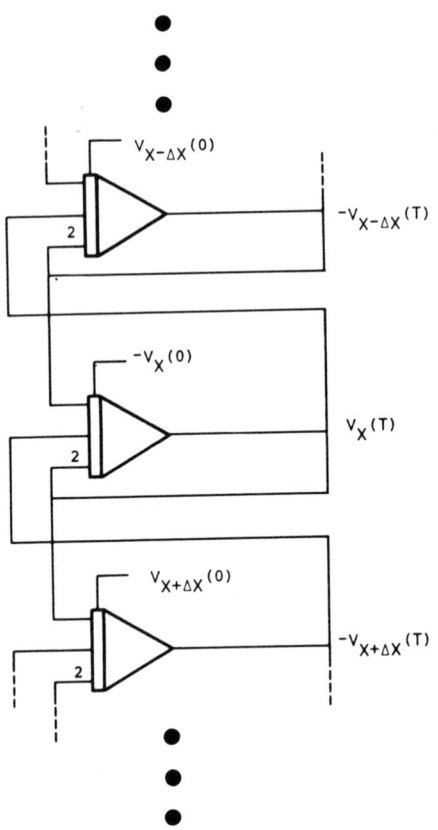

FIGURE 6.15 DSCT SOLUTION OF THE HEAT FLOW EQUATION.

equipment. Thus the accuracy of the approximation depends to a great extent on the number of integrators used for the simulation. With a large number of integrators, the accuracy decreases because of increased numerical difficulties so that, in practice, there is an optimum number of integrators to be used for the best tradeoff between truncation errors and computation errors.

EXAMPLE 6.8

Give an analog computer solution for the DSCT solution of the one dimensional wave equation for a vibrating string

$$\frac{\partial^2 y}{\partial t^2} = a^2 \frac{\partial^2 y}{\partial x^2} \qquad (6.13)$$

with the triangular initial conditions shown in Figure 6.16 and the boundary conditions $y(0,t) = y(1,t) = 0$.

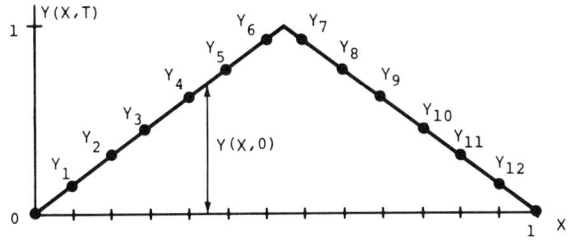

FIGURE 6.16 EXAMPLE 6.8 INITIAL DISPLACEMENT FOR VIBRATING STRING.

SOLUTION:

Taking a discrete approximation for the space variable we have

$$\ddot{y}_n = \left(\frac{a}{\Delta x}\right)^2 (y_{n-1} - 2y_n + y_{n+1}) \qquad (6.18)$$

where $y_n = y(n\Delta x)$. The block diagram for the position $y(x,t)$ at one station $n\Delta x$ corresponding to Equation (6.18) is given in Figure 6.17 where the gain $(a/\Delta x)^2$ has been time scaled out.

If we choose 12 intermediate stations with 13 intervals, $y_0 = y_{13} = 0$ gives the boundary conditions. Since the initial conditions are symmetrical the solution will be symmetrical, requiring only half the

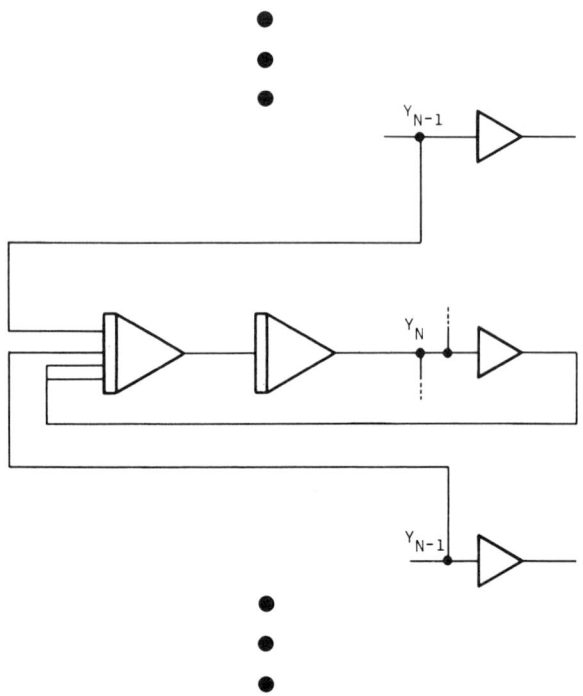

FIGURE 6.17   EXAMPLE 6.8 BLOCK DIAGRAM FOR THE DSCT SOLUTION OF
THE ONE DIMENSIONAL WAVE EQUATION.

components.   Since $y_7 = y_6$, $y_8 = y_5$, and so on, the input of y_7 to y_6 reduces the scaled Equation (6.18) at station 6 to

$$\ddot{y}_6 = (y_5 - 2y_6 + y_7)$$

$$= (y_5 - 2y_6 + y_6)$$

$$= y_5 - y_6$$

as shown in the last set of integrators in Figure 6.17 for y_{n+1}.

Using the above program with the initial conditions

set at $y_n = n/6.5$,†   the simulation was run   and   the position of the vibrating string at each station plotted as a function of time. Figure 6.18 shows the time response of each y_n. ▼

Partial differential equations in the DSCT format

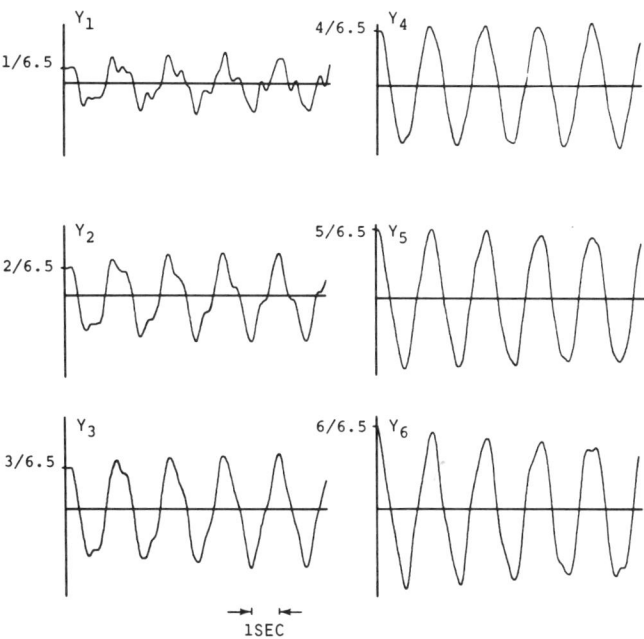

FIGURE 6.18 EXAMPLE 6.8 DSCT SOLUTION OF WAVE OPERATION.

can be solved by the numerical integration of the differential equations obtained by discretizing the space variable. The analog integration is merely replaced by a predictor-corrector method or Runge-Kutta integration, usually with variable step size. The digital simulation languages are well suited for this simula-

---

† Why 6.5, instead of some other number, say 6? Would 10 have been satisfactory? Would the solution have been different? What is the physical meaning of the initial condition $\dot{y}_n = 0$ as shown in Figure 6.17?

tion because the system equations are the same on each variable, so the system description can be accomplished in a loop of the program. (See Problem 6.25.) Also, since the continuous equations are linear the "exact" solution to the DSCT equations can be calculated using the discrete state transmission matrix. (See Problem 6.30.)

CONTINUOUS-SPACE-DISCRETE-TIME
    In the DSCT solution of partial differential equations, the equation is replaced by a set of ordinary differential equations that are integrated in parallel to obtain the solution. The *CONTINUOUS-SPACE-DISCRETE-TIME* CSDT method is similar in approach, but the time variable is discretized, giving a set of ordinary differential equations in the space variable. The resulting equations are then solved serially, stepping through time in the unbounded region as discussed relative to Figure 6.12.

Consider the heat flow equation

$$\frac{\partial^2 v}{\partial x^2} = k \frac{\partial v}{\partial t} \qquad (6.12)$$

Approximating the time derivative by the back difference gives

$$\frac{\partial^2 v_n}{\partial x^2} = \frac{k}{\Delta t}\left[v_n(x) - v_{n-1}(x)\right] \qquad (6.19)$$

where $v_n(x) = v(x,n\Delta t)$. This equation can be integrated over the space interval from $x = 0$ to L. However, Equation (6.19) is not an initial value problem but a TPBVP since the conditions are on $v_n(0) = v(0,n\Delta t)$ and $v_n(L) = v(L,n\Delta t)$. This TPBVP can be solved by shooting solutions to obtain iteratively a best estimate for $v_n(0)$ and a convergent solution before stepping on to the next time increment. The solution of Equation (6.19) requires the storage and playback of the solution for the last time increment

$v_{n-1}(x)$, necessitating memory.

Analog computers are impractical at best for the solution of CSDT systems because of the need of a memory for function playback. Digital computers can solve these problems with numerical integration schemes but the CSDT solution of partial differential equations is most efficient on the hybrid computer. The analog computer runs the solution to the TPBVP in Equation (6.19) while the digital computer controls the operation and performs the function storage and playback. For these problems the hybrid computer is 10 to 100 times faster than comparable digital computers. Thus the per-run cost on the hybrid computer is considerably lower, especially if many runs are to be performed. Vitchnevetsky † has presented a comparison of several methods used for the hybrid computer solution of partial differential equations. We conclude this chapter with an example of the CSDT solution of the wave equation.

EXAMPLE 6.9

Give a CSDT simulation of the one dimensional wave equation

$$\frac{\partial^2 y}{\partial x^2} = a^2 \frac{\partial^2 y}{\partial t^2} \qquad (6.13)$$

subject to the boundary conditions, $y(0,t) = y(1,t) = 0$ and the initial displacement $y(x,0)$ as given in Example 6.8, Figure 6.16

SOLUTION:

Discretizing the time variable we have

---

† See R. Vichnevetsky, "State of the Art in Hybrid Methods for Partial Differential Equations," *Proceedings of the AICA-IFIP International Conference on Hybrid Computation*, Munich, Germany, September 1970, and the extensive references in that paper.

$$\frac{\partial^2 y_n}{\partial x_n^2} = \left(\frac{a}{\Delta t}\right)^2 \left[y_n - 2y_{n-1} + y_{n-2}\right] \quad (6.20)$$

The hybrid computer block diagram for Equation (6.20) is shown in Figure 6.19 where the solution has been

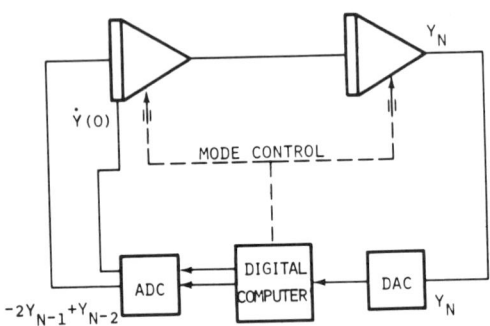

FIGURE 6.19 BLOCK DIAGRAM FOR THE HYBRID COMPUTER SOLUTION OF THE WAVE EQUATION.

time scaled to remove the $(a/\Delta t)^2$ factor so as to have a single space solution run in 2 seconds for convenient XY plots of the solution. Of course the simulation can be run considerably faster (say in 10 – 100 ms per run), the solution stored, and the results plotted off line.

The program is essentially the same as the program used in Example 5.7 to model the human pupil servomechanism. The digital computer controls the mode of the system operation and stores the solutions, playing back not a delayed output but the function $-2y_{n-1} + y_{n-2}$ which is easily calculated from stored values. A program that requires less storage is obtained if the functions are discarded when they are no longer needed. The following program is an abbreviated version of the actual program and uses the machine independent subroutines discussed in the last chapter.

```
C
C     DELTA T=.1    DELTA X=.01
C     SET INITIAL VALUES FOR Y(X,0) AND Y(X,DELTAT)
C
      DO 2 K=1,100
      IY(1,K)=0
    2 IY(2,K)=0
C
C     GUESS THE INITIAL SLOPE.
C
      IS=0
C
C     INCREMENT TIME FROM 0 TO 5 SECONDS.
C
      DO 5 LT=3,52
C
C     SOLVE THE T.P.B.V.P.
C
    6 CALL RESET
C
C     ALLOW 100 MILLISECONDS TO RESET.
C
      CALL DELAY(10000)
      CALL COMPUT
C
C     READ OUT FUNCTION (-2Y(N-1)+Y9 (N-2))
C
      DO 20 LX=3,102
      KY=-2*IY(LT,LX-1)+IY(LT,LX-2)
      CALL DAC(1,KY)
C
C     TAKE SAMPLES OF Y(N)
C
      CALL ADC(1,IY(LT,LX))
      IY(LT,LX)=(16383/2047)*IY(LT,LX)
C
C     DELAY 20 MILLISECONDS FOR NEXT SAMPLE.
C
      CALL DELAY(2000)
   20 CONTINUE
      CALL HOLD
C
C     CHECK FOR ERROR IN END VALUE TO 0.5%
C
      IF(10.GT.IABS(IY(LT,102)) GO TO 5
C
C     CHANGE THE SLOPE ESTIMATE IN TPBVP
C
      IS=IS+.1*IY(LT,102)
      GO TO 6
    5 CONTINUE
```

If the error in the solution to the TPBVP is greater than 10 (or 0.5 percent of full scale 2047) the initial estimate of the slope dy_n/dx is changed by 10 percent of the error. When the TPBVP is solved (that is, the end conditions satisfied), the time increment is stepped.

Figure 6.20 gives plots of the displacement y(x) for several values of time for the scaled problem. ▼

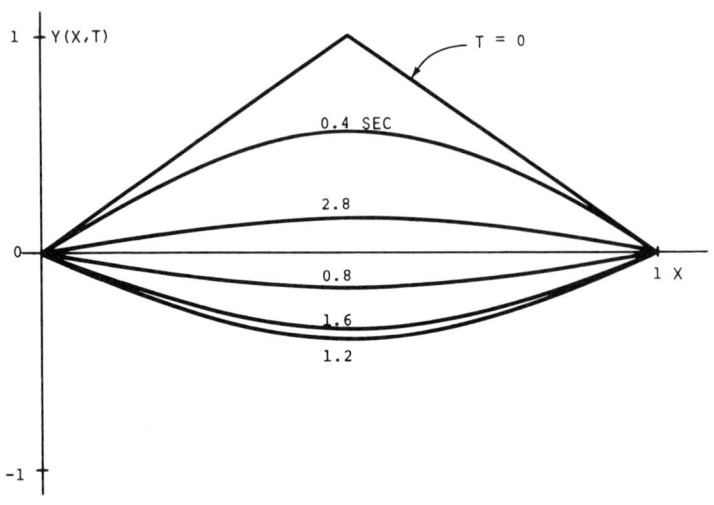

FIGURE 6.20 EXAMPLE 6.9 CSDT SOLUTION OF THE WAVE EQUATION.

If the initial conditions are adjusted to correspond to an initial displacement, as shown in Figure 6.21, the harmonic content of the solution

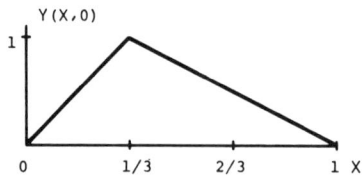

FIGURE 6.21 ANOTHER INITIAL CONDITION ON THE WAVE EQUATION.

becomes apparent.   There are several ways to present the wave motion in both directions.   Perspective and isometric plots are popular.   (See Problem 6.22.)

The CSDT method suffers from the propagation of errors in successive steps.   In particular, for second order equations the error propagation is unstable and errors grow to mask the solution eventually.   One method of solution is to decompose the problem into two first order equations which can be integrated in opposite directions, rendering them stable.   The decomposition technique is of particular interest for the solution of higher order partial differential equations on the hybrid computer. †

PROBLEMS

6.1 Use   the   discrete   state   transition matrix $\Phi(5) = e^{5A}$ to solve the boundary value problem $\dot{A} = AX$ where

$$X(0) = \begin{bmatrix} a \\ 1 \end{bmatrix} \qquad X(5) = \begin{bmatrix} b \\ 0 \end{bmatrix}$$

and A is given by the following matrix.

(a) $\begin{bmatrix} 0 & 1 \\ -5 & -6 \end{bmatrix}$ (b) $\begin{bmatrix} 0 & 5 \\ -5 & -1 \end{bmatrix}$

[Give the initial value a, final value b and plot the solution.]

---

† See Vichnevetsky, R., "The Method of Decomposition for Space Dependent Boundary Value Problems", *Proceedings of the Fourth Annual Princeton Conference on Information Sciences and Systems,* March 1970.

6.2 Use an analog computer to solve the boundary value problem

$$\dddot{y} + 3\ddot{y} + 102\dot{y} + 100y = 0$$

$$y(0) = 1$$

$$\dot{y}(0) = 0$$

$$y(2) = 0$$

Evaluate the initial condition $\ddot{y}(0)$ and plot the solution. [This requires at most 3 computer runs.]

6.3 Use the discrete state transmission matrix obtained in Problem 6.2 to evaluate the final values of $\dot{y}(2)$ and $\ddot{y}(2)$.

6.4 Repeat Problem 5.3 using a trial and error procedure for changing the initial value $\ddot{y}(0)$.

6.5 Solve Problem 5.1 using a trial and error procedure for guessing the initial conditions

(a) use an analog computer and

(b) use a digital simulation. Plot your solution and evaluate the initial conditions.

6.6 (a) Give a scaled analog computer diagram that will solve the following TPBVP by a shooting method.

$$\ddot{y} + (y^2-1)\dot{y} + y = 0 \qquad \begin{cases} y(0) = 3.0 \\ y(5) = 0.0 \end{cases}$$

(b) Evaluate $\dot{y}(0)$ and give a phase plane plot of a sequence of solutions.

(c) Repeat parts (a) and (b) using a digital simulation.

6.7 (a) Sketch a figure like Figure 6.5 to demonstrate the operation of the circuits in Figure 6.6.

(b) Obtain plots from an analog computer to match your results from (a).

6.8 Repeat Problem 5.1 using a finite difference approximation and successive approximation for solution.

6.9 Program Problem 6.8 so that the initial velocity V_0 is adjusted to solve the boundary value problem. Take h = 100 feet, θ = 30°, and d = 250 yards (a good drive for an old professor).

6.10 Solve the TPBVP given in Example 6.2 on an analog computer. Plot y(t), z(t) and w(t) over several cycles on a common time base. Also obtain a sequence of solutions on a single time base to show convergence. Experiment with various values of the gain constant k to obtain varying stability and convergence.

6.11 (a) Give an analog computer diagram that will use a shooting method to adjust ω in the equation

$$\ddot{y} + \omega_n^2 y = 0$$

to satisfy the boundary conditions $\left\{\begin{array}{l} y(0) = 0 \\ \dot{y}(0) = 1 \\ y(1) = -1 \end{array}\right\}$

(b) Obtain plots of the solutions for part (a).

6.12 Give a block diagram that will determine α such that

$$\ddot{y} + 2\dot{y} + y = 1 - 2u(t-\alpha)$$

where $u(t) = \left\{\begin{array}{l} 0 \text{ for } t < 0 \\ 1 \text{ for } t > 0 \end{array}\right\}$

[Hint: Use a time base generator, comparitor and relay to generate $u(t-\alpha)$.]

6.13 (a) Give a finite-difference approximation to solve the TPBVP with boundary conditions at the initial time t_0 and the final time t_f.

$$\ddot{y} + 4y = \sin t \qquad \left\{\begin{array}{l} y(t_0) = A \\ y(t_f) = B \end{array}\right\}$$

(b) Obtain a solution to part (a) for $t_0 = 0$, $t_f = 4$, $A = 0$, and $B = 0$. Estimate $\dot{y}(0)$ and give a plot of the solution.

6.14 (a) Program Example 6.5 for the solution of TPBVP by finite differences and obtain the solution shown in Figure 6.9.

(b) Use an overrelaxation technique as outlined in Chapter 4 to evaluate the value at each grid point by taking the new computed value of y as α_n where

$$\alpha_n = \frac{1}{3.98}[1.9y_{n-1} + 2.1y_{n+1}]$$

Then the overrelaxed value is

$$y_n = y_{n-1} + \omega[\alpha_n - y_{n-1}]$$

Make a plot of the successive values of y_5 as a function of the number of iterations for each weight ω to demonstrate that increasing ω speeds convergence. However, for large ω the iterative scheme is unstable. Estimate the optimum value of ω_{opt} for fastest convergence for this equation.

6.15 Give a recursion relation for a finite difference

solution to LapLace's equation in three dimensions.

$$\frac{\partial^2 v}{\partial x^2} + \frac{\partial^2 v}{\partial y^2} + \frac{\partial^2 v}{\partial z^2} = 0$$

6.16 (a) Obtain a solution to Laplace's equation in the region and boundary conditions given in Figure 6.11.

(b) Investigate the effect of overrelaxation as accomplished in Probelm 6.14 and evaluate ω_{opt}.

6.17 Give a finite difference approximation to solve the wave equation

$$\nabla^2 v = A \frac{\partial^2 v}{\partial t^2}$$

6.18 Solve the one dimensional wave equation on an analog computer.

$$\frac{\partial^2 y}{\partial x^2} = A \frac{\partial^2 y}{\partial t^2}$$

Take A = 1 and the boundary conditions

$$y(0,t) = y(1,t) = 0$$

and the initial displacement as shown in Figure 6.21. Obtain solutions y(x,t) for fixed t and also plot the position of y(1/3,t) as a function of time.

6.19 Devise schemes for setting the boundary values at grid points near the boundary when the boundary does not pass through the grid points. Consider the solutions shown in Figure 6.22 and give an expression for $f_0(c)$ in terms of $f(a)$ and $f(b)$.

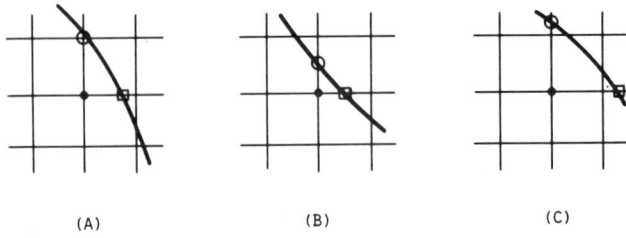

        (A)   (B)   (C)

FIGURE 6.22   PROBLEM 6.19.

6.20 Give an analog computer block diagram that includes logic and mode control to solve the heat flow equation

$$\frac{\partial^2 T}{\partial x^2} = \frac{\partial T}{\partial t} \approx \frac{T_t - T_{t-\Delta t}}{\Delta t}$$

by a CSDT approach having the computer time variable correspond to the problem space variable x. Assume that the initial condition $T(x,0)$ is a given initial condition and $T(0,t) = T(L,t) = T_0$ are the given boundary conditions.

    A tape recorder is available to record the solution $T(x,t)$. Assume that putting the recorder to RESET places it at the beginning of $T(x,t-\Delta t)$ and PLAY plays back $T(x,t-\Delta t)$. The solution for each t must be obtained through some shooting procedure to match up the boundary conditions on the space dimension x.

6.21 Use an analog computer to solve the final value problem

$$\ddot{y} + |\dot{y}|y = 0 \qquad y(5) = 1$$

$$\dot{y}(5) = 0$$

That is, find the initial conditions that correspond to the given final values. [Hint: See Problem 2.38.]

6.22 Find the periodic solutions with period T = 1 to the equation

$$\ddot{y} + 0.1\dot{y} + y^3 = \sin 2\pi t$$

by (a) analog simulation, (b) digital simulation, and (c) hybrid simulation. Plot the time response of the system, with the initial conditions, giving a periodic solution. (See Example 6.4.)

6.23 Write a program for the DSDT solution of the heat flow equation with the conditions given in Example 6.7.

6.24 Answer the questions in the footnote associated with Example 6.8.

6.25 (a) Use a digital simulation language (CSMP, DSL, or your own) to obtain the solution to the wave equation as given in Problem 6.18.

(b) Repeat part (a) using 4th order Runge-Kutta integration for the time variable.

6.26 From the program given in Example 6.9, determine the scale on x, t, and y for the plots shown in Figure 6.20. Assume that the maximum on the initial value of $y(x,0)$ is +1.

6.27 Show that an isometric plot of a function of two variables can be obtained from the transformation on $h(x,y)$ of

$$u = y - x \sin \theta$$

$$v = h(x,y) - x \cos \theta$$

where θ is the angle of perspective.

6.28 (a) Add a damping term to the wave equation in Example 6.9 and obtain the CSCT solution to

$$\frac{\partial^2 y}{\partial x^2} = \frac{\partial^2 y}{\partial t^2} + \alpha \frac{\partial y}{\partial t}$$

for several values of α.

(b) Use CSDT.

6.29 Use the analog computer memory from Example 5.3, connected as a delay as shown in Example 5.4, and the linear data reconstructor as discussed in Example 5.6 as a memory for the CSDT solution of the Wave Equation (6.13). (This problem requires 50 to 100 manhours.)

6.30 In the hybrid computer solution of linear partial differential equations as shown for the wave equation in Figure 6.19 the input to the analog portion of the simulation is the piecewise constant output of a digital-to-analog converter (DAC). Comment on the following conjecture:

"Even though a major application of hybrid computers is for the solution of linear partial differential equations, the entire calculation can be accomplished on the digital computer. Since the output of the DAC is piecewise constant the sampled values into the ADC can be calculated exactly in the digital computer. Therefore the analog portion of the computer is not needed since the direct digital calculation is faster and more accurate!" – T. E. Bullock and A. E. Durling.

7 optimum seeking methods

There are many situations that require the improvement of a parameter value or the selection of a "best" strategy for a problem. In each case we desire in some sense to obtain an optimum solution. In Chapter 4, the overrelaxation factor in the method of successive displacements for the solution of finite difference equations was found to greatly effect the rate of convergence. When the factor is too small, convergence is slow; when it is too large, the algorithm diverges. Some intermediate value yields the fastest convergence. We can call this the optimum value. The solution of a TPBVP requires finding the "best" or optimum values for the parameters of the design. This chapter is devoted to presenting several methods of optimization which provide some insight into this vast area, as well as a few more computational concepts that are conveniently solved by hybrid computation. Details of the methods presented, and many more, are discussed extensively in the excellent book by Wilde and Beighter[11] and in Chapter

9 of Bekey and Karplus[1]

In optimization problems it is desired to obtain the "best" set of parameter values for a given situation. Of course, to obtain a solution we must have a precise definition for what is meant by "best". The first step is to form an *OBJECTIVE FUNCTION* or a *PERFORMANCE INDEX* which is a function of the variables to be optimized and is a measure of the suitability of their values to the desired result. The performance index should have properties that make it realistic from a design point of view and should also have mathematical properties that are convenient for the programming optimization. Of the many ways the objective function can be formulated, probably the most common is the sum of the squares of the difference between the solution having present parameter values, and the desired solution. This sum squared error criterion is motivated by the nice mathematical nature of quadratic functions, as well as the fact that for linear problems the quadratic performance index has a unique minimum (is unimodal) and has all derivatives well defined.

Section 7.1 is an introduction to the concepts of optimization through the minimization of functions of one variable. This treatment is followed by multi-dimensional optimization problems in Section 7.2 by several search techniques and finally the chapter is concluded with some application of optimization methods to TPBVP, system modeling, and parameter optimization.

7.1 ONE DIMENSIONAL OPTIMIZATION

The optimum value of a variable x is that value for which some function of x called the performance index $f(x)$, is maximized (or minimized). An obvious approach is to attack the problem by dividing the interval of investigation, say [a,b], into increments, evaluate $f(x)$ at each point, and pick the best value. This procedure is time consuming and expensive, and suffers from the problems discussed in Chapter 4 for this strategy as applied to solutions of the equation

$f(x) = 0$. Wilde and Beightler present strategies for successively dividing the interval for one dimensional optimizations by golden sections, Fibonacci search, and many others. Here we present a technique that leads conveniently into multidimension optimization for analog, digital and hybrid techniques.

Section 4.1 is concerned with solutions of equations of the form $f(x) = 0$. If $f(x) > 0$ for all x this is equivalent to determining the x for which $f(x)$ is minimum. Consider the functions $f(x)$ and $[f(x)]^2$ shown in Figure 7.1 and the associated Newton-Raphson

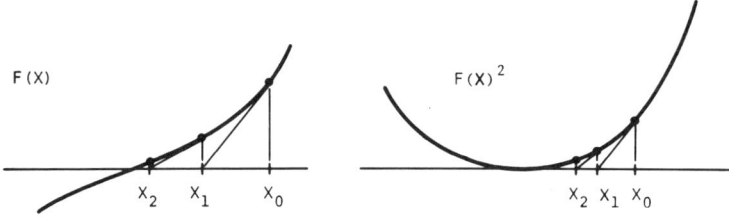

FIGURE 7.1 NEWTON-RAPHSON CONVERGENCE TO $F(x) = 0$ AND $\min[F(x)^2]$.

iteration toward the solution x_0 by the equations

$$x_{n+1} = x_n - \frac{f(x_n)}{\dot{f}(x_n)} \qquad (7.1)$$

for the zero of $f(x)$ and

$$x_{n+1} = x_n - \frac{[f(x_n)]^2}{2f(x_n)\dot{f}(x_n)}$$

$$= x_n - \frac{f(x_n)}{2\dot{f}(x_n)} \qquad (7.2)$$

for the zero of $[f(x)]^2$.

The second procedure depicted in Equation (7.2) and Figure 7.1 is a special case of minimization of a function $f(x)$ where the minimum value $[f(x)]^2 = 0$ is known. If the minimum is not known and we have an initial estimate x_0 we take $\Delta x = x_1 - x_0$ and have the Taylor Series

$$f(x_0 + \Delta x) = f(x_0) + \Delta x \dot{f}(x_0) + \ldots \qquad (7.3)$$

Searching for zeros of $f(x) = 0$ by a Newton-Raphson search we set $f(x_0 + \Delta x) = 0$ and solve for x_1 where $x_1 = x_0 + \Delta x$. However, in general, we do not know the minimum so we just choose a Δx so that $f(x)$ is decreased, $f(x_0 + \Delta x) < f(x_0)$, or $f(x_0 + \Delta x) - f(x_0) < 0$, and the converse if we desire the maximum. Basically we do not know how far to go but at least we can go downhill on each step.

Based only on first derivative information we have

$$f(x_0 + \Delta x) - f(x_0) = \Delta x \dot{f}(x_0) < 0$$

Thus we desire $\Delta x \dot{f}(x_0) < 0$ which is accomplished by taking

$$x_{n+1} = x_n - \alpha \dot{f}(x_n) \qquad (7.4)$$

where $\alpha > 0$. When $\dot{f}(x_n)$ is negative, Δx is positive and $x_{n+1} > x_n$, and for $\dot{f}(x_n)$ positive $x_{n+1} < x_n$. The obvious statement is that when $\dot{f}(x)$ is positive, increase x to increase $f(x)$ and decrease x to decrease $f(x)$. The equivalent statements for negative derivatives exist also.

If the increment factor α is small, convergence is slow and expensive. However, a large α may jump x to values beyond the minimum or even to values where $f(x_{n+1})$ is greater than $f(x_n)$. A convenient method is to start at an initial guess x, with $\alpha = 1$. If $f(x_1) < f(x_0)$ increase α for the next iteration and compare $f(x_2)$ and $f(x_1)$. After each successful decrease of $f(x_n)$ α is increased, perhaps by a factor 2. When α is too large, so that $f(x)$ is not decreasing as desired, α is decreased, by perhaps a factor of 0.1 or 0.2, and the iteration continued. Of course, after an unsuccessful iteration the procedure is restarted with the new   at the past point. To maximize a function, Equation (7.4) becomes

$$x_{n+1} = x_n + \alpha \dot{f}(x_n) \qquad (7.5)$$

where $\alpha > 0$ and the same procedures apply.

If the evaluation of $\dot{f}(x)$ is difficult or time consuming an alternate method, sometimes called the neighboring grid method, is obtained by taking a step in x_0 to obtain $f(x_0 + \delta)$ for some δ, say $\delta = 1$. Then if $f(x_0 + \delta) < f(x_0)$ we take $x_1 = x_0 + \delta$ and continue increasing δ with each successful step. If $f(x_0 + \delta) > f(x_0)$, check $f(x_0 - \delta)$. If both $f(x_0 + \delta)$ and $f(x_0 - \delta)$ are greater than $f(x_0)$ for $\delta = 1$, decrease k by a factor of, say 0.1.

EXAMPLE 7.1

Give a flow chart for the minimization of a scalar function $f(x)$ by the neighboring grid method.

SOLUTION:

See Figure 7.2. The inequalities in blocks 1 and 2 are used to obtain the best current value of x. δ_{min} is the criterion for the accuracy required in x to terminate calculation. ▼

7.2 MULTIDIMENSIONAL OPTIMIZATION

The one dimensional procedure of the last section can be applied to the solution of multidimensional optimization problems by extension to vector notation. Suppose we desire to maximize a scalar valued function $F(x)$ of the n dimension vector augment x having elements $x_1, \ldots x_n$. Starting from the initial guess x_0, we desire to increment x into more promising territory where $F(x)$ is larger. We do not know where the maximum is so, like climbing a mountain in the

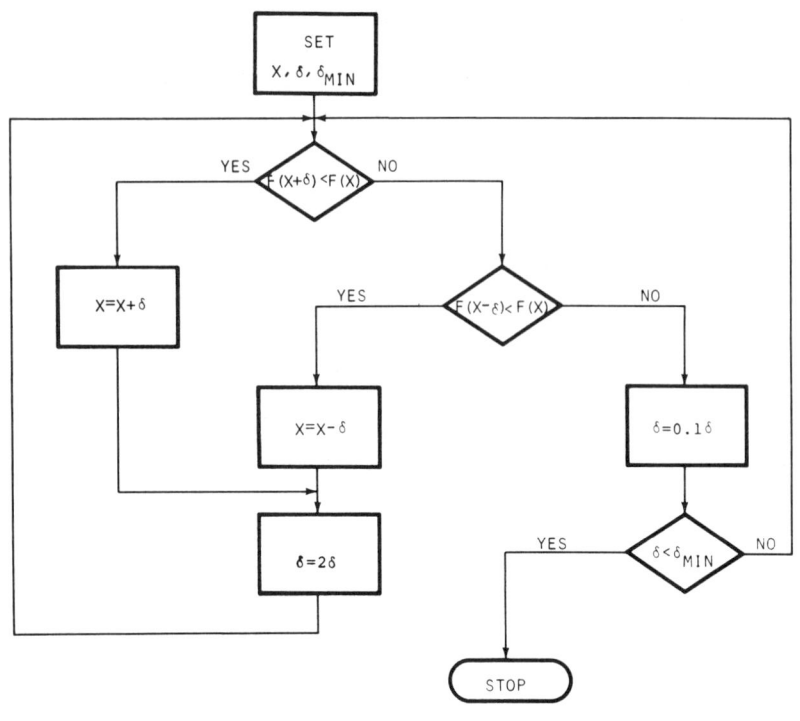

FIGURE 7.2 EXAMPLE 7.1 ALGORITHM FOR THE MINIMIZATION
OF A ONE DIMENSIONAL FUNCTION F(X).

fog, the best we can do is to step off in an uphill
direction.   The direction of *STEEPEST ASCENT* is along
the maximum directional derivative, or the gradient
$(\text{grad } F(x) = \nabla F(x))$ where $\text{grad } F(x)$ is the vector

$$\text{GRAD } F(X) = \nabla F(X) = \begin{bmatrix} \dfrac{\partial F(X)}{\partial x_1} \\[2ex] \dfrac{\partial F(x)}{\partial x_2} \\[1ex] \cdot \\ \cdot \\ \cdot \\[1ex] \dfrac{\partial F(X)}{\partial x_n} \end{bmatrix}$$

which has components denoting the rate of change of $F(x)$ as a function of each of the components of x. Extending the scalar case of the last section we take the new point x_1 as

$$X_1 = X_0 + \alpha \nabla F(X_0) \qquad (7.6)$$

Starting with $\alpha = 1$ we can increase a after successful steps and decrease α when Equation (7.6) produces a new x_1 such that $F(x_n) < F(x_{n+1})$ in which case we return to x_{n-1} and try again with the smaller step length. To minimize $F(x)$ we step x in the direction of the negative gradient or *STEEPEST DESCENT*

$$X_{n+1} = X_n - \alpha \nabla F(X_n) \qquad (7.7)$$

in a similar manner, attempting to descend the surface $F(x)$ as rapidly as possible.

The following two dimensional example demonstrates the method of a gradient search by casting the solution of algebraic equations as an optimization problem.

EXAMPLE 7.2

The system of algebraic equations

$$x_1^2 + x_2 = 2$$
$$\qquad \qquad \qquad (7.8)$$
$$x_1 - x_2^2 = 0$$

has solution $x_1 = x_2 = 1$. Write the equations in the form $F(x) = 0$ such that $F(x) > 0$ and obtain the solution by minimizing $F(x)$ with a discrete gradient search.

SOLUTION:

Equation (7.8) can be written

$$f_1(X) = x_1^2 + x_2 - 2 = 0$$

$$f_2(X) = x_1 - x_2^2 = 0$$

It is convenient to choose the performance function to be minimized as the sum of the squares of $f_1(x)$ and $f_2(x)$ since this results in a nonnegative function with well defined derivatives.

$$F(X) = f_1^2(X) + f_1^2(X)$$

$$(x_1^2 + x_2 - 2)^2 + (x_1 - x_2^2)^2 \qquad (7.9)$$

We desire that x be such that $F(x) = 0$ where Equations (7.8) are satisfied, so we are looking for that x which minimizes $F(x)$.

Figure 7.3 shows contours of constant values of $F(x) = K$ which of course are not known when solving an optimization problem. They are presented here to demonstrate the convergence of the gradient algorithm. The contours of constant $F(x)$ indicate that there is also a solution $(F(x) = 0)$ in the vicinity of $x_1 = 2$, $x_2 = -1$. Also there is a local minimum of $F(x)$ but not $F(x) = 0$ near $x_1 = -1$, $x_2 = 0$. We will see below that the existence of multiple minima create difficulties in all optimization procedures.

The gradient of $F(x)$ is the vector function of (x_1 and x_2)

$$\nabla F(X) = \begin{bmatrix} \dfrac{\partial F(X)}{\partial x_1} \\[2em] \dfrac{\partial F(X)}{\partial x_2} \end{bmatrix} = \begin{bmatrix} 2(x_1^2 + x_2 - 2)2x_1 + 2(x_1 - x_2^2) \\[2em] 2(x_1^2 + x_2 - 2) - 2(x_1 - x_2^2)2x_2 \end{bmatrix}$$

$$(7.10)$$

To minimize $F(x)$ we substitue Equation (7.10) into Equation (7.7) and choose initial guesses for x_1, x_2

and α. Then the values of x_1 and x_2 are iterated by the relations,

$$x_1 (\text{new}) = x_1 - \alpha [4x_1 (x_1^2 + x_2 - 2) + 2(x_1 - x_2^2)]$$

$$x_2 (\text{new}) = x_2 - \alpha [2(x_1^2 + x_2 - 2) - 2(x_1 - x_2^2) 2x_2]$$

$$(7.11)$$

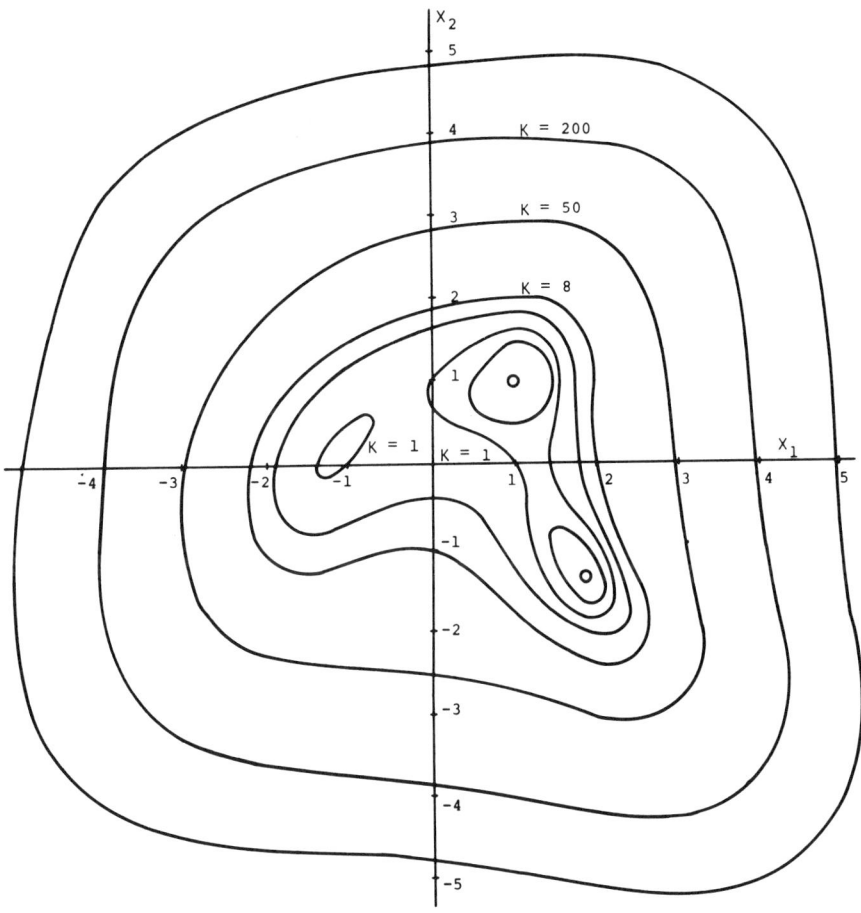

FIGURE 7.3 EXAMPLE 7.2 CONTOURS OF $F(X) = (x_1^2 + x_2 - 2)^2 + (x_1 - x_2^2)^2 = K$.

where the values of x_1 and x_2 on the right side are
the past values. We now have the minimization
algorithm shown in the flow chart of Figure 7.4. The
program is straightforward and left as an exercise.

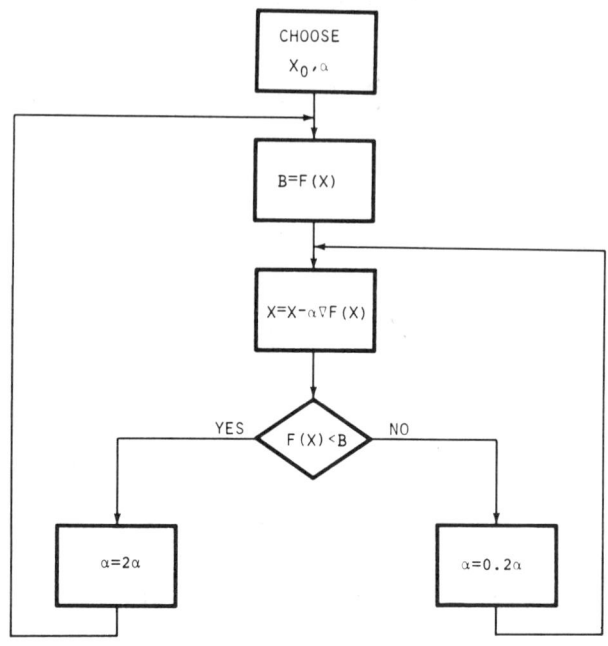

FIGURE 7.4 EXAMPLE 7.2 FLOW CHART FOR THE GRADIENT MINIMIZATION OF F(X).

Figure 7.5 shows several convergent paths for this
example using the above strategy. ▼

The algorithm must be terminated when the objective
function $F(x)$ is sufficiently close to the minimum, δ
is less than a predetermined value δ_{min}, or the number
of iterations becomes excessive. For general optimization problems the desired value of the objective
function is not known as it is in Example 7.2. Therefore, it is necessary either to maximize or to
minimize $F(x)$ without knowing the optimum value. Any
termination strategy must contain a criterion for

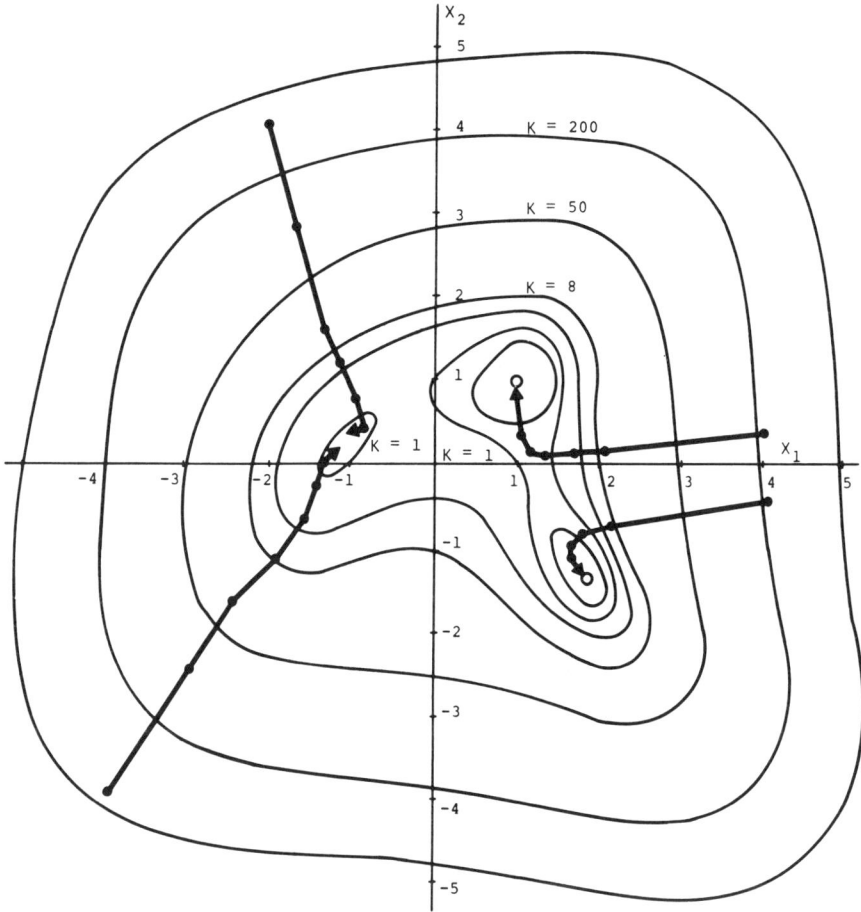

FIGURE 7.5 EXAMPLE 7.2 GRADIENT OPTIMIZATION.

determining that the solution is indeed the maximum or minimum desired. This strategy usually involves constant monitoring of the changes in δ, $\times$ and $F(\times)$ as well as an experimental method for investigating the nature of the stationary point obtained to ascertain that it is indeed a maximum or minimum. This can be accomplished by using a higher order method, by including second derivative terms, or by a pattern type of search as discussed below.

If it is difficult or impossible to calculate the gradient of the objective function, $\nabla F(x)$ can be approximated by finite differences of the form

$$\frac{\partial F(X_0)}{\partial x_i} \approx \frac{F(X_0 + \delta_i) - F(X_0)}{\Delta} \qquad (7.12)$$

This finite difference procedure entails evaluation of $F(x)$ in the vicinity of x_0 for increment changes along each element of x and does not require any gradient information. Thus it is just a search technique based on neighboring points. The neighboring point strategy using Equation (7.12) and the gradient concept results in the iteration

$$X_{n+1} = X_n + \alpha \begin{bmatrix} \frac{1}{\delta_1}[F(X_n + \delta_1) - F(X_n)] \\ \frac{1}{\delta_2}[F(X_n + \delta_2) - F(X_n)] \\ \cdot \\ \cdot \\ \cdot \\ \frac{1}{\delta_n}[F(X_n + \delta_n) - F(X_n)] \end{bmatrix}$$

$$(7.13)$$

Gradient techniques have a tendency to oscillate and converge slowly, particularly as the solution approaches the "optimum" value. The method of *DEFLECTED GRADIENTS* (or *CONJUGATE GRADIENTS*) developed by Fletcher and Powell is presented nicely in Wilde and Beightler,[11] and has accelerated convergence as the optimum is approached. Having *QUADRATIC CONVERGENCE* the error at one step is proportional to the square of the error at the last step. Observe in Figure 7.1 that the Newton-Raphson minimization has quadratic convergence for $f(x)$ but not for $[f(x)]^2$. The Fletcher-Powell algorithm will converge in exactly n steps for an n dimensional problem if the objective function is quadratic and all calculations are exact. Since accurate gradient computations are required, deflected gradient methods are most applicable to situations where computation of the gradient is accurate and not costly. These methods are outside the scope of this text and the reader is referred to the references for further detail.

In many cases there are regions in which the objective function has widely varying characters, consisting of flat spots, steep walls and ridges which frequently create convergence difficulties for the gradient methods. Figure 7.6 shows constant contours for an objective function of two variables that contains a ridge and typical gradient trajectories.

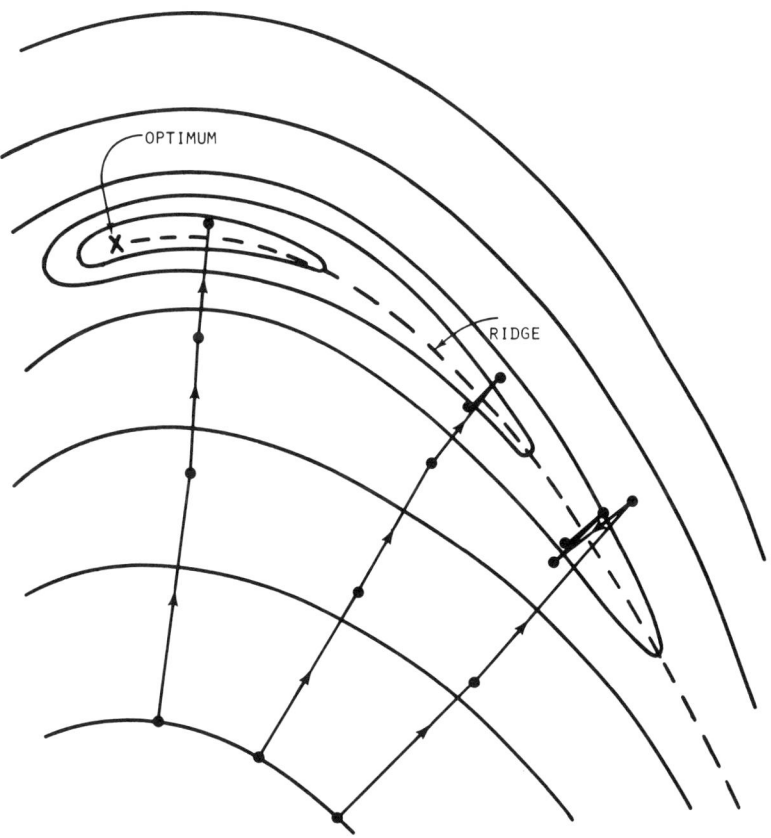

FIGURE 7.6 GRADIENT TRAJECTORIES IN THE PRESENCE OF A RIDGE.

The trajectories obtained by Equations (7.7) or (7.13) converge rapidly to the ridge and then oscillate across the ridge, making little headway toward the optimum. The same difficulty is encountered in optimizations that minimize in one dimension at a time since diagonal moves are ruled out. Extensions of the neighboring grid and finite difference methods are the so-called *RIDGE FOLLOWING* and *PATTERN SEARCH* methods which are particularly applicable when computation of the gradient is costly, since they depend only upon evaluations of the objective function. These methods and many others comprise a set of rules to be performed on past and present calculations of the objective function $F(x)$ to obtain the next value for x, hopefully closer to the optimum.

PATTERN SEARCH METHODS

Successive one dimensional minimization as shown in Figure 7.7 can be made significantly more efficient by the addition of intermediate logic based on past calculations. For example, in the search started at p_0 in Figure 7.7, $F(x)$ is minimized in x_2 holding x_1 constant to determine the new point p_2, from which

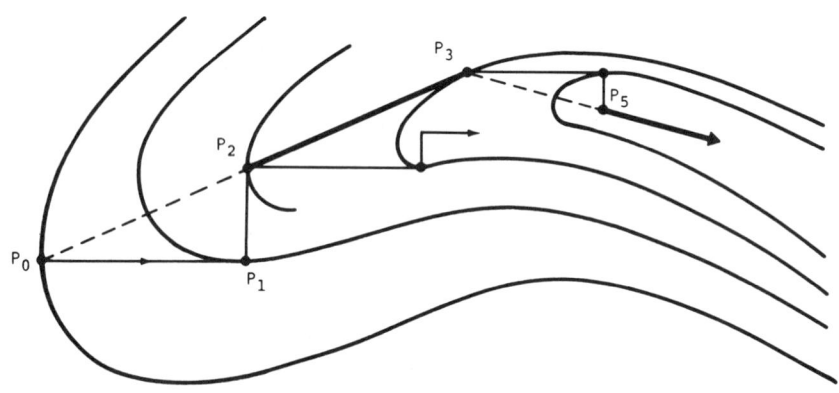

FIGURE 7.7 BEGINNER'S PATTERN SEARCH.

$F(x)$ is minimized in x_1 holding x_2 constant to determine the next point p_2. Now rather than repeat this procedure, which usually performs poorly and stops on diagonal ridges, a promising direction seems to lie in extrapolating along the line between p_0 and p_2 - again a linear search but in a better direction than either x_1 or x_2 alone. This is one of the many strategies of a pattern search.

It is not necessary to perform the one dimensional minimization (maximization) to get a more promising direction. A small increment step in each variable merely gives an estimate for the gradient direction. Note, however, how the pattern search follows the ridge toward the optimum.

To generalize the above procedure to n dimensions it is required to minimize the objective function in each dimension before taking off in the direction $p_n - p_0$. The evaluation of n independent minimizations to obtain the new promising direction is, of course, inefficient. Wilde and Beightler present several pattern search methods which are efficient algorithms for following ridges. In particular, the method of Hooke and Jeeves provides an adaptive technique for changing the distance taken along the new direction based on past experience so that the steps grow with repeated success. Failure to increase the objective function results in a reduced step size and hopefully the calculation of a new direction.

For complex optimization problems in high dimensions more sophisticated logical strategies for optimization must be employed. Furthermore, if the system being optimized or modeled is easily represented on the analog computer or contains hardware (actual equipment), the simulation portion is most economically performed on the analog computer. Thus it is frequently desirable to do a portion of the calculation by analog computation and the arithmetic and/or logical computation by digital computation. The hybrid computer is particularly adept at these more complex problems.

Optimization techniques generally have difficulty in the vicinity of flat spots, where the gradient is approximately zero, and in the presence of ridges. As observed above the method of gradients perform poorly near ridges, stopping before reaching the optimum. However, it avoids the stationary points, called saddle points, which are neither maxima nor minima. Thus the gradient method used in conjunction with some ridge following technique appears worthy of investigation. (See Problems 7.10 and 7.11.)

When an optimization search is terminated because of an excessive number of iterations, lack of further change in the objective function, or failure of the parameter values to change, an ending strategy must be initiated to ascertain whether or not the method has terminated at an optimum point. This requires some functional evaluations in the vicinity of the stationary point. One such strategy is to evaluate the objective function along a circle (or n dimensional sphere) around the stationary point by a one dimensional search, as discussed in Section 6.1. Then the direction yielding the maximum (minimum) value can be investigated in detail. If, indeed, the stationary point is optimum, the objective function will be uniformly inferior to the value of the starting point. If however, the stationary point is a saddle point the objective function will go through maximum and minimum values while traversing the circle and there may be several directions to investigate. If the stationary point is merely an inflection along a ridge only one direction is obtained. Clearly it is necessary to determine the nature of the stationary point before concluding that the optimum has been reached.

Finally, if an optimum value is obtained from any of the search schemes known, there is no guarantee that it is other than merely a local minimum (maximum). One common way to avoid overlooking a true minimum in the presence of several local minima is to start the search from regions far from the minima already obtained and hope that the search goes to another minima. After several minima are found, that value which yields the smallest objective function is taken as the true minimum or optimum. Occasionally it is possible to design the objective function so that it has only a single extremum which eliminates the problem of uniqueness of the optimum.

7.3 PARAMETER OPTIMIZATION

Functional approximation, parameter optimization, and two point boundary value problems can be solved by the optimization procedures of the last section. Application to these areas make it clear that in many cases it is impractical or impossible to obtain an analytic expression for the gradient of the objective function since its value may depend upon the parameters being adjusted in a complex nonlinear manner.

The characteristic value problem discussed in Example 6.3 for the adjustment of ε in the van der Pol equation is an optimization problem that can be stated as follows. Choose ε in the equation

$$\ddot{y} + \varepsilon(y^2-1)\dot{y} + y = 0$$

to best satisfy the boundary conditions

$$y(0) = 0, \quad \dot{y}(0) = 0.5 \quad \text{and} \quad y(2) = 2.0$$

where the parameter value ε can be adjusted by any optimization scheme. In Example 6.3 and the other TPBVPs discussed the optimization has been performed by adjusting the parameter proportional to the error by a gain factor K which corresponds to the factor α in the gradient or pattern search.

In general, the set of parameters to be ajusted is a column vector

$$\beta = \begin{bmatrix} \beta_1 \\ \beta_2 \\ \cdot \\ \cdot \\ \cdot \\ \beta_k \end{bmatrix} \tag{7.14}$$

and the system equations can be written in state variable form as

$$\dot{X} = F(X,U,\beta) \qquad (7.15)$$

The system response $x(t,\beta)$ is a function of the vector of parameters β. It is desired that we pick the best set of parameters β to optimize the performance of the system. "Best" and "optimize" are defined for each problem by the specification of the performance index. A mathematically convenient performance index is the integral square error of the system response as it deviates from the desired response. If the desired response is specified as $y_D(t)$ then we can take for the objective function

$$I(\beta) = \int_0^T [y_D(t) - y(t,\beta)]^2 dt \qquad (7.16)$$

and we adjust β by an optimization strategy to insure that

$$I(\beta_n) < I(\beta_{n-1})$$

EXAMPLE 7.3

Give a block diagram to adjust the parameter β in the equation

$$\ddot{y} + \beta y^3 = u(t) \qquad \begin{array}{l} y(0) = 0 \\ \dot{y}(0) = 0 \end{array} \qquad (7.17)$$

to give the desired step response

$$y_D(t) = \begin{cases} 0 & t < 2 \\ 1 & t > 2 \end{cases}$$

SOLUTION:

First, we cannot run the solution for all time so we must choose an interval over which to optimize, say $t = 0$ to $t = 10$. A convenient objective function is the integral square error, so we will minimize

$$F(\beta) = \int_0^{10} [y_D(t) - y(t)]^2 dt \qquad (7.18)$$

Figure 7.8 gives a block diagram both for the generation $F(\beta)$, which is easily adapted to analog, digital or hybrid simulation, and for adjustment of β to decrease $F(\beta)$ on the next run. [See Problem 7.15.] A reasonable neighboring grid strategy is to start β at some initial value, say $\beta_0 = 1$, and evaluate the index $F(\beta_0)$. Then evaluate $F(\beta_1)$ where $\beta_1 = \beta_0 + \Delta\beta$ for, say $\Delta\beta = 0.1$. If $F(\beta_1) > F(\beta_0)$, take $\beta_1 = \beta_0 - \Delta\beta$. After each successful decrease in $F(\beta_n)$ increase $\Delta\beta$ by a factor of 2 in an attempt to speed up

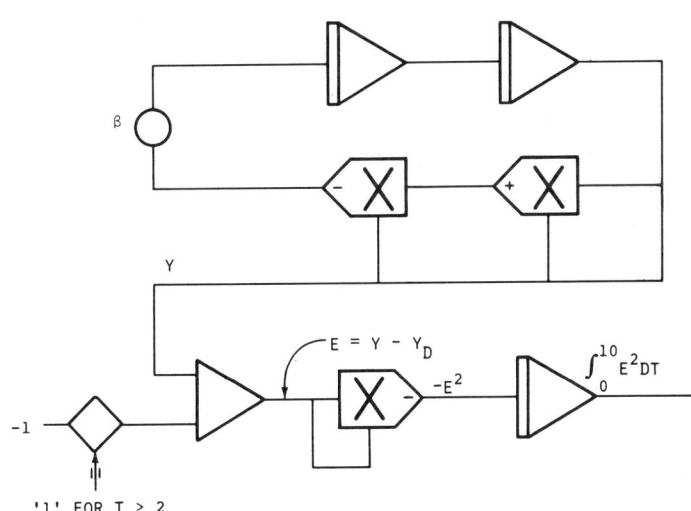

FIGURE 7.8 EVALUATING INTEGRAL SQUARE ERROR FOR PARAMETER OPTIMIZATION.

convergence.   If $F(\beta_n) > F(\beta_{n-1})$ and $\Delta\beta$ is decreased
twice, change the sign of $\Delta\beta$ and restart in the   other
direction. ▼

In the example above, as in any pattern search, the
gradient is never calculated directly.   All of the
optimization performed is based on calculations of
$F(\beta)$. It is frequently possible to evaluate the
gradient directly from the definition of $F(\beta)$.   In
these cases the analytical calculations of   the
gradient can greatly improve convergence since the
optimization can proceed in the direction of steepest
descent based on exact gradient calculations.

As with all functional optimization schemes there
is no guarantee that the minimum obeained is a   global
minimum.   This chapter is concluded with the analysis
of two parameter optimization problems to indicate the
difficulty with multiple extrema.

EXAMPLE 7.4 †

The identification of discrete nonlinear systems
can be performed by a parameter optimization based on
the input-output record for a discrete system which is
to be modeled.   Assume a form for the system like

$$H(z) = \frac{\alpha}{1 + \beta z^{-1}} \qquad (7.19)$$

Then the input sequence $\{x_n\}$ can be put into Equation
(7.19) by the recursion relation

$$y_n = -\beta y_{n-1} + \alpha x_n \qquad (7.20)$$

and the response $\{y_n\}$ can be compared with the system
output record.   The parameters α and β are then
adjusted to minimize the deviation of the model
response $\{y_n\}$ and the known system output $-\{y_{Dn}\}$ A
reasonable error criterion is the sum square error
over the available record in the form

---

† S.   Goldberg and A. Durling, "A Computational
Algorithm for the Identification of Nonlinear
Systems," *Journal of the Franklin Institute,* Vol. 291,
No. 6, June 1971.

$$E = \sum_{k=0}^{N} (y_k - y_{D_k})^2 \qquad (7.21)$$

[See Problem 7.17.]

An input record of 200 Gaussian random numbers was applied to a system having pulse transfer function

$$H(z) = \frac{0.5}{1 + 0.75z^{-1}} \qquad (7.22)$$

to generate an input–output record. The optimization problem to minimize the sum squared error is then as shown in Figure 7.9.

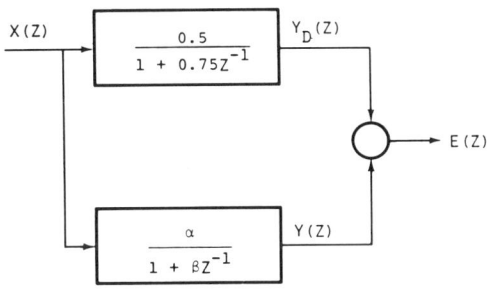

FIGURE 7.9 EXAMPLE 7.4 PARAMETER OPTIMIZATION.

Figure 7.10 shows contours of constant sum square error E given by Equation (7.21) for this input–output record. Clearly the error contours are well defined and essentially any optimization scheme will obtain the correct values for α and β. Problem 7.18 requests the optimization of α and β to best model the system.

If the recursion relation is nonlinear the error contours are not so nice and we may encounter convergence difficulties. Figure 7.11 shows the error contours for the system shown. The nonlinear recursion relation

$$v_k = y_k + 2y_k^2$$

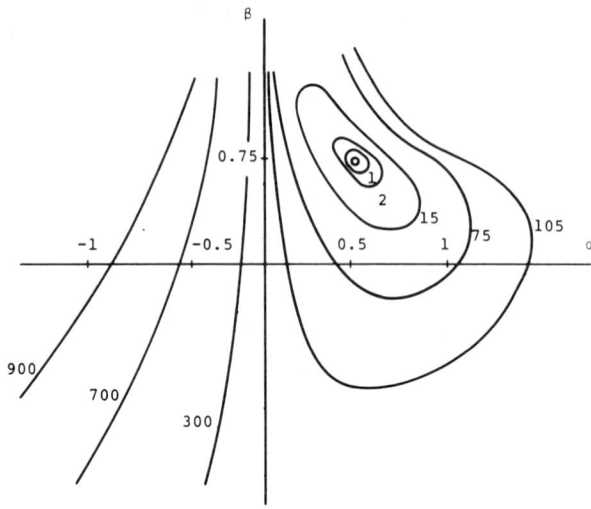

FIGURE 7.10   ERROR CONTOURS FOR THE MODEL IN FIGURE 7.9.

is assumed to be known and the error contours are plotted as a function of α and β. Note that in this case there are several local minima in the α - β plane and the one that an optimization procedure converges to depends on the initial guess for the parameters.

Once a local minimum has been obtained in a optimization problem the search can be reinitiated, starting from random values of the parameter vector β. Presumably, if all initial guesses result in the same optimum value of β, then this is the global minimum. This is of course not guaranteed.

The random selection of initial conditions seems like an expensive proposition but the selection can be biased as additional information is obtained about the character of F(β). In fact, there are several optimization methods based on a random step at each increment of the solution. ▼

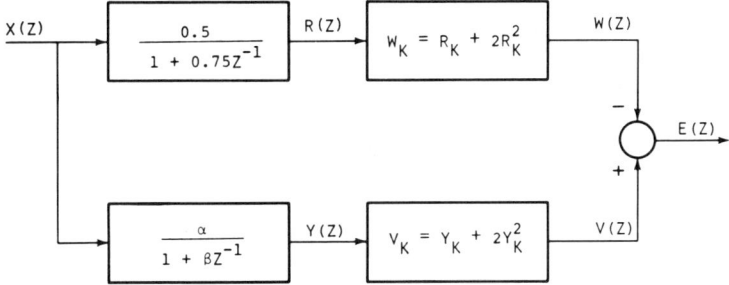

(A) NONLINEAR SYSTEM AND MODEL

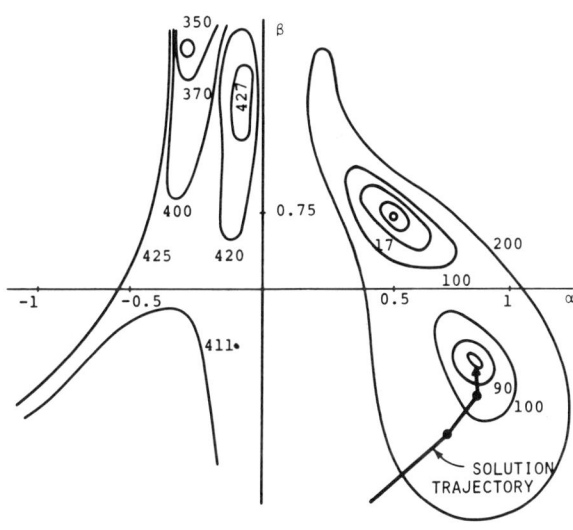

(B) ERROR CONTOURS AND SOLUTION TRAJECTORY

FIGURE 7.11   ERROR CONTOURS FOR A NONLINEAR SYSTEM AND MODEL.

PROBLEMS

7.1 Write a program that will minimize the following
     functions by a sequence of one dimensional mini-

mizations along successive variables.

(a) $F(x,y) = x^2 + y^2$

(b) $F(x,y) = x^2 + (y-1)^2$

(c) $F(x) = x_1^2 + x_2^2 + x_3^2 + 10x_4^2$

Start the program at the initial point (1,1) for (a) and (b), and at (1,1,1) for (c), and give a table of the iteration number, the variable values, and the value of the functions.

7.2   Write a program for the gradient minimization of the following functions using an adaptive gain adjustment similar to Example 7.3. Give a plot of the gradient trajectories starting from the point x = y = 1.

(a) $F(x,y) = 10x^2 - 40x - 10y + y^2 + 65$

(b) $F(x,y) = (e^{-x} + x^2)(y^2 + (y-x)^2)$

7.3   Use a gradient search technique to solve the following equations

(a) $x^2 + 2xy + e^{-x} - 4 = 0$

(b) $3xy - 2y^2 - x = 0$

7.4   Obtain the minimum of the functions given in Example 7.2 with a program that starts at the given starting point and continues in the direction of the gradient at that point in a one dimensional minimization. A new gradient is computed and the one dimensional minimizations along that line performed. Show trajectories of this "optimum gradient" search.

7.5   Implement the pattern search shown in Figure 7.7 to obtain the minimum of the functions given in Problem 7.2.

7.6 (a) Use the digital-analog simulation language to simulate the system

$$\ddot{y} + 2\zeta\omega_n\dot{y} + \omega_n^2 y = x(t)$$

and use an optimization procedure to adjust ζ to best approximate the step response $y(t) = u(t)$ in the interval $t = 0$ to 5 for $\omega_n = 1$. You might take the objective function

$$F(\zeta) = \int_0^5 (y(t)-1)^2 dt$$

Obtain a plot of the response for several values of the damping ration ζ, including the optimum, and tabulate or plot $F(\zeta)$ versus ζ. How many iterations were required to obtain your desired accuracy?

(b) Adapt the program in part (a) to adjust ω_n with $\zeta = 0.4$ to minimize the integral objective function. Sketch and discuss your results.

7.7 Write a FORTRAN subprogram that finds whether or not a stationary point is the optimum value of an objective function by a one dimensional search around a circle centered at the starting point. Apply the program to the stationary points $(x,y) = (0,0)$ and $(0,1)$ of the function

$$F(x,y) = \frac{y^2 + 1}{x^2 + 1}$$

and have the program make a decision about the nature of the point.

7.8 Write a program that will supply data for the plotting of contours of constant value $F(x,y) = K$

for various K based on the gradient search. The line perpendicular to the gradient is tangent to the contour at that point. Use a gradient search to obtain (x,y) for the desired K. Then go along the tangent to trace out the contour. When the value of $F(x,y)$ deviates from K excessively take a step in the gradient direction to get back to the contour. Use the program to plot the contours shown in Figure 7.3.

7.9 Wilde[10] presents a poor man's ridge follower originally proposed by Mugele. The plan is to increment each of the components from x_0 independently. If excursions in both directions for all components fail to produce a better value for $F(x)$, then the best two points (other than x itself) are picked out, say a_1 and a_2. Then the point $((a_1 + a_2)/2)$ is tried and if $F((a_1+a_2)/2)$ is better than $F(x_0)$ the successive search is resumed. If not, a quadratic interpolation is performed between the values of $F(x)$ at $x = a_1$, a_2, and $((a_1+a_2)/2)$ and the minimum of this quadratic is tried next. Verify that the minimum of the quadratic is at $a_3 = (1-\alpha)a_1 + \alpha a_2$ where

$$\alpha = \frac{F(a_1) - F(a_2)}{2[2F((a_1+a_2)/2) - F(a_1) - F(a_2)]}$$

7.10 Use the gradient method to minimize the function given in Problem 7.7 and plot gradient trajectories in the vicinity of the saddle point at $x = 0$, $y = 0$ to demonstrate that the gradient method avoids the saddle point thereby circumventing a difficulty of other methods.

7.11 Write a program that will minimize (maximize) by the method of gradients with the modification that when the search is terminated due to lack of improvement the poor man's ridge follower of Problem 7.9 is initiated to follow the ridge if there is one. Use the program to minimize the Rosenbrock test function

$$F(x_1,x_2) = 100(x_2-x_1^2)^2 + (1-x_1)^2$$

starting from the point $x_1 = -1$, $x_2 = 0$.

This test function, used by Rosenbrock to test a ridge following method with rotating coordinates, is used as a test function to compare several optimization procedures in Wilde and Beightler[11].

7.12 Use the program in Problem 7.11 to plot gradient trajectories from several starting points for the minimization of the function

$$F(x,y) = [100(y-x^2)^2 + (1-x)^2][(2-x)^2 + y^2]$$

Obtain sufficient trajectories to demonstrate the nature of the convergence to the two minima $x = y = 1$ and $x = 2$, $y = 0$.

7.13 To find the real roots of a polynomial $y(x)$ we can use an optimization technique to minimize $F(x) = [y(x)]^2$. Write a digital computer program to determine the complex roots of a polynomial

$$y(x) = u(x) + jv(x)$$

where u and v are real functions of the complex variable x. If you choose an objective function

$$F(x) = [u(x)]^2 + [v(x)]^2$$

then the $x_0 = \alpha_0 + j\beta_0$ which yields $F(x_0) = 0$ is a desired solution.

Use the program to find a solution to the equation $x^2 + 2x + 100 = 0$.

7.14 If the coefficients of a polynomial are real the roots are real or appear in complex conjugate pairs. Adapt the program in Problem 7.13 to divide out real roots automatically, and quadratic terms when complex roots are obtained. Use this program to locate the roots of the polynomials

(a) $x^3 - 1$

(b) $x^6 - 1$

(c) $x^4 + 10x^3 + 11x^2 + 10x + 10$

7.15 (a) Complete the block diagram for Example 7.3.

   (b) Perform the simulation and evaluate the optimum β.

7.16 Cast Example 6.4 as an optimization problem in which it is desired to minimize

$$F(a,b) = [y(0)-y(T)]^2 + [\dot{y}(0)-\dot{y}(T)]^2$$

and perform the optimization.

7.17 Show that if y_k is given by Equation (7.20) in Example 7.4 and the error criterion is the sum square error, Equation (7.21) then the gradient elements $\partial E/\partial \alpha$ and $\partial E/\partial \beta$ can be calculated directly from the response since

$$\frac{\partial E}{\partial \alpha} = \frac{\partial}{\partial \alpha} \sum e^2 = \frac{1}{2} \sum e \frac{\partial e}{\partial \alpha}$$

7.18 Generate an input-output record for the system in Figure 7.9 and perform an optimization to select α and β. Discuss the convergence properties of your algorithm.

7.19 Repeat Problem 7.18 for the system in Figure 7.11. Choose initial conditions to converge to at least two different minima.

7.20 (a) The steady state solution to the vector differential equation

$$\dot{X} + AX = B$$

is that X which satisfies the algebraic equation $AX = B$.

Define the error vector

$$E = AX - B$$

and the scalar objective function

$$F(X) = \frac{1}{2} e^T e = \frac{1}{2} \sum_{i=1}^{n} e_i^2$$

Show that the algebraic equations $AX = B$ can then be solved by a gradient search, where the gradient of $F(X)$ is given by the expression

$$[\text{jth element of } \nabla F] = \sum_{i=1}^{n} \frac{\partial e_i}{\partial x_j} \sum_{k=1}^{n} \frac{\partial e_i}{\partial x_k} x_k - b_k$$

(b) Give a hybrid computer block diagram to implement the result of part (a) for the solution of linear algebraic equations. [Note: See Bekey and Karplus[1], Section 9.4 for further details.]

references

1. G. A. Bekey and W. J. Karplus, *Hybrid Computation*, John Wiley & Sons, Inc., New York, 1968.

2. M. Cuénod and A. Durling, *A Discrete-Time Approach for System Analysis*, Academic Press, New York, 1969.

3. V. N. Faddeeva, *Computational Methods of Linear Algebra*, Dover Publications, Inc., New York, 1959.

4. G. E. Forsythe and C. B. Moler, *Computer Solution of Linear Algebraic Systems*, Prentice-Hall, Inc., Englewood Cliffs, N. J., 1967.

5. G. E. Forsythe and W. R. Wassow, *Finite Difference Methods for Partial Differential Equations*, John Wiley & Sons, Inc., New York, 1960.

6. F. B. Hildebrand, *Finite Difference Equations and Simulations*, Prentice-Hall, Inc., Englewood Cliffs, N. J., 1968.

7. G. A. Korn and T. M. Korn, *Electronic Analog and Hybrid Computers*, McGraw-Hill, Inc., New York, 1964.

8. F. F. Kuo and J. F. Kaiser, *System Analysis by Digital Computer*, John Wiley & Sons, Inc., New York, 1966.

9. A. Ralston, *A First Course in Numerical Analysis*, McGraw-Hill, Inc., New York, 1965.

10. D. J. Wilde, *Optimum Seeking Methods*, Prentice-Hall, Inc., Englewood Cliffs, N. J., 1964.

11. D. J. Wilde and C. S. Beightler, *Foundations of Optimization*, Prentice-Hall, Inc., Englewood Cliffs, N. J., 1967.

appendix

```
C
C     MAIN PROGRAM FOR THE SOLUTION OF VECTOR DIFFERENTIAL
C     EQUATIONS OF THE FORM XDOT=F(X) USING FOURTH ORDER
C     RUNGE-KUTTA INTEGRATION.
C
C     PURPOSE
C       TO DEMONSTRATE THE INTEGRATION OF STATE EQUATIONS
C       FOR SYSTEM SIMULATION.
C
C     USAGE
C       OCCASIONAL AND SPORADIC
C
      DIMENSION X(2,1),XDOT(2,1),TIMEO(101),Y(101),YDOT(101)
      DATA TIMEO,Y,YDOT/303*0.0/
      DATA N/2/,TAU/0.01/,FINTIM/10.0/,NUMOUT/100/
C
C     INITIALIZATION STATEMENTS.
C
      NSTEPS=FINTIM/TAU
      OUTDEL=FINTIM/NUMOUT
      TIMOUT=OUTDEL-0.5*TAU
C
C     INITIALIZE THE STATE VARIABLE X.
C
      X(1,1)=0.0
      X(2,1)=1.0
C
C     STORE THE INITIAL VALUES OF THE ARRAYS USED IN PLOTTING.
C
      NN=1
      YDOT(1)=X(2,1)
C
C     SIMULATION LOOP...TO STATEMENT # 100.
C
      DO 100 K=1,NSTEPS
      TIME=K*TAU
      CALL RK4(TAU,N,X)
C
C     IS IT TIME TO STORE?
C
      IF(TIME.LT.TIMOUT) GO TO 100
      NN=NN+1
      TIMEO(NN)=TIME
      Y(NN)=X(1,1)
      YDOT(NN)=X(2,1)
      TIMOUT=NN*OUTDEL-0.5*TAU
      IF(NN.GE.(NUMOUT+1)) GO TO 999
  100 CONTINUE
  999 CALL PLOT(TIMEO,Y,51)
      CALL PLOT(TIMEO,YDOT,51)
      CALL XYPLT(Y,YDOT,NN)
      STOP
      END
```

```
      SUBROUTINE SYSTEM(H,N,X,XDOT)
      DIMENSION X(N,1),XDOT(N,1)
      DATA EPSILN/2.0/
      M=N-1
      DO 1 I=1,M
    1 XDOT(I,1)=X(I+1,1)
C
C
C     INTRODUCE HERE THE EQUATION FOR THE PARTICULAR SYSTEM.
C
      XDOT(N,1)=EPSILN*(1.0-X(1,1)**2)*X(2,1)-X(1,1)
      RETURN
      END



      SUBROUTINE RK4(H,N,X)
      REAL K(10,1)
      DIMENSION X(N,1),ACCU(10,1),XU(10,1)
      DATA ACCU,K,XU/30*0.0/
      CALL SYSTEM(H,N,X,K)
      CALL MTX(0.0,ACCU,1.0,K,ACCU,N,1)
      DO 1 I=2,3
      CALL MTX(1.0,X,H/2.,K,XU,N,1)
      CALL MTX(1.0,ACCU,2.0,K,ACCU,N,1)
    1 CONTINUE
      CALL MTX(1.0,X,H,K,XU,N,1)
      CALL SYSTEM(H,N,XU,K)
      CALL MTX(1.0,ACCU,1.0,K,ACCU,N,1)
      CALL MTX(1.0,X,H/6.,ACCU,X,N,1)
      RETURN
      END



      SUBROUTINE MTX(CA,A,CB,B,C,N,M)
      DIMENSION A(N,M),B(N,M),C(N,M)
      DO 1 I=1,N
      DO 1 J=1,M
    1 C(I,J)=CA*A(I,J)+CB*B(I,J)
      RETURN
      END
```

```
C
C      SUBROUTINE PLOT
C
C      PURPOSE
C         TO PLOT A FUNCTION F(T) VS T WITH THE MAGNITUDE OF F
C         PLOTTED HORIZONTALLY ACROSS AN OUTPUT PAGE AND EACH
C         VALUE OF T REPRESENTED BY A LINE OF OUTPUT. PLOT IS
C         8 INCHES WIDE BY INDEFINITE LENGTH. 50 OUTPUT POINTS
C         GIVES AN 8.5 BY 11 INCH PLOT.
C
C      USAGE
C         CALL PLOT(T,Y,N)
C
C      DESCRIPTION OF PARAMETERS
C         T - ARRAY OF CONSECUTIVE VALUES OF THE INDEPENDENT
C             VARIABLE TO BE PLOTTED. EACH VALUE IS REPRESENTED
C             BY A LINE OF OUTPUT.
C         Y - ARRAY OF THE FUNCTIONAL VALUES CORRESPONDING
C             TO EACH OF THE CONSECUTIVE T VALUES(I.E.Y(T))
C             EACH VALUE OF Y IS REPRESENTED BY A ROW OF
C             ASTERISKS IN THE OUTPUT GRAPH.
C         N - THE NUMBER OF CONSECUTIVE Y & T ARRAY VALUES TO BE
C             PLOTTED. THE FIRST N VALUES OF Y TO BE PLOTTED.
C
C      REMARKS
C         THE OUTPUT VALUES ARE AUTOMATICALLY SCALED WITH THE
C         SCALE FACTOR PRINTED ABOVE THE GRAPH. ADJUST YOUR
C         JOB CONTROL LANGUAGE TO OVERRIDE THE STANDARD 50 LINES
C         PER PAGE AUTOMATIC SKIP MECHANISM IF N IS GREATER THAN 50.
C
       SUBROUTINE PLOT(T,F,N)
       INTEGER PLINE(61),ASTRX,AXIS,BLANK,PLUS,VDASH
       DIMENSION T(N),F(N),DISP(3),DY(11),SCALE(3),SFRCT(5)
       DATA ASTRX,AXIS,BLANK,PLUS,VDASH/'*','I',' ','+','I'/
       DATA SFRCT/1.,2.5,5.,7.5,10./,DISP/1.5,61.5,31.5/,SCALE/2*60.,30./
C
C      NORMALIZATION SCHEME.
C
       FMAX=F(1)
       FMIN=F(1)
       DO 4 I=1,N
       IF(F(I)-FMAX)2,2,1
     1 FMAX=F(I)
     2 IF(F(I)-FMIN)3,4,4
     3 FMIN=F(I)
     4 CONTINUE
       IF(ABS(FMAX)-ABS(FMIN))6,5,5
     5 DIV=ABS(FMAX)
       GO TO 7
     6 DIV=ABS(FMIN)
     7 NEXP=IFIX(ALOG(DIV)/ALOG(10.))
       IF(DIV.LT.1.)NEXP=NEXP-1
       PP=10.**NEXP
       P=10.**(-NEXP)
       FRACT=DIV*P
       DO 8 I=1,5
       DIV=SFRCT(I)*PP
       IF(FRACT-SFRCT(I))9,9,8
     8 CONTINUE
     9 WRITE(6,10)FMAX,FMIN
    10 FORMAT('1'/' MAXIMUM = ',G15.7/' MINIMUM = ',G15.7/)
```

```
C
C    INDEX THE APPROPRIATE PLOTTER SUBSECTION.
C
         WRITE(6,11)PP
   11 FORMAT(T38,'(OUTPUT SCALED BY: ',1PE7.1,')')
         IF(FMIN)12,20,20
   12 IF(FMAX)25,25,30
   20 INDEX=1
         DY(1)=0.
         YINC=DIV/10.
         WRITE(6,21)
   21 FORMAT(T21,'0 + =====>')
         GO TO 100
   25 INDEX=2
         DY(1)=-DIV*P
         YINC=DIV/10.
         WRITE(6,26)
   26 FORMAT(T72,'<===== - 0')
         GO TO 100
   30 INDEX=3
         DY(1)=-DIV*P
         YINC=DIV/5.
         WRITE(6,31)
   31 FORMAT(T42,'<===== - 0 + =====>')
  100 DO 110 I=2,11
  110 DY(I)=DY(I-1)+YINC*P
         WRITE(6,120)DY
  120 FORMAT(16X,11F6.1)
         WRITE(6,125)
  125 FORMAT('    TIME        OUTPUT',T21,'+',10('-----+'))
         DO 190 J=1,N
         I=IFIX((F(J)/DIV)*SCALE(INDEX)+DISP(INDEX))
         DO 135 L=1,60,6
         DO 130 K=1,5
  130 PLINE(L+K)=BLANK
  135 PLINE(L)=VDASH
         PLINE(61)=VDASH
C
C    BRANCH TO APPROPRIATE SUBSECTION./
C
         GO TO (140,150,160),INDEX
  140 DO 145 L=1,I
  145 PLINE(L)=ASTRX
         PLINE(1)=AXIS
         GO TO 170
  150 DO 155 L=I,60
  155 PLINE(L)=ASTRX
         PLINE(61)=AXIS
         GO TO 170
  160 IF(I-31)165,169,167
  165 DO 166 L=I,30
  166 PLINE(L)=ASTRX
         GO TO 169
  167 DO 168 L=32,I
  168 PLINE(L)=ASTRX
  169 PLINE(31)=AXIS
  170 PLINE(I)=PLUS
         FP=F(J)*P
         WRITE(6,180)T(J),FP,PLINE
  180 FORMAT(' ',G10.3,1X,F7.3,1X,61A1)
  190 CONTINUE
         WRITE(6,195)
  195 FORMAT(20X,'|',10('_____|'))
         RETURN
         END
```

```
C
C
C      SUBROUTINE XYPLT
C
C      PURPOSE
C         TO PLOT TWO FUNCTIONS X(T) & Y(T) OF AN INDEPENDENT
C         VARIABLE T VS EACH OTHER WITH X(T) AS ABSCISSA AND
C         Y(T) AS ORDINATE.THE GRAPH PRODUCED FITS IN 8.5 BY 11
C         INCH PAGE.
C
C      USAGE
C         CALL XYPLT(X,Y,N)
C
C      DESCRIPTION OF PARAMETERS
C         X - ARRAY OF X(T) FUNCTIONAL VALUES.
C         Y - ARRAY OF Y(T) FUNCTIONAL VALUES CORRESPONDING
C             TO THE SAME T VALUES USED IN GENERATING X(T).
C         N - THE NUMBER OF X & Y VALUES TO BE PLOTTED.
C             THE FIRST N VALUES ARE PLOTTED.
C
C      REMARKS
C         THE FIRST TEN VALUES OF X & Y ARE REPRESENTED BY THE
C         NUMBERS 0 THRU 9 IN THE GRAPH TO SIGNIFY DIRECTION.
C
       SUBROUTINE XYPLT(X,Y,N)
       INTEGER GRAPH(81,51), ASTRX,BLANK,CENTR,PLUS,VDASH
       DIMENSION X(N),Y(N),NUM(10),SFRCT(5)
       DATA GRAPH/4131*' '/,SFRCT/1.,2.5,5.,7.5,10./
       DATA ASTRX,BLANK,CENTR,MINUS,PLUS,VDASH/'*',' ','0','-','+','|'/
       DATA NUM/'0','1','2','3','4','5','6','7','8','9'/
C
C      NORMALIZATION SCHEME.
C
       XMAX=X(1)
       XMIN=X(1)
       YMAX=Y(1)
       YMIN=Y(1)
       DO 9 I=2,N
       IF(X(I)-XMAX)3,3,2
     2 XMAX=X(I)
     3 IF(X(I)-XMIN)4,5,5
     4 XMIN=X(I)
     5 IF(Y(I)-YMAX)7,7,6
     6 YMAX=Y(I)
     7 IF(Y(I)-YMIN)8,9,9
     8 YMIN=Y(I)
     9 CONTINUE
       IF(ABS(XMAX)-ABS(XMIN))10,11,11
    10 DIVX=ABS(XMIN)
       GO TO 12
    11 DIVX=ABS(XMAX)
    12 IF(ABS(YMAX)-ABS(YMIN))13,14,14
    13 DIVY=ABS(YMIN)
       GO TO 15
    14 DIVY=ABS(YMAX)
    15 NX=IFIX(ALOG(DIVX)/ALOG(10.))
       NY=IFIX(ALOG(DIVY)/ALOG(10.))
       FX=DIVX/10**NX
       FY=DIVY/10**NY
       DO 16 I=1,5
       DIVX=SFRCT(I)*10**NX
       IF(FX-SFRCT(I))17,17,16
```

```
   16 CONTINUE
   17 DO 18 I=1,5
      DIVY=SFRCT(I)*10**NY
      IF(FY-SFRCT(I))19,19,18
   18 CONTINUE
   19 SIZEX=DIVX/5.
      SIZEY=DIVY/5.
C
C    BUILD GRAPH'S AXIS.
C
      DO 24 I=1,80,8
      DO 20 J=1,51,25
   20 GRAPH(I,J)=PLUS
      DO 24 K=1,7
      DO 21 J=1,51,25
   21 GRAPH(I+K,J)=MINUS
   24 CONTINUE
      DO 25 J=1,51,25
   25 GRAPH(81,J)=PLUS
      DO 30 I=1,50,5
      DO 27 J=1,81,40
   27 GRAPH(J,I)=PLUS
      DO 30 K=1,4
      DO 28 J=1,81,40
   28 GRAPH(J,K+I)=VDASH
   30 CONTINUE
      DO 35 J=1,81,40
   35 GRAPH(J,51)=PLUS
      GRAPH(41,26)=CENTR
C
C    START PHASE PORTRAIT.
C
      DO 50 K=1,N
      I=IFIX((X(K)/DIVX)*40. + 41.5)
      J=IFIX((-Y(K)/DIVY)*25. + 26.5)
      IF(K-10)40,40,45
   40 GRAPH(I,J)=NUM(K)
      GO TO 50
   45 GRAPH(I,J)=ASTRX
   50 CONTINUE
      WRITE(6,60)XMAX,YMAX,XMIN,YMIN
   60 FORMAT('1','MAXIMUM X VALUE =',G15.7,T41,'MAXIMUM Y VALUE =',G15.7
     .        /' MINIMUM X VALUE =',G15.7,T41,'MINIMUM Y VALUE =',G15.7 )
      WRITE(6,65)SIZEX,SIZEY
   65 FORMAT('2','X SCALE : ',G15.7,' PER MAJOR HORIZONTAL DIVISION.'
     .        /' Y SCALE : ',G15.7,' PER MAJOR  VERTICAL  DIVISION.')
      WRITE(6,70)((GRAPH(I,J),I=1,81),J=1,51)
   70 FORMAT(1X,81A1)
      RETURN
      END
```

```
MAXIMUM =    1.680097
MINIMUM =   -2.032362

                                     (OUTPUT SCALED BY: 1.0E 00)
                                        (===== - 0 + =====)
                    -2.5  -2.0  -1.5  -1.0  -0.5   0.0   0.5   1.0   1.5   2.0   2.5
  TIME    OUTPUT  +-----+-----+-----+-----+-----+-----+-----+-----+-----+-----+
  0.0      0.0    |     |     |     |     |     +     |     |     |     |     |
  0.100    0.110  |     |     |     |     |     |+    |     |     |     |     |
  0.200    0.242  |     |     |     |     |     |**+  |     |     |     |     |
  0.300    0.399  |     |     |     |     |     |****+|     |     |     |     |
  0.400    0.581  |     |     |     |     |     |******+     |     |     |     |
  0.500    0.782  |     |     |     |     |     |********+   |     |     |     |
  0.600    0.992  |     |     |     |     |     |***********+|     |     |     |
  0.700    1.194  |     |     |     |     |     |*************+    |     |     |
  0.800    1.369  |     |     |     |     |     |***************+  |     |     |
  0.900    1.505  |     |     |     |     |     |*****************+|     |     |
  1.00     1.598  |     |     |     |     |     |******************+     |     |
  1.10     1.653  |     |     |     |     |     |*******************+    |     |
  1.20     1.677  |     |     |     |     |     |*******************+    |     |
  1.30     1.680  |     |     |     |     |     |*******************+    |     |
  1.40     1.668  |     |     |     |     |     |*******************+    |     |
  1.50     1.645  |     |     |     |     |     |*******************+    |     |
  1.60     1.615  |     |     |     |     |     |*******************+    |     |
  1.70     1.579  |     |     |     |     |     |******************+     |     |
  1.80     1.539  |     |     |     |     |     |*****************+|     |     |
  1.90     1.495  |     |     |     |     |     |*****************+|     |     |
  2.00     1.447  |     |     |     |     |     |****************+ |     |     |
  2.10     1.396  |     |     |     |     |     |****************+ |     |     |
  2.20     1.341  |     |     |     |     |     |***************+  |     |     |
  2.30     1.281  |     |     |     |     |     |**************+   |     |     |
  2.40     1.217  |     |     |     |     |     |**************+   |     |     |
  2.50     1.147  |     |     |     |     |     |*************+    |     |     |
  2.60     1.071  |     |     |     |     |     |************+    |     |     |
  2.70     0.986  |     |     |     |     |     |***********+     |     |     |
  2.80     0.890  |     |     |     |     |     |**********+|     |     |     |
  2.90     0.782  |     |     |     |     |     |********+  |     |     |     |
  3.00     0.657  |     |     |     |     |     |*******+   |     |     |     |
  3.10     0.511  |     |     |     |     |     |******+    |     |     |     |
  3.20     0.336  |     |     |     |     |     |***+  |    |     |     |     |
  3.30     0.123  |     |     |     |     |     |+     |    |     |     |     |
  3.40    -0.136  |     |     |     |     |   +*|     |    |     |     |     |
  3.50    -0.449  |     |     |     |     |+****|     |    |     |     |     |
  3.60    -0.810  |     |     |    +**********|     |    |     |     |     |
  3.70    -1.191  |     |     |  +*************|***|     |    |     |     |     |
  3.80    -1.533  |     |    +***************|****|     |    |     |     |     |
  3.90    -1.785  |   +***************|*********|     |    |     |     |     |
  4.00    -1.936  |+****************|**********|     |    |     |     |     |
  4.10    -2.008  +*****************|**********|     |    |     |     |     |
  4.20    -2.032  +*****************|***********|     |    |     |     |     |
  4.30    -2.030  +*****************|***********|     |    |     |     |     |
  4.40    -2.013  +*****************|**********|     |    |     |     |     |
  4.50    -1.989  +*****************|*********|     |    |     |     |     |
  4.60    -1.960  +****************|*********|     |    |     |     |     |
  4.70    -1.929  |+***************|*********|     |    |     |     |     |
  4.80    -1.896  |+***************|********|     |    |     |     |     |
  4.90    -1.862  | +**************|********|     |    |     |     |     |
  5.00    -1.826  | +**************|********|     |    |     |     |     |
                  |_____|_____|_____|_____|_____|_____|_____|_____|_____|_____|
```

```
MAXIMUM =    2.088953
MINIMUM =   -3.793948
```

```
                                   (OUTPUT SCALED BY: 1.0E 00)
                                     <===== - 0 + =====>
                    -5.0  -4.0  -3.0  -2.0  -1.0   0.0   1.0   2.0   3.0   4.0   5.0
       TIME   OUTPUT +----+----+----+----+----+----+----+----+----+----+
       0.0    1.000  |    |    |    |    |    |*****+    |    |    |    |
       0.100  1.214  |    |    |    |    |    |******+   |    |    |    |
       0.200  1.453  |    |    |    |    |    |********+  |    |    |    |
       0.300  1.704  |    |    |    |    |    |*********+ |    |    |    |
       0.400  1.933  |    |    |    |    |    |************+  |    |    |
       0.500  2.084  |    |    |    |    |    |*************+ |    |    |
       0.600  2.089  |    |    |    |    |    |*************+ |    |    |
       0.700  1.906  |    |    |    |    |    |***********+|  |    |    |
       0.800  1.557  |    |    |    |    |    |*********+  |    |    |
       0.900  1.127  |    |    |    |    |    |*******+    |    |    |    |
       1.00   0.712  |    |    |    |    |    |****+   |    |    |    |
       1.10   0.371  |    |    |    |    |    |**+  |    |    |    |    |
       1.20   0.117  |    |    |    |    |    |+    |    |    |    |    |
       1.30  -0.061  |    |    |    |    |    +|   |    |    |    |    |
       1.40  -0.184  |    |    |    |    |   +|   |    |    |    |    |
       1.50  -0.271  |    |    |    |    |   +*|   |    |    |    |    |
       1.60  -0.334  |    |    |    |    |   +*|   |    |    |    |    |
       1.70  -0.383  |    |    |    |    |   +*|   |    |    |    |    |
       1.80  -0.423  |    |    |    |    |   +**|   |    |    |    |    |
       1.90  -0.460  |    |    |    |    |   +**|   |    |    |    |    |
       2.00  -0.496  |    |    |    |    |   +**|   |    |    |    |    |
       2.10  -0.533  |    |    |    |    |   +**|   |    |    |    |    |
       2.20  -0.573  |    |    |    |    |   +**|   |    |    |    |    |
       2.30  -0.619  |    |    |    |    |  +***|   |    |    |    |    |
       2.40  -0.671  |    |    |    |    |  +***|   |    |    |    |    |
       2.50  -0.733  |    |    |    |    |  +***|   |    |    |    |    |
       2.60  -0.808  |    |    |    |    | +****|   |    |    |    |    |
       2.70  -0.900  |    |    |    |    | |+****|   |    |    |    |    |
       2.80  -1.016  |    |    |    |    |+*****|   |    |    |    |    |
       2.90  -1.164  |    |    |    |    +*******|   |    |    |    |    |
       3.00  -1.355  |    |    |    |   +*******|   |    |    |    |    |
       3.10  -1.605  |    |    |    |  +*********|   |    |    |    |    |
       3.20  -1.933  |    |    |    | +***********|   |    |    |    |    |
       3.30  -2.356  |    |    |   +**************|   |    |    |    |    |
       3.40  -2.872  |    |    |+***************** |   |    |    |    |    |
       3.50  -3.416  |    | +******************** |   |    |    |    |    |
       3.60  -3.794  | +***********************|   |    |    |    |    |
       3.70  -3.704  |  +**********************|   |    |    |    |    |
       3.80  -2.999  |    | +******************|   |    |    |    |    |
       3.90  -1.961  |    |    |  +************|   |    |    |    |    |
       4.00  -1.038  |    |    |    |    +*****|   |    |    |    |    |
       4.10  -0.423  |    |    |    |    |   +**+|   |    |    |    |    |
       4.20  -0.072  |    |    |    |    |    +|   |    |    |    |    |
       4.30   0.115  |    |    |    |    |    |+   |    |    |    |    |
       4.40   0.215  |    |    |    |    |    |+   |    |    |    |    |
       4.50   0.269  |    |    |    |    |    |**  |    |    |    |    |
       4.60   0.301  |    |    |    |    |    |**  |    |    |    |    |
       4.70   0.322  |    |    |    |    |    |**  |    |    |    |    |
       4.80   0.338  |    |    |    |    |    |**  |    |    |    |    |
       4.90   0.351  |    |    |    |    |    |*+  |    |    |    |    |
       5.00   0.364  |    |    |    |    |    |*+  |    |    |    |    |
                     +----+----+----+----+----+----+----+----+----+----+
```

MAXIMUM X VALUE =   2.034571   MAXIMUM Y VALUE =   3.835082
MINIMUM X VALUE = -2.032362   MINIMUM Y VALUE = -3.793948
X SCALE :   0.5000000   PER MAJOR HORIZONTAL DIVISION.
Y SCALE :   1.000000   PER MAJOR VERTICAL DIVISION.

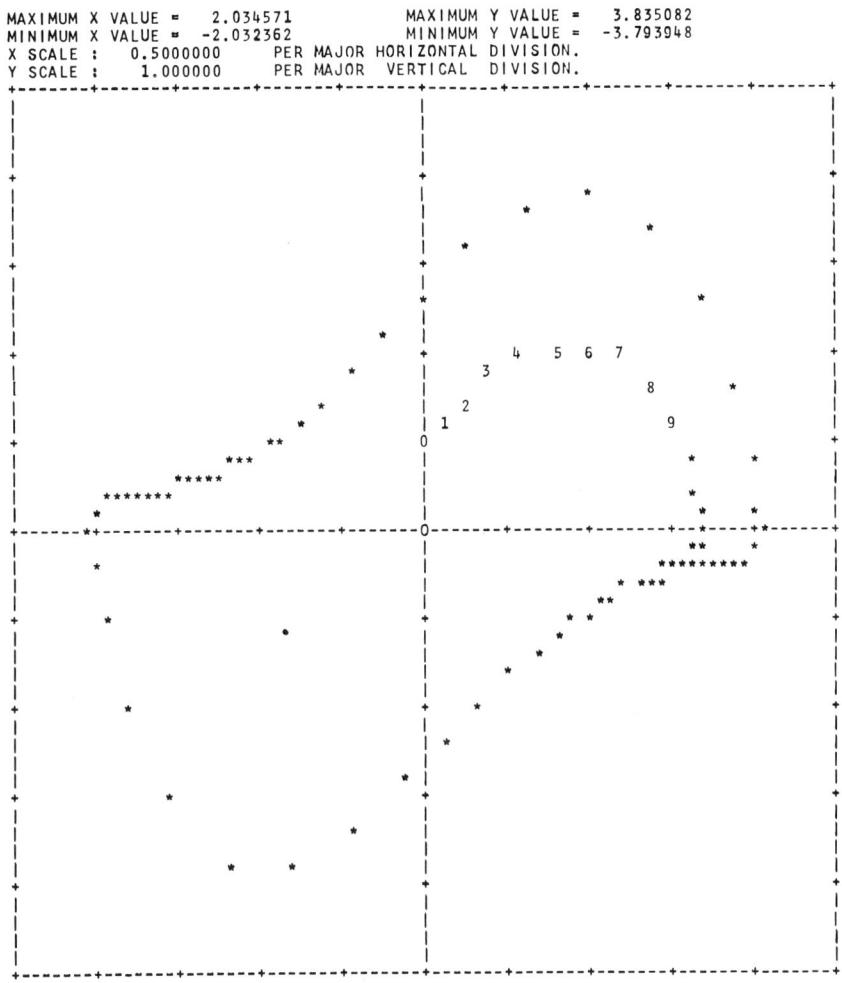

index